Seductions of Place

Seductions of Place explores questions at the crossroads of contemporary issues in travel and tourism, human geography, and the complex cultural, political, and economic activities at stake in touristed landscapes, as a result of processes of globalization. The seductiveness of touristed landscapes is simultaneously local and global, as traveled places are formed and reworked by the activities of diverse, mobile people, in their desires to experience situated, sensuous qualities of difference.

The initial chapters link the existing literature on mobility and travel with current theoretical perspectives in human geography. They include a theorization of the concepts of place and landscape, and the importance of tourism in the world economy, to move research on tourism to a central position in geographic research. Chapter contributions include analysis of the representational character of landscape and the built environment; historic constructions of place seduction; the importance of class, racial, and gender dimensions of place; how mobility and the seduction of place orient identity formation; the role of tourism in the world economy; local tourism economies in relation to the global economy; and the environmental impacts of tourism economies. Throughout the book, examples are given from urban and environmental touristed landscapes, from major world cities to tropical islands.

Seductions of Place assesses travel and tourism as simultaneously cultural and economic processes, through ideas about place seduction and the formation of landscapes. This approach emerges from the new significance of tourism as the largest industry in the world economy and from increased international mobility. The book's broad approach will garner interest from social scientists and humanists alike, who are interested in contemporary debates about place studies, mobility, and the located realities of globalization.

Carolyn Cartier is Associate Professor of Geography at the University of Southern California.

Alan A. Lew is Professor and Chair of the Department of Geography, Planning and Recreation at Northern Arizona University.

CRITICAL GEOGRAPHIES

Edited by **Tracey Skelton**, Lecturer in Geography, Loughborough University, and **Gill Valentine**, Professor of Geography, The University of Sheffield.

This series offers cutting-edge research organised into three themes of concepts, scale and transformations. It is aimed at upper-level undergraduates, research students and academics and will facilitate inter-disciplinary engagement between geography and other social sciences. It provides a forum for the innovative and vibrant debates which span the broad spectrum of this discipline.

Seductions of Place

Geographical perspectives on
globalization and touristed landscapes

**Edited by Carolyn Cartier and
Alan A. Lew**

Routledge
Taylor & Francis Group

LONDON AND NEW YORK

First published 2005 by Routledge
2 Park Square, Milton Park, Abingdon, Oxon OX14 4RN

Simultaneously published in the USA and Canada
by Routledge
270 Madison Ave., New York NY 10016

Routledge is an imprint of the Taylor & Francis Group

Editorial material and selection © 2005 Carolyn Cartier and Alan A. Lew
Individual chapters © 2005 the contributors

Typeset in Perpetua by
Taylor & Francis Books
Printed and bound in Great Britain by
TJ International Ltd, Padstow, Cornwall

British Library Cataloguing in Publication Data
A catalogue record for this book is available from the British Library

Library of Congress Cataloging in Publication Data
A catalog record for this book has been requested

ISBN 0–415–19218–8 (hbk)
ISBN 0–415–19219–6 (pbk)

Contents

Illustrations

Figures

Tables

Contributors

Jeff Baldwin is a Doctoral Candidate in Geography at the University of Oregon.

Carolyn Cartier is Associate Professor of Geography and Coordinator of the Urban and Global Studies Initiative at the University of Southern California.

T.C. Chang is Associate Professor of Geography at the National University of Singapore.

David Crouch is Professor of Cultural Geography, Tourism and Leisure at Derby University, UK, and Royal Geographical Society Visiting Professor of Geography and Tourism at the University of Karlstad, Sweden.

Jon Goss is Professor of Geography at the University of Hawai'i, Mānoa.

C. Michael Hall is Professor and Head of the Department of Tourism, University of Otago, Dunedin, New Zealand.

Anne-Marie d'Hauteserre is Senior Lecturer and Tourism Program Coordinator in the Department of Geography at the University of Waikato, New Zealand.

Shirlena Huang is Associate Professor of Geography and Head of the Department of Geography at the National University of Singapore.

Alan A. Lew is Professor of Geography and Chair of the Department of Geography, Planning and Recreation at Northern Arizona University.

Dean MacCannell is Professor of Landscape Architecture and Dean of the School of Landscape Architecture at the University of California, Davis.

Mary G. McDonald is Associate Professor of Geography at the University of Hawai'i, Mānoa.

Fung Mei Sarah Li is a Doctoral Candidate in the Tourism Programme at Murdoch University, Western Australia.

Claudio Minca is Chair in Geography at the University of Newcastle, UK.

Janet Henshall Momsen is Professor of Geography at the University of California, Davis.

Tim Oakes is Associate Professor of Geography at the University of Colorado, Boulder.

Ginger Smith is Associate Dean, College of Professional Studies, and Associate Professor of Tourism Studies, School of Business and Public Management, The George Washington University, Washington, DC.

Trevor H. B. Sofield holds the Foundation Professorial Chair in the Tourism Programme at the University of Tasmania, and is also Adjunct Professor in the Tourism Programme at Murdoch University, Western Australia.

Margaret Byrne Swain is Associate Adjunct Professor of Anthropology and Co-Director of the Women's Resources and Research Center at the University of California, Davis.

Ning Wang is Professor of Sociology at Zhongshan University (Sun Yat-Sen University), Guangzhou, People's Republic of China.

Alan Wong is Lecturer in the School of Hotel and Tourism Management at the Hong Kong Polytechnic University.

Acknowledgments

Seductions of Place, an interdisciplinary project addressing the challenges of transcending two fields of scholarly research, tourism studies and cultural geography, presents perspectives on understanding the cultural–economic conditions of places through tourism, in which aspects of tourism figure as a complex lens through which to assess multiple conditions of landscape formation. The scope of the project absorbed considerable time and text space. We are consequently especially grateful to Routledge and our editorial team in the London office for their support, patience and encouragement. In particular we thank Andrew Mould, our editor, and Ann Michael, Anna Somerville, Barbara Duke, Faye Kaliszczak and Marianne Bulman, who guided the production of the book. Our institutional appreciation also extends to our departments, at the University of Southern California and Northern Arizona University, to the National University of Singapore, where we initially conceived the project, and to the Association of American Geographers, at whose meetings many of our authors initially presented papers that form chapters of this book.

Our contributing authors worked generously to adopt the vision we sketched and the rather unorthodox approach to subject matter we encouraged: write about places where you have lived. We sought perspectives from scholars who think critically about cultural geography and tourism and who have experienced the places and palimpsests of landscapes wrought by tourism's presence across the paths of daily life, and because we are committed to the position David Crouch takes, in the first conceptual essay of the book, that the practice of tourism is regularly inseparable from local leisure and its embodied contexts. We are especially grateful to Dean MacCannell for sharing interests with us, and as his own critical work on culture and tourism verges into its fourth decade of relevance and inspiration.

We acknowledge too people whose support made the book possible in particular ways. I owe in part my interests in leisure/tourism to an intimate group of family and friends who shared with me the districts and cultures of San Francisco. I have never known a better booster of the City than my mother, who grew up there as Helen Boller. She allowed only two other cities (New York and Paris) into the astral orbit of her own, and I finally got the point after leaving the Bay Area to take up academic jobs. My friends also sheltered me on return furloughs from the labor migration, especially Bob Burnett and Kim Rhody, who shared the lures of the city not to mention tickets to 49ers playoff games. Yet this project gelled thousands of miles away during a research visit and discussions with Alan in Singapore, where the distinctively accomplished geography faculty works largely on local issues. For me, research typically means boarding a 747, but there, along the margin of the South China Sea, I realized I could write about home.

Alan would like to thank his administrative assistant, Debbie Martin, for her ongoing support of his research efforts; his graduate assistants for their help with classes while this project was going on; his children Lauren, a budding scholar in her own right, Chynna, and Skyland, for allowing their Dad space to work at home; and the constant and devoted support of his wife, Mable. In geography at USC, doctoral student Jacqueline Holzer ably assisted with editorial matters on chapter drafts and references, and Billie Shotlow supervised documents deliveries between Los Angeles and London. Ou ultimate thanks are, again, to our contributors for sharing the motivation to work at the interstices of fields, the reliable sites of new questions–if not paradigms.

Carolyn Cartier, Los Angeles
Alan A. Lew, Flagstaff, Arizona

1   Introduction

Touristed landscapes/seductions of place

Carolyn Cartier

Sometimes it seems like nearly any subject of difference or sensory distinction becomes the object of tourism. In Americus, Georgia, Habitat for Humanity International opened the Global Village and Discovery Center, featuring 'authentic' slum housing from around the world (except tents and shopping carts of the US homeless). In Newfoundland, Canada, local boat operators lobby to maintain iceberg tourism in the face of commercial 'berg harvesters who make vodka from the frozen waters, claiming its purity moderates hangovers. China's Guangdong Province, historically famed for culinary indulgences, issued a ban on the leisure practice of eating exotic animals after the SARS epidemic. In the US Midwest, Europeans sign on for twister tours, pack into minivans and race hundreds of miles around the Plains states in hopes of catching sight of a class 5 tornado. *Wired* reports on

> the hacker tourist [who] ventures forth across the wide… meatspace of three continents [encountering] the exotic manhole villagers of Thailand, the U-turn tunnelers of the Nile Delta, the cable nomads of Lan Tao Island, the slack control wizards of Chelmsford, and subterranean ex-telegraphers of Cornwall and other previously unknown and unchronicled folks, [all engaged in] laying the longest wire on earth.
>
> (Stephenson 1996)

During the course of this project, we realized from reactions to its main title that people thought the book might be about such forms of extreme and unique tourism. Or that it would focus on places like the night markets of Bangkok, or the Rosse Buurt of Amsterdam. The subject is really more fundamental: what places, and formative processes of place, generate and sustain significant desire, what are their material landscape qualities, and how should we theorize and narrate their conditions? What we learned is that these places, their landscapes, and even their histories, are dynamic and contested, changing in relation to transformations in society and economy. These are places of complexity, in some cases landscape palimpsests of great historical depth, and whose draw owes to multiple sites of possible experience and sensory encounter.

The project's broad conceptual interest draws closer engagements between tourism studies and the heart of innovations in contemporary human geography. We are interested in lines of intersection between tourism and questions raised in the new cultural geography, and with a vision toward economic geography in its related 'cultural turns.' In the way that tourism, as a set of service industries, is the largest in the world economy, tourism implicates a full range of questions about culture and political economy in an era of globalization, and so must embrace issues beyond its traditionally more empirical areas of inquiry. Thus the wider ideational scope of the project explores questions at the crossroads of contemporary issues in cultural studies of travel as well as economies of tourism, theoretical currents as a consequence of the poststructural shift in social theory, and the complex cultural, political, economic, and environmental conditions at stake in places as a result of processes of globalization. Situating the ideas in the larger arena of globalization reflects too the rise of so-called mass tourism with the restructuring of the world economy in the emergent period of late or disorganized capitalism, since the 1970s. In other words, we have also consciously written against the historical stereotypes that work on tourism geography generally has been less developed than other fields (perhaps because its conceptual orientation stopped short of giving full play to its array of contingent processual issues and complexities; or because its practitioners were accused of succumbing to the lure of travel!). Two central ideas orient the project and suggest these complexities: the tourist*ed* landscape and seductions of place. Landscape and place, two of geography's fundamental concepts, are closely interrelated and also carry their own contested intellectual histories.

The tourist*ed* landscape

The idea of landscape typically concerns visual qualities of landscapes and their representations, in designs, plans, paintings, and imaginaries about how landscapes *should* look (cf. Casey 2002, Cosgrove and Daniels 1988, Meinig 1979, Mitchell 2002, Rose 1993, Schama 1995), no less in the mind of the tourist. This project recognizes the significance of the visual, yet is more interested in uncovering its formative conditions and ways the visual can 'mystify' political ideologies and relations of production (Cosgrove 1984a). So we are not transfixed by the visual and, moreover, like David Crouch (1999b), see the need to move beyond emphasis on textual representations (e.g. Barnes and Duncan 1992, Duncan and Duncan 1988a, Duncan 1990, Duncan and Ley 1993) to favor more varied sensibilities in explaining emplaced experiences. We are interested in the complexity of landscape formation, in who designs landscape and why; in landscape experience as a multi-sensory, located subjectivity, including memories about it; and in impacts on landscape, understood through perspectives on nature–society relations.

This scope of interest points to examination of places whose larger, a priori significance arguably initiates desire to experience, tour, travel, and explore, rather than those where tourism economies have been explicitly created. Such tourist*ed* landscapes are found less in 'tourist towns', theme parks, and 'holiday destinations'; their sites are more likely in cities and regions of diverse purposes and meanings, and natural environments, whose integrity, as beaches, mountains, rivers, and oceans, fundamentally orient their geographies. So we use 'touristed' to signal that tourists significantly patronize these landscapes but that their formation has not fundamentally owed to the culture and economy of those who pass through. In these ways, touristed landscapes, and as places, represent an array of experiences and goals acted out by diverse people in locales that are subject to tourism but which are also places of historic and integral meaning, where 'leisure/tourism' (Crouch 1999a) economies are also local economies, and where people are engaged in diverse aspects of daily life. In a move relatively uncommon in contemporary cultural geography (but see Olwing 2002), we include both urban and 'natural' environments in this project, suggesting that work on landscape, and especially the touristed landscape, reintegrate critically how nature–society relations, whether material or as geographical imaginaries, remain central to processes of landscape formation.

The idea of the tourist*ed* landscape then concerns the possibilities of understanding landscape as toured and lived, places visited by their own residents, the dialectic of moving in and out of 'being a tourist.' Touristed landscapes are about complexity of different people doing different things, locals and visitors, sojourners and residents, locals becoming visitors, sojourners becoming residents, residents 'being tourists,' travelers denying being tourists: resident part-time tourists, tourists working hard to fit in as if locals. These pivoting and juxtaposed positions take inspiration from Dean MacCannell's (1976a) recognition that "we are all tourists." And so the project recognizes the messiness of tourism as category of activity, experience, and economy. In the touristed landscape people occupy simultaneous or sequential if sometimes conflicted positions of orientation toward landscape experience and place consumption. This kind of landscape necessarily reflects histories of travel and mobility, relations between local, national, and global economies, the possibilities for different identity positions, and the environmental contexts, built and natural, of places and sites. While landscape studies in the new cultural geography have taken account of different and contested identity dimensions in landscape formation as well as their representational qualities, including perspectives on race, ethnicity, class, gender, sex, and sexuality (cf. Jackson 1989, D. Mitchell 2000), the touristed landscape consciously seeks to complicate these positions by signaling multiple and shifting points of view in the context of leisure economy production and consumption.

Landscape and place

In these landscapes, and the places they constitute and represent, interest in experiencing them reflects aspects of desire as well as multiple positions of sensory engagement, attraction, and legibility—ways in which landscapes can be read, imagined, and experienced, from diverse points of view and positions of orientation. As Lucy Lippard (1999: 52) has put it, "Finally, after the pictorial seduction, people flock to places not because of their beauty but because of their promise." This landscape is the place where locals and visitors negotiate identity, seeking renewal or exploration, the possibilities of alterity and liminal identity shift. Understanding experiences and meanings of the touristed landscape depends in part, then, on understanding subject positions and subject formations of the touring, the toured, and those who would work at being both or neither, and from one moment or place to the next. In such contexts, ideas people hold about places substantially inform identity formation, human agency, and questions of subjectivity.

Our emphasis on the multiple conditions of landscape reflects the concept of place, and its different theorizations and intellectual lineages in social theory and phenomenological views. Where landscape concerns distinctive forms, features, and assemblages of natural and built environments, and their representations, the concept of place is differentiated space, broadly encompassing diverse aspects of locales (Agnew 1987). It sometimes stands in as a synonym for landscape, but is really the broader concept; landscape is also place, landscapes make up places, and places may include diverse landscapes. We focus on touristed landscapes because the mobile subject tends to seek particular scenes of leisure/tourism activity, while emphasis on place reflects geographical imaginaries, how people think about destinations, and from multiple points of perspective and distance.

Comparative conceptualizations of place lend greater interpretive meaning to place/landscape desire and experience. Contemporary interpretations of place theorize it in the context of its processes of formation through social relations: how space is transformed by human activity and how those activities connect to diverse other places. The idea of place as "social relations stretched out" and the "spatial reach of social relations" (Massey 1994, 1995) is compelling in a contemporary world buffeted by global forces; place becomes dynamic, contested, and multiple in its symbolic qualities and representative identity positions. This view of place challenges traditional views that interpret place as simply fixed and located, and in relation primarily to humanistic ideas about sentiment, meaning, and place attachment, as in 'sense of place.' But such differing views may be reasonably transcended if we understand place through contemporary theory on the body in geography and philosophy.

Feminist geography has made the discipline's substantial contributions to theorizing the body, including the fundamentally emplaced conditions of embodiment (e.g. Duncan 1996, McDowell 1999, Nast and Pile 1998, Pile 1996) and how sensory positionality influences perceptions of landscape, through critiques of the male gaze (e.g. Nash 1996, Rose 1993). These issues have also been the subject of so-called representational geographies (Thrift 1996). Contemporary philosophers have developed the concept of place in relation to the body as the ontological basis of human existence (Casey 1996, 1997, Malpas 1999). For example, Malpas (1999: 176) finds identity formation and human subjectivity as "necessarily embedded in place, and in spatialised, embodied activity." This view interrelates concepts of agency, spatiality, and experience, in which embodiment is "one's extended, differentiated location in space…[and] essential to the possibility of agency and so to experience and thought" (133). It follows that to be embodied is to be emplaced. Historically constituted, these relations are both phenomenological and material, embracing the spatial reach of social relations.

Seductions of place

Following our understanding of place, place seductions must be situated subjectivities and emplaced experiences, inviting encounters with touristed landscapes, scenes where people seek particular aspects of attraction, desire, and possibilities for liminal experience. Seductiveness of place differs for visitors and residents, across genders and sexualities, and so touristed landscapes are places simultaneously perceived, formed, and reworked by activities of diverse people and their uneven ties to arenas of the local/regional/national/global. The basis for seduction lies in multiple positions of legibility.

But the psychology of seduction is also about illegibility: *mundus vult decipi* (the world wants to be deceived). Sensory modes beyond the visual may be more elusive, qualities that are aural, haptic, flavorful, olfactory. What stimulates these senses may be fleeting; we might own the visual environment via the gaze, but sounds, tastes, smells have their temporal limits. Seduction's psychological orientation also asks us to consider contradictions of tourist imagining, anticipation, and memory, which suggest its tensions and illegibility. In the words of Jane Miller (1991: 21), "the word seduction presides somewhat willfully over chains and clusters of meanings which are contradictory, tautologically driven and embedded in and issuing out of a number of worlds and variegated accounts and visions of those worlds." Seduction's tension works through discourses and their interpretations, and how we map them onto the material world of landscapes, in seeking access to desirable place characteristics. We might also treat seduction as a form of knowledge, as awareness or promise of the potential for experience. Tension is implicit in this potential. It is the seduction of place, implying the unknowns of the journey, embodied movement, to travel to a place, encounter its landscapes, and open up to its possibilities of experience.

In seduction…it is somehow the manifest discourse, the most 'superficial' aspect of discourse, which acts upon the underlying prohibition (conscious or unconscious) in order to nullify it and to substitute for it the charms and traps of appearances. Appearances, which are not at all frivolous, are the site of play and chance taking the site of a passion for diversion—to seduce signs is here far more important than the emergence of any truth.

(Baudrillard 1988c: 149)

Among contemporary social theorists, Jean Baudrillard substantially considers seduction and as a basis for his critiques of structural theory. Baudrillard's seduction engages the surface, highlighting the postmodern interpretation of visual spectacle and prefiguring his interest in the 'hyperreal.' As Norbert Bolz (1998: 1) writes, "bring on the beautiful illusions. Yet simulacra, as such would have it, don't actually deceive: they seduce." And if seduction is a surface effect, the analytical complement is to "look for the meaning of the surface, as well as the meaning within the surface." As if it is all surface. In *Seduction*, Baudrillard's (1990) central subjects are in some ways predictable, based in aspects of heterosexual attraction and desire and the male gaze. Critiques of the work have zeroed in on its patriarchal priorities and scopophilia, even "an attack on feminism" (Kellner 1989: 143). However, in the context of social relations, Baudrillard proposes seduction as a form of information and alternative to production and symbolic exchange as primary orienting social forces. We cull from such broader suggestions to pursue different ways seduction works out in sensory/embodied/emplaced possibilities.

Baudrillard's apparent preferred sense of seduction is nostalgic, before the era of the hyperreal and the simulacrum, i.e. "the aura of secrecy produced by weightless, artificial signs." We know that the built environment at the basis of his signs is Las Vegas:

No charm, no seduction in all this. Seduction is elsewhere, in Italy, in certain landscapes that have become paintings, as culturalized and refined in their design as the cities and museums that house them. Circumscribed, traced-out, highly seductive spaces where meaning, at these heights of luxury, has finally become adornment. It is exactly the reverse here: there is no seduction, but there is absolute fascination—the fascination of the very disappearance of the aesthetic and critical forms of life in the irradiation of the objectless neutrality.

(Baudrillard 1988a: 124)

Here, Baudrillard himself appears seduced by imaginations of authenticity, constructed memories of a sensory historic culture, before the industrial era compelled regimented social time, before hyperreal surfaces subsumed refined substance. Claudio Minca's essay on the 'two Bellagios' demonstrates how Italian landscapes continue to convey this kind of 'traditional' seduction for the global professional class. But such nostalgia for past regimes can neither define seduction nor claim its authenticity; seduction must be continually constituted, socially, spatially, and discursively, reflecting transformative changes in culture, society, and economy.

Among a wider range of subjects, Baudrillard (1990) agrees that the symbolic elements of seduction must vary broadly. These include wealth and power, and imaginations about them, in political space: "Since Machiavelli politicians have perhaps always known that the mastery of a *simulated* space is the source of power, that the political is not a *real* activity or space, but a simulation model, whose manifestations are simply achieved effects" (Baudrillard 1988c: 158). Among the essays, those by Margaret Byrne Swain and Trevor Sofield and Sarah Fung Mei Li both show how the state in China reworks representations of touristed, symbolic space in order to support state ideological perspectives on national identity and nation-building.

We can map diverse touristed landscapes onto Baudrillard's (1990) themes. Symbolic elements of seduction are also to be found in places that represent luck, fortune, and speculation (from temples to gambling houses), landscapes that challenge ('natural' environments, for urbanites; Everest as tourism) as well as in sites that are new and spectacular (the 'Bilbao effect'). Such places present to visitors the lure of potential, the tension of possible outcomes. Secrets and enigmatic places are seductive too: archaeological landscapes from the edges of history, the Egyptian pyramids, Mayan capitals, the mounds of imperial graves that mark China's millennial history, still unopened, too many to excavate. Baudrillard also finds seduction in the symbolic order of things, collections and their arrangement (but not in collecting itself—the antithesis of seduction), implicating museums, libraries, and archives of all types. This is a concentration effect, which only urban institutions can produce and accommodate. The seduction of technology is "cool seduction," "the 'narcissistic' spell of electronic and information systems, the cold attraction of the terminals and mediums that we have become, surrounded as we are by consoles, isolated and seduced by their manipulation" (162). Here too tension locates, in what MacCannell (this volume) identifies as "the triangle…that connects keyboard, eyes, and monitor," "the most meaningful space in Silicon Valley." *The Ecstasy of Communication* (Baudrillard 1988b) continues the theme of cool seduction, in how we surrender to the seductive power of mass media, whether via the TV, the movie screen, the computer monitor—the full range of surfaces.

In *The Consumer Society*, seduction finds its extended role in the continued growth of services, which results not only in ever increasing specialization but in increasing personalization. We find this "smallest margin of difference" (Baudrillard 1998: 87–98) commonly in leisure/tourism services, such as in an advertising campaign for Princess Cruises—"personal choice cruising"—in which the female voice-over on the TV commercial begins line after line with, "I want…," concluding with "I want personal choice cruising." This logic of personalization "*ushers in the reign of differentiation*" (89, italics in original) in which products of mass consumption are apparently tailored for the individual, thereby encouraging endless seduction for limitless consumption. This strategy is working for the cruise industry, which, as Janet Momsen reports in her essay on the Caribbean, is distinctively increasing capacity.

The consumer society's increasing individualization takes especially embodied forms, in all the ways that personalization implies. But "The narcissism of the individual in consumer society *is not an enjoyment of singularity;* it is *a refraction of collective features*" (Baudrillard 1998: 95). So we learn, as Ginger Smith's essay shows, that international body-oriented tourism is one of the fastest growing niche markets. Finally, Baudrillard (151–8) finds any remaining democracy of the consumer society in the myth of leisure as 'free time,' which, he notes, is especially promoted by the corporate sun, sand, and sea vacation industry. Given leisure time's simultaneous evaporation for the majority of the world's working population and its increasing proliferation of services for the professional classes, leisure may be the ultimate seduction of time/space.

The essays

Interventions

This project set up a number of interrelated intellectual problems: how to reread landscape as tourist*ed*; ways of retheorizing the relationship between tourists and the toured, 'travelers' and 'locals,' and who is a 'tourist'; what constitutes place seduction (as opposed to sense of place) and landscape desire; and the context of these meanings, processes, and formations in an era of (dramatically uneven) globalization. We also sought to assess the global tourism economy, its empirical contexts as well as travel trends, to ground the significance of travel as an economic activity. In the face of the industry's enormity, we have wondered at mainstream geography's relative lack of engagement with 'tourism geography'; but the greater dilemma is that economic activity constituting tourism and travel is especially difficult to define and measure: the umbrella industry involves a full range of service industries not all directly involved in 'travel' and patronized by diverse people, not all 'tourists.' This problem—identifying what is 'tourist'— emerges as a central theme in this first part of the book. Its four essays explore the limits of prevailing assumptions and ideas about who are tourists and travelers, what they do, what and how they desire and consume, what mobility means, and how tourism and travel register in economic activity. The conceptual interventions in these chapters coalesce around refining and redefining ideas about the tourist and tourist activity. Whether in regard to agency, activity or practice, consumption or economic impacts, their authors converge on rethinking 'tourism' at the interface of mobility/movement, sensory experience, and consumption. So at more abstract levels of conceptualization, central processes constituting 'tourism' are sometimes about (local) movement(s) and not always about conventional ideas of travel and mobility, they are profoundly multi-sensory and need not depend on the 'tourist gaze' (cf. Urry 1990, 2002), and, in the consumption of goods, services, experiences, and landscapes, they transcend the tourist/traveler/resident dilemma.

What are the realities of living through the tourist experience, of being a tourist in place/space and in the landscape? David Crouch briefly evaluates then puts aside accepted ideas about 'tourists' as a category of subjects and agents, and assumptions about the power of the industry to construct tourist consumption. He instead asks us to proceed in terms of people 'doing tourism,' a process of self-reflective encounter in space, with people, and while negotiating expectations, experiences, and desire. Concerned with nuanced aspects of the tourist's exploratory performativity (to wander, to hear an interesting sound and turn this way instead of that), Crouch sees tourism desire as a process of seductive encounter, how the tourist 'flirts with space.' In the process of movement and consuming space, Crouch pursues strategies of desire not through the symbolic power of 'signs' but through a substantially more complex process of seduction, a personal, embodied semiotics of doing diverse things, creating 'tourism in process.' He moves beyond geographies of representation to geographies of embodied experience and practice—'doing tourism'—through at least four dimensions: multi-sensuousness, intersubjectivity, expressive embodiment, and subjectivity as process. Such dimensionality arguably provides the conceptual ground for pursuing a more empirically refined experience of tourism as practice and process.

Of course the problem of the subject remains at the forefront among important theoretical issues emerging from the postmodernism/feminism debates of the 1980s, and the chapters by both David Crouch and Tim Oakes bring feminist readings to questions of subjectivity. Place context is the focus of Oakes' precise rereading of modern tourist subjectivity, as originally theorized by MacCannell (1976) in terms of the mobile traveler in search of staged authenticity and meaning. Oakes' vehicle is *The Sheltering Sky*, in which, in the course of a long journey across North Africa, a woman embraces an unpredictable sexual encounter as a seduction of place, in the region and home of the 'other.' The interpretation suggests how we might look carefully at a dialectical subjectivity-in-formation vis-à-vis encounters in place and dynamic place formation. Here, seductions are reflexively constituted placed encounters between traveler and other, the essence of modern subjectivity.

In the next essay Jon Goss theorizes the 'tourist mode of consumption' in which the signature commodity, the souvenir, embodies meanings of place and landscape that are distinctively estranged from their historical and geographical origins. Here we find the potential for allegory, in which, materially, the souvenir is modernity's refiguring of the commodity and its capacity to effect memory and nostalgic desire for lost people and places, yet ever symbolically bound to located landscapes of departed persons and ruined pasts. Goss reflects on historic land-scapes in Hawai'i, about war, battle, fallen kings, and the loss of indigenous culture. Landscape allegories of these sites tell about the life journey, inevitably wasting away toward death, but, as the contemporary landscape fundamentally portrays, demonstrating the ultimate lifepath seduction—the possibility for resurrection. Goss reminds us that even typically framed liminal and ludic land-scapes have their consumptive histories and geographies, a fundamental seduction of place in which the symbolic landscape tears at awareness of the sensory limits of mortality.

The first part concludes with an assessment of the world tourism industry, an umbrella category of diverse service industries. Ginger Smith sees the global tourism economy as a microcosm and reflection of the world economy in the information age, as technology, information, and capital converge in new and increased flows of information, people, and things. Smith presents a framework for understanding tourism through an 'integrative model of international rela-tions,' which embeds complex networks of tourist services in the global economy. The discussion considers tourism in relation to the expansion of finan-cial services, which parallels the evolution of so-called mass tourism, and in relation to impacts of wars and terrorism, especially September 11. Rather than simply damping leisure travel, we can see how events of 2001 effect shifts in types of tourist activity toward achieving more intimate connections with people/places and health and well-being, especially medical tourism. Smith's analysis also considers the lack of substantial scholarship on the 'industry': it is conceptually difficult to study systematically because its services accounting, from various modes of transport to restaurants, makes no distinction between visitors and residents. Thus we have the 'travel and leisure industry'—and another reason why we are all tourists.

The city

The city is arguably the most reliable place of touristed landscapes: it is the scene of complexity where identifying the bounds of what is 'tourist' culture and economy remains elusive because of its embeddedness in local life. In the urban scene the transition from work to leisure takes immediate forms, in moments and as gestured from one turn to the next, on the street, in restaurants, bars, clubs, the 'dressing rooms' of department stores, through a variety of commercialized personal services; in essence, consumption of all types. Yet as Ning Wang notes, except for some major (tourist) cities, such as London and Paris, cities in general, and with a few exceptions (e.g. Ashworth 1990, Judd and Fainstein 1999, Law 2002), have not reliably been the focus of substantial tourism inquiry. Even if we could shift the focus of tourism research to the city, what we are interested in are multi-valent perspectives, ones that recognize the city as both the place of "escape attempts" (Cohen and Taylor 1992) and the "place of escape" (Amin and Thrift 2002).

From a historic perspective on cultural economy, the seduction of the city is about surplus, excess, in a word, luxury, and the conditions and contradictions of modernity it engenders. The city is the vortex for the concentration of things and affords the most extreme views of consumption. For Sombart (1967: 27), the pursuit of luxury was centrally at stake in the evolution of early modern capitalism and urbanization; "the early development of the cities [was] based on the concentration of consumption in them." Each century had its cities of luxury consumption. In sixteenth-century Europe, "it seems that Madrid, next to Rome, was the first modern city to have an appreciable influx of pleasure-bent foreign visitors" (*ibid.*). And Sombart (1967: 60) draws for us too direct connection to the sensory: "All personal luxury springs from purely sensuous pleasure. Anything that charms the eye, the ear, the nose, the palate, or the touch, tends to find an ever more perfect expression in objects of daily use." Then there is consumption's psychological operative: how the thrill of consumption peaks subsequent to acquisition and then downtrends, 'the half-life of consumer satisfaction.' Consumption is a widely promoted and sanctioned addiction. Each of the chapters on the city deals with aspects of modernity's cultural economy, capital accumulation and investment, and consumption.

In the 1990s the 'Silicon Valley' boomed again with the dot.com phenomenon, attracting tens of thousands more tech professionals from around the world. A large group of engineers from India formed one new prosperous community; on another end of the migrant spectrum, landlords, impelled by sky-rocketing housing costs, divided up three-bedroom tract houses into warrens of rooms, found in Sunnyvale and occupied by Chinese renters in the double digits (Goodell 2000). Silicon Valley in the second half of the 1990s was an ultimate seduction of place, one recent version of the modern dream of getting rich quick. But international tourists to this center of the world economy do not have an easy time of it: it is not clear where to go. We can see the dilemma at Stanford University's oval, where busloads of Asian tourists regularly debouche and assess the photo options. Large institutional buildings loom above manicured lawns; though it is true that engineers in the early industry were on the Stanford faculty, nothing about the scene suggests the *garragiste* origins of the region's computing and electronics industries, and the university long antedated the industry. But tourists on tour typically need to be brought to monumental sites for their inherent ability to convey significance. And so as Dean MacCannell so perceptively negotiates, 'place,' in the usual sense of significant geographies of recognition, is absent from Silicon Valley. Instead, he argues, the region's signature industry, of indisputably global proportions, values primarily powers of miniaturization and speed—and as valued ultimately in money. How these values transfer to the landscape are, by conventional modes of landscape assessment, unreliable and illegible. We can see strip malls, 'monster houses,' and campus-style office parks, but the Valley does not present a *public* landscape of wealth and power. Instead, the seduction of miniaturization and speed have led to a culture space of the computer monitor, in which people relate to technology, its images, and imaginaries, which, in turn, have arguably influenced a new 'extreme leisure' of alternative 'sports' distinguished by rapid, quirky rhythms, always seeking 'the edge.' Significant disposable income is directed toward the requisite equipment and the newest model, not unlike the Valley's mantra, what is 'the next big thing'?

Las Vegas might be an all too predictable place in which to assess urban tourism, but how does it represent the touristed landscape, in which the landscape is the scene of interrelations between locals and residents, a place of historic and differentiated meaning? Isn't the Las Vegas condition largely a one-way flow, of tourists to the town, resources to the desert, the most ersatz city in the United States—if not the world? Claudio Minca authorizes a distinctive reading of the touristed landscape by revealing a dialectic of place formation between the authentic and the inspired copy, the real and representation of the representation: Bellagio, Italy and the Bellagio hotel development, the most successful resort in Las Vegas. How can it be that the hotel put the *place* on the world map of the international professional class? Italians have had to cope with the Bellagio (hotel) making a small lake town of historic villas into a world-recognized symbol of elite leisure experience—its developers even marginalizing the authentic in the process by visiting the town, gaining knowledge of place from local planners, and then colonizing the website, www.bellagio.com (not unlike uncompensated corporate appropriation of indigenous resources). Minca critically evaluates the Bellagini response to the phenomenon, which understands how especially American tourists, having absorbed the Bellagio-sign in imaginaries of status travel, come to Bellagio to see themselves within the space of a 'post-A' lifestyle in which having made it is performed through a distinctively staged, leisurely leisure; they sit around and observe themselves, hardly interacting at all. No wonder when tourists visit the real thing they do not know how to engage: the *hotel*-in-creating-Bellagio has utterly de-localized connections to place-centered subject formation, creating instead scenes for playing out a commodified global lifestyle in the space of a controlled, privatized *qua*-place.

Recent scholarship on 'the tourist city' (Judd and Fainstein 1999) recognizes tourism's service industry assemblage as a basis for attracting investment capital. The bulk of capital flow in the neoliberal era is between world cities, and it has been the role of urban elites to attract investment for high-profile projects—as if cities themselves are engaged in a global competition for capital, status, and power. Michael Hall's ideas about seduction of place in Melbourne and Sydney reflect these processes and the ways that the state–capital alliance seeks to 'seduce' capital and engage in 'urban imaging' and 'branding' of the city. In both cities, the state is promoting new leisure-oriented waterfront development, replacing industrial and low-income landscapes. Through targeted projects, including Sydney's successful bid for the 2000 Summer Olympics, Hall shows how the state has sought to reimage not only particular landscapes, attracting both tourists and locals, but how it has also worked to promote lifestyle, in which local places defer to a kind of global template of daily life consumption for the professional class. The question remains, how sustainable is such constructed, packaged, and branded 'seduction'?

China's urban tradition is the world's longest, and its cities are repositories of ancient history as well as centers of globalizing modernity. Ning Wang's chapter on China's southern city of Guangzhou, known as 'one step ahead in China under reform,' demonstrates how the modernizing city itself is a place of significant attraction, as well as a showcase for China's modernization project. Wang describes the dramatic changes in the city, where, in the early 1980s, the first new world-class hotel became, even for locals, a significant touristed site/sight. Now a major high-rise city, Guangzhou has become an accessible place of consumer desire—a veritable landscape of spectacular shopping malls—and also a magnet for visitors and migrants from rural areas. Wang argues for the importance of treating the city as a regular focus of tourism analysis, and offers one model for doing so by assessing frequency of tourist visits to Guangzhou's historic sites as a baseline for determining what kinds of landscapes draw people to the city.

San Francisco may be 'everybody's favorite city,' but accounting for it is another story. My chapter on San Francisco assesses the complexity of the touristed landscape in this city of superlatives, an archetypical seduction of place with a fully elaborated touristic complex integrated in diverse districts of the city. The analysis recovers both ideational and material aspects of San Francisco's evolution, and reads its landscapes through perspectives on the maritime cultural economy, values of urban environmentalism, and cultures of masculinity and the gay community. The essay argues that explaining the city's transhistorical draw depends on cutting a wide arc across the local social formation, developed here as the 'spectacle of environment,' and its relation to what is elsewhere in the United States labeled the 'left coast.' This magnetic economic region and its globally distinctive industries combine with a culture of alternative and leading-edge urban lifestyle practices to produce a place seduction that simultaneously represents the height of normative desire economically and anything but normative lifestyle boundaries culturally and politically. This cultural economic tension lends to all those going to San Francisco the psychological opportunity for individual participation in the discourses of difference and greatness which also shape the city's leisure/tourism narratives.

The beach

Seduction of the beach is about the sensory allure of 'sun, sand, and sea,' and more. It is about the interface of continents and oceans, the potential fear of being faced with the enormity of a water planet, the seduction of being at ends, of earth, land, and the imprint of urban life. It is about essences, of sand as the earth's constituent or residual matter, as it dissolves into the shallows of oceanic vastness. It is about the simultaneity of the sensory: sound of the waves, the sight of the surging water continually resurfacing the beach, the feel of the sand on the body, the smell of the marine, the taste of salt in the air. Sun hats, umbrellas, lounge chairs, bathing suits, buckets, shovels, surfboards, these are things used to negotiate sensory being at the juncture of the extra-terrestrial; they distract and make us focus on how to do leisure in a zone historically feared. (We are drawn to edges: the narrative power of looking over the cliff, the roof of a tall building; wanting a seat at the side of the pool, the corner of a dining room, a corner office.) The beach, in tropical and temperate zones, is arguably the most seductively powerful, yet accessible, natural site. (We might argue for the seductive power of tops of mountains, as pilgrimage sites, reaching them as feats of triumph, but the mountaintop, in its extreme in the Himalayas, is a moment for a few and there is nothing subtle about it.) The world population is distributed and settled most densely in coastal regions, for reasons that have been historically practical and economic as well as they have also been in more recent times cultural and political.

But the potential democracy of the beach is contested. In any number of urban coastal regions the oceanfront site is an elite address; critical planning histories document 'innovations' that kept urban public transit from reaching the beach, both in New York City (overpasses too low for buses) and in Los Angeles (ripping out the east–west street-car lines). Along the California coast, the most expensive housing is in Santa Barbara (after the 2001 stock market and dot.com downturns impacted San Francisco and environs), and planners there have found "the middle class is being driven out of an entire region…a danger…eventually for the entire California coast" (Overend 2003). This perspective from southern California is also an indicator of the price put on comparatively high residential environmental quality in the coastal zone. In one of the ultimate landscape gestures, Bertrand Delanoe, Mayor of Paris, has reminded us that the annual sun, sand, and sea vacation is not the 'mass' event that word's connotations imply; he has conceived the 'Paris-*plage*' for those whose economy precludes France's August vacation ritual. Its second annual iteration required 3,000 tons of sand hauled in to form the month-long 'urban beach' on two sides of the Seine, from the Tuilleries Gardens to Ile St. Louis. A huge success in Paris, this recovery of environment in the city has caught on around Europe, as Berlin, Budapest, and smaller cities around France have all embraced the ephemeral urban beach (Chambon 2003).

Then we have islands. If the beachscape is the ultimately seductive natural environment, then the island, the oceanic island, is that essence reduced, concentrated, in mythic form. Here seduction lies in a kind of tension, in inaccessibility, the island as refuge combined with its distance from metropolitan centers. One geographical history of modernity, in both Western and Asian traditions, is overcoming fear of the sea and crossing oceans (Cartier 2001: 41–5, Corbin 1994, Lencek and Bosker 1998); islands made that navigation ever more possible. In the tourist imagination, the island is the ultimate beach (even as the geomorphology of so many island coasts precludes substantial beach formation). The seduction of the island landscape, even more than the beachscape, is much more about myth than reality.

Most countries have their coasts and their islands; some countries are islands. Yet for islands at some distance from their metropolitan capitals, we see how countries-as-colonizers have had particular interests in islands, as windows on marine resources, as places of tropical plantation economies, and especially as remote extensions of sovereign power. The Hawaiian Islands have served this purpose for the United States, France has French Polynesia, the Caribbean was a kind of multi-colonial sea. These 'colonial window boxes,' past and present, are also crucibles of historic world political economy, places that remind us how the era of exploration and discovery turned into long-term territorial imperialism. All four essays in this part on the beach are also about islands.

For many American tourists, the US State of Hawai'i is the most desirable island destination: the environment is exotic and the Hawaiians are interestingly different but the currency of communication is English and the US dollar is the common denominator. It is like a reality theme park. It could be closer, but otherwise the Hawaiian vacation provides for many tourists a seamless transition. Mary McDonald's essay on Hawai'i inverts this equation by focusing on what we don't expect: Japanese wedding tourism. The Japanese tourist wedding in Hawai'i—in traditional Western style, colored white, with all the accessories—is the very one that North American couples are trying to escape. McDonald theorizes the growth of Japanese wedding tourism through desire for ritual fulfillment, enacted in liminal time–space. Her discriminating reading of the packaged, transnational corporate wedding economy reveals how these staged rituals are overwriting local meanings of place, even becoming NIMBY problems in residential landscapes while government officials are often invested in the lucrative industry.

Many islands are worthy of their legendary narratives: Bali for an extraordinary culture complex and its continuity in the face of mass tourism, Cuba for its adherence to socialist principles, Jamaica for the Rastafarian scene and its impact on globalizing pop culture. But arguably no island in the contemporary world has a myth like Tahiti. If Tahiti remains the iconic paradisical island in the touring imagination, Anne-Marie d'Hauteserre's account of the Society Islands reveals the power of myth to endure: hardly anybody goes there. Distant from all major population centers and its economy underwritten by France, Tahiti has a minimally developed tourism industry, strong local culture, and a certain sophistication about its political economic position in the zone of France's nuclear testing program. D'Hauteserre argues that Tahiti's myth is kept alive in part by tourism industry use of its metaphors to produce desire for sun, sand, and sea tourism worldwide.

Janet Momsen's essay on touristed landscapes of the Caribbean reorients the tourist gaze: visitors to the Caribbean seek views of the sea. Examining 'seascapes' shows how visitors are so rarely concerned with island societies and islands at large that resorts have increasingly become enclaves on the beach, and cruise tourism, i.e. just looking at islands in the sea, has increased. Momsen also disentangles tourist industry narratives about the Caribbean from local issues and perspectives, revealing for us alternative Caribbeans: where local male escorts serve women tourists, scenarios in which tourism density has already exceeded island resource capacities, and how environmental NGOs are seeking to regain public access to the beach. 'The Caribbean' is a leading world tourist brand but its seduction as buying power is being curtailed by environmental limits of the region.

The final essay on the beach presents a microcosm of the Caribbean condition through a political economy/ecology interpretation of the landscapes of resort development on Antigua. Jeff Baldwin's reading of Antiguan political economy demonstrates how even this long-established island destination is also the site of major enclave resort development, facilitated by government leaders and their alliances with transnational capital—as the state deploys discourses about jobs, race, and nationalism to deflect attention from their accumulation project. Baldwin foregrounds too the impacts of development on local communities, including sand mining and changes in mangrove subsistence systems, and how dissent and resistance emerge in calypso and protest actions. Passionate concern with place here lies in local concern for living landscapes.

'The Orient'

'The Orient,' as a complex and multi-sited regional place seduction, from the 'Middle East' (West Asia) to the 'Far East' (East Asia), is likely the outstanding Western construction of a regional cultural alterity in the modern period. If we think of cultural alterity in Clastres' (1994: 46) terms of a system of "inferiority on a hierarchical axis," the Orient is Europe's nearest 'other' (the Occident/the Orient); its claims on history, from philosophy, religion, and belief systems to economic and technological development, cities, and urban life, put the Orient at the top of any hierarchy the European world could construct. So the seduction of the Orient is about 'otherness' but also substantially about difference—desirable difference—a seduction of place constituted in desirability to know and command the Orient, to assert superiority over its superior traditions, and to have it by having its things, especially its luxury goods. The commodity trade in 'oriental' luxuries of the China trade, the silks, the tea, the porcelain, and ultimately the opium, engendered transcontinental trade routes before the era of sail turned the Pacific into the 'Spanish Lake' (Spate 1979), inspired colonial botanizing, transformed European domestic rituals, and caused several major wars. Through it all—the 'era of exploration and discovery' otherwise known as colonialism—it was Asia, the 'East Indies,' that centrally motivated the drive that laid the foundation for globalizing world trade. From the perspective of China, it was indeed this imperialist project and the desire for its products that transformed the country's nineteenth century into a series of unequal treaties and treaty ports, derailing it from certainly different paths of modernization and development. In this world order, China's contemporary rise in the international economy is overdue.

In a world regional map of 'othering,' ethnocentricity, cultural essentialism, and racism, Asian gets reduced to 'Chinese.' This is one version of Said's (1978: 3) "Orientalism as a Western style for dominating, restructuring, and having authority over the Orient." Historically, all maritime world powers were bent on access to the China trade. The size of China and its population, its worldwide diaspora, and now its prominence in the world economy together put China at the forefront of what is 'Asian' in the geographical imagination. The visibility of 'Chinatowns,' their apparently generic toponym (even as their Chinese names are different), and their role as tourism districts contribute to constructing common notions about 'Chineseness' and Chinese landscapes. For example, in Antigua the local nationalist anti-development discourse constructs as 'Chinese' a development project capitalized by a Malaysian company. Such constructions of regional images also suggest how ideas about regions work as geographical imaginaries, how people think of distant places: the greater the distance, the more likely ideas about places are imagined at larger scales, just as international travelers are identified as citizens of particular countries when abroad. 'America' is not frequently identified as America within the United States. 'Chinese' people in diaspora differentiate Chinese ethnicity based on dialects, provinces, and other such local characteristics. The regional place seduction also works by foregrounding alternative geographies lacking definitive political boundaries: 'the Orient' is more a regional idea than a defined region. Each of the four essays in this part examines aspects of 'orientalism' as the ways ideas about Asian landscapes and cultural practices have been constructed, contested, and negotiated. They all also deal with 'China' or 'Chineseness,' but not with any focus on tourism to China; instead the questions are about ways state and society interpret and reimage 'Chinese' touristed landscapes, dimensions of China's 'internal orientalism' (Schein 1997) in the production of ethnic tourist sites, and the significance of the idea of China as a place of essential encounter for overseas Chinese.

The greatest tourists in China may be local and national: domestic tourism surged in China under reform as people gained new opportunities for widespread travel. Many of China's touristed landscapes represent particular goals of the industrialization and modernization drive, and its own citizens are interested to see them. Margaret Byrne Swain's study of tourism at the Stone Forest National Park and the local Sani people in Yunnan Province exemplifies the local state's developmental strategies in borderland regions. Like practices of colonial 'othering,' the state sexualized the myth of the Sani heroine Ashima and feminized the Sani to promote commercialized desire and popularity of the park. The site is an internationally significant karst landscape, but the state's representations are more interested in anthropomorphizing its rock formations in this region of 'exotic' peoples for tourism development.

China's outstanding iconographic monument, touristed or not, is the Great Wall. But as Trevor Sofield and Sarah Fung Mei Li reveal, the Great Wall is different than its constructed history, and it isn't even one wall. But Western writers have represented it as a singular monumental structure, thereby reducing its history and producing an orientalism of domination in which the condition of an Asian place is redefined from the perspective of a Western gaze; it is this one wall most international tourists expect to see. The modern Chinese state has also found the constructed idea of the unitary wall a useful symbol of national history and identity, representing the greatness of the Chinese people. Sofield and Li adopt a consciously postmodern perspective to both theoretically and metaphori-cally deconstruct their project: the singular wall comes down and instead the multiplicity of wall projects, their histories and meanings, rise instead.

Modernization of historic landscapes was practically a state obsession in Singapore until the city began to lose some of its cultural appeal, and in the process, its international tourism market. The case in point was Bugis Street, an outdoor nightspot famed for drag queens on parade, an utterly sensory seduction of place, a touristed landscape of fully embodied practice. It was well known internationally but a rather debauched scene and so the ever-moralizing state shut it down in the name of construction for a Mass Rapid Transit station. Then in the early 1990s the government sanctioned rebuilding a Bugis Street site near the original, including commemorative signboards and occasional workers dressed like transvestites. Singaporeans had had enough. When the state sought to trans-form the historic Chinatown into a multi-themed tourist district, the public responded with incisive criticism: for one, the plan was 'too Chinese.' The essay by T. C. Chang and Shirlena Huang analyzes the role of the state and industry in 'tourism imaging' and 'place enhancement' schemes for Singapore. The case of Chinatown was part of an otherwise successful tourism imaging strategy, 'New Asia-Singapore,' which has sought to image Singaporean landscapes for the global tourist while meeting the challenge of maintaining their appeal for domestic leisure. Like Hall's assessment of Melbourne and Sydney, Chang and Huang find the possibilities for seduction in commercialized place imaging promoted by government and tourism industry capital.

In the final quarter of the twentieth century overseas Chinese could travel to China for the first time in decades, and many sought to visit historic family homes and ancestral villages. The chapter by Alan Lew and Alan Wong examines the draw of experiencing the motherland as 'existential tourism,' in which travelers actively negotiate a dialectical subjectivity-in-formation through encounters along the journey. Their analysis looks at what is 'Chineseness' and the need to explore this aspect of personal identity by immersing in the 'homeland.' Lew and Wong conceptualize the analysis through intersecting perspectives on modernity and migration, and the role of social structures in maintaining ties over space and time, testing these perspectives with empirical information on patterns of over-seas Chinese visits to China. In the end, this chapter also explores the dilemmas of not being 'Chinese enough.' This is one version of the ultimate seduction of travel: the opportunity to explore the liminal and consequent emplaced and embodied contexts of sensory difference.

Part I

Interventions

In the areas set aside for leisure, the body regains a certain right to use, a right which is half imaginary and half real, and which does not go beyond an illusory 'culture of the body,' an imitation of natural life. Nevertheless, even a reinstatement of the body's rights that remains unfulfilled effectively calls for a corresponding restoration of desire and pleasure. The fact is that consumption satisfies needs, and that leisure and desire, even if they are united only in a representation of space (in which everyday life is put in brackets and temporarily replaced by a different, richer, simpler and more normal life), are indeed brought into conjunction; consequently, needs and desires come into opposition with each other. Specific needs have specific objects. Desire, on the other hand, has no particular object, except for a space where it has full play: a beach, a place of festivity, the space of a dream.

Henri Lefebvre, The Production of Space (1991: 353)

2 Flirting with space

Tourism geographies as sensuous/expressive practice

David Crouch

In this chapter I consider tourism as something people do in spatial encounters. The discussion continues previous work on tourist practices and experiences (Crouch 1999a, 2001), focusing on the subjective and embodied content of tourist practice as a form of seduction, which I term here flirting with space. First I consider the range of debates on tourism and seduction. Central to this is the power of the gaze, constituted in sightseeing, in tourism geographies and sociologies as it relates to the engagement of the tourist in making sense of tourism, but as observer, detached or self-aware, consuming—prioritizing—signs. I also make a brief consideration of what I consider to be an extreme version of what tourism is: Baudrillard's distracting journey, and tourism promoted as a seduction process (Rojek 1995). Arguably people are lured to go touring, enticed to particular cultures, sites and sights across the world through the tourism industry and numerous agencies who collude in this—writers, film-makers, advertisers and governments themselves. Desire is engaged, perhaps produced, in this process. Spaces and the cultural use/content and context are important in these processes and can be used to entrap desire.

We can make the argument that prevailing explanations of tourism and tourism's relation to power and identity over-emphasize external constructions of what makes tourism. The thesis developed here suggests that people as human subjects practise and perform places, in a perhaps chaotic relation to themselves, each other, and diverse cultural contexts. Considering what tourists do in relation to space, metaphorically and materially, contributes to the critical re-examination of the human subject temporarily a tourist. Central to this thesis is that places are lived, played, given anxiety, encountered (Crouch 1999b).

Recent developments in cultural geography suggest grounds for rethinking the tourist, desire and space. Notwithstanding the investment power of tourism and its roles in constructing contemporary identities, the power of industry to construct and constitute tourists' consumption desires and the meanings that people make from them may be exaggerated; pre-figured meaning may be disrupted by the way people practise tourism and its spaces. Tourist desire and seduction can, I argue, be considered in terms of subjectivities. What the tourist does is a process of seductive encounter: where the tourist flirts with space in ongoing practice, where space is performed rather than merely arrived at as the product of seduction by others. Identities are constituted, temporally, in the flow of these practices and in their relation to contexts. This points to a theoretical approach towards 'doing tourism'.

Seductions

Baudrillard (1981: 85) argued the importance of "strategies of desire" through which consumers'—pace tourists'—needs are mobilized and provoked, their nascent interests captured in a process of consumption before consumption. These strategies, he argued, consist of the signs through which the value of products are conveyed in the process of seduction. For tourism, these signs may include brochures of destinations, their power displayed and systematized through their display and communication. The symbiotic power shifts from the objects themselves to their circulation in representations, their fuller consumption dominated by their sign-value, their value invested in anticipation (see also MacCannell 1976). But Crang's (1996a) critique has called into question the familiar claim that meaning is composed of free-floating signifiers from which the world, according to Baudrillard, is made sense as being somewhat detached from what people seem to do, detached from any in depth investigation of lay geographies.

MacCannell (1976) has argued powerfully for acknowledgement of the signification of tourist sites as locations of the authentic, i.e. where people could—and want to—feel their lives authenticated, anxiously seeking the reflection of something authentic into their own lives as the man in the mirror. Building his argument on the interpretation of tourism as sightseeing, MacCannell discusses how the baggage of visual material is constructed to deliver the tourist what is understood to be authentic, even though this may be a staged authenticity, achieved through the skilful manipulation of artefacts and signs. Indeed, artefacts, even constructed ones, become translated into commodities for tourism, privileged as 'markers'. He calls this constructed recognition: "Sightseers have the capacity to recognise sights by transforming them into one of their markers…not permitted to attach the…marker to the sight according to his own method of recognition" (MacCannell 1999: 123–4). Tourists' photographs, too, are programmed by the markers; the 'Kodakization' of tourism, clearly marked sites from which to see, direction of view, even framing in a circuit of signification. This contributes to an interpretation of the alienated modern individual, anxious for identity. Such framing of space, landscape, heritage as 'static' content remains powerful in the way in which places are inured in tourist seduction by contemporary capitalism—the business sector and its marketing (Selwyn 1996).

The rich and varied body of work exploring how 'tourist spaces' have historically been, as increasingly they are today, tends to explore how they are deeply implicated in the formation of identities. In the British case, few historical examinations of the shaping of the identities and dispositions of the upper and upper-middle classes would be complete without exploring the influences of the Grand Tour and travel more generally, including travel to classical Mediterranean sites (Inglis 2000). Likewise no examination of the shaping of more contemporary identities could be framed without extensive reference to, for example, sites in the Caribbean, the Indian sub-continent, and/or once again (although in a rather different way) the crowded shores of the Mediterranean.

But the notion of tourist gazing has developed through an increasingly subjective vein (Urry 1990, 1995, Crawshaw and Urry 1997). This more self-reflective tourist, or, better, human subject doing tourism, is less duped than aware, less desperately needing identity than using tourism in the negotiation of identity. Whilst Urry stays with an emphasis on the visual and its evolution through increasing mobility, easily related to aspects of tourism (2000), he is increasingly attuned to the incompleteness of vision alone (2002). Critically for Lash and Urry (1994), this 'making use' of signs 'out there' became a focus for the mental reflexivity of the subject's own life in a practice of postmodern subjectivities. Following Baudrillard's desire, this might suggest that people come to respond to their life through the provided spectoral seduction. Like Cohen and Taylor, however, Lash and Urry (1994) argue a subject-centred and less dependent seduction, the former in terms of even a reverie of self-seduction in an escape, in an everyday sense of wanting change, adjustment, rediscovery, in emotion, identity and relations.

The burden of debate so far has concerned the shift to mental reflexivity where the available contextual signification may be subjectively, perhaps ironically, made sense of by the individual. In contradistinction to a 'floating signs' perspective, it may be claimed that human subjects live their lives, live space, encounter each other, work through their own lives in a negotiation of the world around them/us.

In order to develop this deconstruction of the sign-seduction further, by way of departure I explore the possibility that desire rests at least in part in people's feelings and actions, conveyed, figured and developed in their embodied self, mind/body, body/mind, inadequately presentable in the language (Crouch 2001). Such a perspective is oriented, I argue, to the increasingly humanistic postmodernism emerging from across the disciplines (Crouch 2000, 2001). Such a perspective orients more to a sympathetic, though not uncritical, subject, in Levinas's terms, unfolding, negotiating and contesting rather than constantly finding the world always unstable, empty of identity and overwhelming (Levinas 1961). Moreover, we may problematize the privileging of vision accorded in these critiques of tourism and its modernisms. Indeed it would seem to be over-reductive to interpret tourism as sightseeing. The moments of sightseeing on any tourist experience are fleeting, not necessarily essential (in terms of meaning) and need to be understood in terms of a much wider, more complex practice. Indeed, Foucault, whose work on vision is very informative to Urry's powerful discourse on the gaze, speaks of vision as much more complicated than a detached, surveilling gaze, and one more of multi-dimensionality. These issues are pursued further in the following section.

Subjective seductions

The following passages explore the processes of human subjects themselves and their practice of desire, expressive sensualities and poetics, and their constitution of spaces or landscapes of desire. This discussion develops from the early work on so-called "non-representative geographies" by Thrift (1996, 1997). In these terms, embodiment is considered less as inscription, although that is not overlooked, and more as practice and the constitution of subjects' knowledge in a process that is pre-discursive–discursive–pre-discursive (Crouch 2001). Of course, bodies, and places/spaces bear inscription, but may not be totally pre-inscribed upon meaning.

The body as object provides a significant object of desire through which tourists are themselves seduced to visit places, and places can be inscribed with significance through the representation of bodies (Selwyn 1996). Like bodies, space becomes inscribed with significance, and both are powerfully deployed in tourism. Spaces may be signified in the visual depiction and representation of the body—lying on the beach as a symbol of what to do, what it feels like, to be a tourist; similarly, bodies may be signified in relation to places as in the club culture of Ibiza (Game 1991). However, we can also consider the 'other side' of bodies and spaces, i.e. as ways we engage and consume places through our own embodied agency. Towards this end, I have developed ideas about 'embodied semiotics' (Crouch 2001) based on Game's (1991) work on materialist semiotics.

Social practices do not exist outside the domain of the semiotic, i.e. the practice and production of meaning. However, Shotter (1993) argues that meaning and significance are also constituted in practice as practical knowledge, practical ontology. Articulated meanings are developed and circulated in everyday discourse. Such discourses can become the instruments and effects of power. Everyday practices, such as in tourism and the refraction of tourism, and everyday life, in mutual relation (thereby avoiding the familiar academic fetishization of tourism as essentially experience of difference, a bubble of practice detached from the rest of life itself, an experience of the other) (Crouch and Ravenscroft 1995), can thereby be significant in the development and enactment of power relations. The problem of tourism annexization has been compounded by the separation of 'leisure' and 'tourism' as wholly distinct arenas of human practice, knowledge and signification (Crouch 1999a). The cultural contexts for the subject as temporary tourist may be far more complex than the tourist brochure and the sightseeing positioning of the 'tourist' map. Temporarily doing tourism involves a complex patina of practices. This suggests that there may be tensions between and amongst practices and contexts, not simply polarized but commingled.

Potentially, tourism practice relies on a discourse on contemporary identities that is complex and may not be wholly pre-figured, but which re-figures, constructs, constitutes. Tourists may be the object of seduction but the ways in which tourists make sense of places, including their cultures, are made through encountering in a practice, or performance, of cultural complexity. Furthermore, the human complexity of this process may include more subjective relations between tourists' and not-tourists' 'hosts' [sic] than normally considered.

I argue, then, that space and landscape, as appropriated in tourism seduction and (encountered by) the tourist, are not adequately theorized as text (Duncan and Duncan 1988a). Space and landscape instead may be considered as constructed through practice in a tension with context but not pre-figured by context. That 'context' may be significantly re-figured in the process (Crouch and Matless 1996). With Barnett (1999) we may problematize whether practice itself may not be considered a 'context' through which the world, ourselves and our identities in relation to others are made sense of. Thus, emerging from this attempt to reposition our approach to seduction in tourism, this discussion shifts focus from the brochure and tourist sites/sights to focus more firmly upon the subjective practice involved in doing tourism. To consider process like this acknowledges connections, disruption, multiple routes of semiotics and sources of power and its enactment. Subjective sensuous/expressive embodied identities and their semiotics may be sometimes softer, possibly as powerful, more complex, and their positioning in people's everyday lives and identities deserves exploration. The ways in which space/landscape are given meaning through practices, and performances, of being a tourist, lent to constitute meaning in things, actions, relations and identities, demands a closer 'look' at the encounter. Cultural contexts during a period of doing tourism are more than contained in the circumscribed hand-resource, the sightseeing map or brochure, the entrance to the hotel or the framed spectacle of the heritage site (sight). Doing tourism involves a much more complex set of practices than sightseeing alone. I explore some of these nuanced, subtle and powerful contexts through a discussion of embodiment theories.

Practices of seduction: flirting with space

Seduction in tourism can be developed through an acknowledgement of the agentive and dynamic role the tourist plays in a process that is many-dimensioned. The person doing tourism is, it is suggested, a subjective individual who, rather than being merely a recipient of other-produced signs and operating in particular contexts not of her own making, participates in those contexts. She metaphorically produces landscape (sometimes also transforming landscape materially, even if temporarily), refigures pre-figured context and content, and makes space in a process of space-ing. Thus I explore the subject, the tourist, figuring and practising her own embodied semiotics. These theoretical orientations have been introduced elsewhere and the present chapter takes these orientations further, with particular theorization of the tourist and space—or rather the space the tourist practises (Crouch 1999a, 2000, 2001, Crouch and Matless 1996, Crouch and Ravenscroft 1995).

Central to this discussion is the notion that tourism is an encounter, between and amongst several things: people, people and space, people as embodied and socialized subjects; in relation to contexts in which people find tourism: between expectations and experience, and desire. Whilst Lash and Urry (1994) embarked upon a critical consideration of the tourist as active subject involved with mental reflexivity, there is an increasing demand that we do not delimit the human subject to mind dis-embodied (Csordas 1990, 1994). Equally, rather than take the human subject as merely sensate, this discussion explores the ground of the subject making sense (incoherently, temporally, in flow) of touring cultures and spaces. Tourism necessarily becomes verbalized in our shift to consider the tourist as active subject, making sense. Our selves are always becoming, never resolved, momentarily felt to be so, and then moving on. In this flow our 'making sense' can be called lay geography. Lay geographical knowledge is thus akin to Shotter's (1993) ontological knowledge: the importance of practice in making sense of place, space, life, different worlds we live in, is constituted in part as a practical knowledge, a practical ontology of 'learning through' and 'making sense by.' This takes us to the content and process of that doing. Doing tourism, touring, being the tourist, encountering space, it is suggested, amounts to more than a mental engagement and reflexivity; instead, it is all of these and embodied encounter.

Several components of practice, outlined in this way, are interrelated, relational, coincident, mutually informed and informing and so on. Artificially dis-embedding them in order to examine their distinctive facets rather than separateness, we identify multi-sensuousness/sensuosity/uality and its attendant multi-dimensionality, expressivity, intersubjective dimensions of embodied practice, and the poetics of body-practice (Crouch 2001, Pile and Thrift 1995, Thrift 1996, 1997, Thrift and Dewsbury 2000). These facets are parts of the developing 'non-representational geographies' that do not refute the significance of representations in 'making sense' but seek to critique and contest the limits of pre-figured representational geographies and their privileging of representations in relation to these subjective properties and practices. Non-representational geographies argue that practices constitute senses of the real, that representation is part of presentation. Of course, we do not act in detachment from contexts, but nor do we act attached; rather we act as semi-detached, semi-attached, articulated (Crouch and Matless 1996).

One of the facets of embodied practice is to understand our contact with the world as a multi-dimensional encounter, rather than a front-on 'view.' As Merleau-Ponty (1962) argued, the world is all around us, we are engulfed by it. The individual moves, speaks and practises space through and in relation to the body and its body-space, the immediate material and metaphorical space of doing things. Such multi-dimensional encounter combines with the multi-sensual inter-play of the bodily senses, and of course includes visuality (Urry 1990, 2000). The individual is engaged; spaces are felt, heard, and so on. Moreover, seeing is a complex faculty that is not restricted to the gaze alone; it can be caring and mutual, haptic, multi-focal.

Second, tourism practice is intersubjective. Whilst it is familiar to consider tourism as individualistic, sometimes solitary, even self-centred, this is only a partial story. Visiting great locations, viewing monuments and museums, spending time in a campsite, walking along the beach, all these usually happen with, or at least amongst, people. Encountering, tourism is made amongst others whom we may or may not know, intersubjectively and expressively. By our own presence we have an influence on others, on their space and on their practice of that space, and vice versa, often considered as negative, as source of conflicts, but such a position overlooks its positive potential. We may be in this way open to each other, and so things we are doing, places we wander across, feel different. "To see the other is not to have an inner representation of her. It is not to have her as an object of thought—although this is possible. It is to be-with-her" (Crossley 1995: 57). The popular or lay geography of the tourist is one which is influenced through her relations with others, as Shotter argues (1993); through the practice of shared body-space that space becomes transformed as social. Our bodies are informed in ways that are more than experienced alone, even amongst others we may not 'know', such as 'the crowd,' and in shaping space, places become different (Nielsen 1995). The feeling of being together can animate the space. Sociality, then, includes 'closeness' of shared activity, as a proxemic tribe, and participation in a loose crowd of, for example, tourists, or football supporters (Maffesoli 1996). People also go on holiday with friends, make friends on holiday and see them later. In all these instances, the presence of people may not detract from the experience (Crawshaw and Urry 1997), but, even if people are not spoken with, can be part of the practice (Birkeland 1999). Our movements and stillnesses are made in relation to them, in relation to an absence of others, as well as solitarily. The movement of others, and ourselves in their relation, complicates and nuances this process through which we make more sense of the space (if confounded in the process).

In moving around, in watching, the body is expressive, our third dimension. The body is active because it extends and connects amongst people and the mate-rial geography of places (Radley 1995, 1996). Of course this can happen in terms of the inscription of culture on the body surface—the wearing of adornments in tourism experiences—at a disco, surfing, or at a Balinese festival—but also as a means through which to express oneself, as 'being', enjoying life, 'making fun,' or not. Thereby the 'fun' of tourism may be a means of being in the world, of reaching and engaging the world, a medium through which it is enjoyed, and the subject declares herself within that world. Dance, white-water rafting and other 'adventure tourism' (Cloke and Perkins 1998), but also more mundane tourism such as camping and coach-touring—moments of emerging from the bus to stand and stare—express not only systems of signs but expressions of feeling, subjectivity in the world, and our unique personalities. Spatial practices provide means through which the individual can express emotional relationships with others and make sense of what is there (Wearing and Wearing 1996). People who in particular time/space are not being tourists also make these practices.

Fourth, the human subject doing tourism can be poetic. Kristeva (1996) suggests that the body's existential immediacy or felt multi-dimensional relationship with the world can provoke imagination and its translation into different feelings and spatial intimacies. Thus bodily drives are never totally repressed. The subject remains always in process, expressed in movement of the body as often wilful, in voice, rhythm and laughter (Sharpe 1999). De Certeau (1984) explores poetic encounters and the limitless imaginative translation of space in a playful, child-like free-thinking, acknowledging the possibly romantic practice that the body can engage and provoke through its potential to disrupt mental rationality through bodily encounter. Walking is not merely mechanical purpose, objectivized in health, but subtle and expressive bodily practice, with both feet and the whole self, through which space can be felt intimately and imaginatively. As emotion is located within the body (Lyon and Barbolet 1994), being a tourist may enable playfulness, an opportunity in which the world can be experienced as a child does. 'Being there' in playful practice can overflow boundaries of rationality and objectivity in an embodied rather than cerebral game, in joie de vivre, encountering deep feelings as well as surface play (de Certeau 1984, Game 1991). This pre-discursivity is not detached from the discursive process of lay geographies but is bound in relation to them, as those sensuous/expressive encounters are worked in dynamic subjective reflexivity. The distinctions of this playfulness between tourist and 'not-tourist' may not be complete, as de Certeau's practice, or flows, of everyday life suggest.

In these complex ways the tourist encounters space, and achieves, unevenly, incompletely, a feeling of doing. The (temporal) 'result' is a patina that flows with encounters and moments and responds to the 'excess of the body' (Dewsbury 2000) that persistently practises. These multiple fragments of practice as discussed so far can each be termed performativity, a term that demands our closer attention to the lack of intentionality and rationality through which we make our way, doing tourism. In turn concern with performativity attends to the nuances of embodiment rather than linear making sense and dominating perspectival vision and gaze.

Thereby our contact with the world can be considered a multi-dimensional encounter and can express emotion and relationship with it, through intersubjective body-communication (Crossley 1996, Goffman 1967, Merleau-Ponty 1962, Radley 1995, Thrift 1997). The individual moves, speaks and experiences space through and in relation to the body and its body-space, the immediate metaphorical and material space of does things. That feeling of being can be practised as 'feeling of doing,' a means of grasping the world and making sense of what it feels like, the 'feel' at once and mutually physical and in a process of 'making sense' (Harré and Gillet 1994). The world may be felt and discovered, partly (re)constituted through this process, and 'made sense.' The body is rendered involved in the world in which it extends itself metaphorically, transforming the space, flirting with space.

Space, place, identity: processes of engagement and encounter

To wander, to turn, to bend, during a visit to a heritage site, a beach and so on, to be still, is to encounter space through different senses, to be aware of space surrounding rather than as detached, 'out there,' 'over there,' in the spectacle. The space of the spectacle is multiply co-terminus with these other dimensions and apprehensions of space. In these movements or stillness we convey our feelings of ourselves, others, the space itself, and the complex and nuanced readings we receive, multi-dimensionally, inform how we feel. In these complex multi-directions we may express delight, unease, in ways of juxtapositioning the body. The meaning of the self shifts in the process, and the meaning of spaces for us shifts as they are apprehended in each dimension (metaphorically too) anew. The expressive 'journey' can be also intersubjective. 'Making sense' does not, then, equate to making clear rationality but rather working our way through things, spaces, relations. Less as another 'layer' than another spontaneity amongst these others, our movements and turns, rests and excitements can be also emotional and poetic. Together with an acknowledgement of the potential for micro-politics in negotiation of the self, the complexity potentially disrupts the linear connection between the power of tourism signs and the human subject in relation to constructing and constituting identities in contemporary culture

Approaching practice from the bundle of embodiment theories provides a complementary means through which it may be possible to get closer to the constitution of spatialities. Space may be considered a site of subjective exploration, creating space in a way that challenges a priori assumptions as to what 'context' is and how it works—or is worked—in practice. Places can serve as contexts and constraints, but also as creative possibilities in relation to human practice and its imagination, which can be positive and negative. Contexts may be problematized by investigating processes of making sense of space through practice.

We may 'look' at spaces and seek out those inscriptions (Lash and Urry 1994) and practice may travel along pre-figured channels unless disrupted by ideology (Macnaghten and Urry 1998: 211). However, rather than spatialities being prefigured in representations and contexts through which human subjects encounter space, Clifford (1997: 54) and de Certeau (1984) suggest that they may be developed in practice, ontologically, discursively made sense of and felt in an embodied way. It is through these plural elements of embodiment, practice, knowledge and spatialities that the following ethnographic investigation is developed to explore the processes of lay geographies. This analysis considers identities as made inter-subjectively, and ontological knowledge as constituted practically and dynamically. Crang (1997) has argued for example that photographs taken on holiday can become artefacts performed through acts of talking and circulation, through which identities can be negotiated. Thus places and their uses in processes of working identities can be constantly (re)produced. Self-interpreting individuals (Taylor 1989), and through rhetorical exchange, work out and negotiate through embodied practice and in social relations in a construction of distinctive situational meanings (Billig 1991). It is likely that these processes work in ways that are distinctive by gender, ethnicity, class and age.

Semi-detaching tourists from a hegemonic sign-discourse offers potential multiple points for connecting discourses concerning responsibility in tourism. Empowerment, amongst and for tourists and not-tourists, may be more significant in these processes than authenticity, in terms of power taken, realized in people's lives, and negotiated. This suggests a new ground for the politics of tourism.

An overlooked dimension of politics in tourism practice concerns the potential micro-politics of fashioning human-centred identities and positions in relation to the world in terms of apparently mundane everyday activities, and through the ways people encounter the world, including themselves, negotiating their lives (Domosh 1998, Young 1990). These practice-informed negotiations of identity and space are significantly political at personal and interpersonal levels. Thus the present discussion is located at a point of tension between, and within, structure and agency, representations and practices, and acknowledges the presence of each in the other. Such self-identity searching is of course part of the argued desire for self-identity and its lack in the contemporary world 'outside the tourism bubble.' However, this argument is not tourism as a separate part of life, but part of it (see Desforges 2000). Perhaps what is happening is a more creative, subjective and personal use of sign-resources; their deployment is less a crisis of authenticity than a negotiation, however fraught and unsuccessful, of empowerment in one's own identity. Doing tourism may sometimes be and take place in a space/time of liminality, but so do many other moments of our lives (Cohen and Taylor 1992).

Embodiment may therefore provide ground on and through which we can develop a grasp of how the human subject copes, manages, explores and negotiates her world. Following from this, there may be room for a more agentive approach to negotiating identity; such an interpretation has already been developed in terms of mental reflexivity (Lash and Urry 1994). Engaging embodied subjective practice enlarges the process through which reflexivity may occur, that thereby becomes discursive, pre-discursive, discursive and so on (Crouch 2001). Moreover, such an inclusive understanding of practice disrupts the grounds for linear connection between signs and their 'reading'.

Tourism brochures, magazines and other media influence, but the individual, as human being, interacts materially and metaphorically with numerous other contexts for making sense of being a tourist (Game 1991). If we consider cultures visited in tourism as felt bodily (Csordas 1994, Haraway 1991), culture is lived, worked, made meaningful, developed, refigured, felt, laughed over, practised, through lived experience; not mere 'background' or 'setting' in a tourism trip [sic], or marker of something else. Tourists' subjectivities become less a separate reflection but more a practical involvement. Embodiment offers a conduit through which we make our sense of heritage, culture, landscape, as we are involved in each of these, we feel them, work our emotions through them. These features become, even if for a moment, part of our lives; such encounters become part of life when as tourists we go home and become transferred to the everyday.

Considering the physicality of moments of time enables us to attend to a range of embodied practices. It is in the flow of time-related moments of encounter with the self and others, artefacts and memory, that lay geography is constituted. Thus time/space attention helps us to take a step in understanding "discursive choreography of performative behaviour" (Gren 2001: 222). By developing our interpretation through embodied non-representational geographies (which includes encounters with representations) we can pay closer attention to the apprehension of materiality and how that material world is made sense.

Thus heritage, great views and intimate corners, adventure and theme parks are brought into our lives and may not remain detached from our own identities. Indeed, Bachelard (1994) argues that to an extent we encounter these in much more diverse, nuanced and potentially intimate ways. Even in grand sites, the tourist performs in close encounters. This means that artefacts, heritage features, landscapes are embodied, socially, through our relations with each other. Materiality and metaphor are worked together, as Radley (1990: 50) argues, "artefacts and the fabricated environment are also there as a tangible expression of the basis from which one remembers, the material aspect of the setting which justifies the memories so constructed". Sharing body-space provides means through which the value of being together in tourism—in Ibiza discos or eco-tourism treks—can be enacted and understood (Maffesoli 1996, Malbon 1998). These places are 'performed' as enactments of identity, relations, the self, but also embodied and made sense of through and with others (Crossley 1996). This is important in terms of memory and the 'role' of artefacts, including other human subjects in shared body-space.

From these perspectives, the simple distinction between the human subject doing tourism and not doing so becomes much less clear (Clifford 1997). Moments of doing tourism, and making sense of identities and practices in tourism and so-called 'everyday life,' are not mutually sealed but mutually embedded and interactive (Couldry 2000). Such fragments of practice may constitute the identity they each make of the space they come to know, revisit and refigure, as embodied subjects. Each materiality may be encountered differently between different tourists. The situation of the tourist is evidently culturally complex; and relations—with the physical world, contexts and each other, as well as the self—are negotiated.

People are participants, actively, in the seduction process. Wanting to do something in tourism, and doing it, are complexly interwoven—not separate. As the tourist anticipates, walking further on, she feels already the embodied encounter; back home, looking at the images, the physicality of the encounter is recalled; in each, she is mentally reflexive on what it all means, how it all feels. As she steps forward to walk further on, she feels the space around her, emotionally and materially, remembering perhaps a film; she thinks the path goes forward, and others reassure her; she is uncertain, notices something else, she turns. Throughout she is performing a process of seduction. She flirts with space (her identity, memory, each unfolding and closing), actively constitutes the seduction of this space.

It is possible, then, to return to the brochure, the pictures, site markers and other contexts of tourism. For example, in doing tourism, the physical exhilaration depicted in images is exchanged for, at least refigured in, the immediacy of practice. The picture in the brochure may become a reminder of encounter, directly or indirectly, or an anticipation. In the flow of practice, these dimensions—anticipation, reflection, experience and recall—are more complex than linear notions of events. Instead of being identified in relation to the structure of its present by reference to its history, context and heritage, tourism, as lay geography presented here, becomes much more uncertain, refigured; figured anew.

The seductive power of tourist site promotion and the tourist's encounter with space/place are processes of tourist identity formation; dynamic subjective processes through which identities are (re)figured in space. The person as tourist realigns the 'marker' or sign in relation to these consumption practices, its contents then crumpled and torn, its meanings shredded through a relation of who we are and what we do (Game 1991). Then one of the challenges for cultural and tourism geographies is to engage interrelations and contestations of subjective negotiation of 'context' and 'practice' involved in this dynamic process. Focusing our interpretation on sight and seeing over-limits our realm of investigation. In focusing [sic] on the total human subject, the privileging of vision is displaced into the complexity of human practice. The general lack of ethnographic investigation that follows the human subjects from home and back, their doing tourism in the middle, is significant in itself, and should become among subjects of study.

Conclusions

Working beyond semiotics as pre-figured and only reflective of underlying structures, and towards acknowledgement of embodied semiotics, enables us to break
the linear relationship between sign-reading and making sense, ontological
knowledge and lay geographies, in a process of refiguring apparently pre-figured
knowing and meaning (Crouch and Matless 1996, Crouch 2001). It also enables
us to get closer to the processes through which human subjects negotiate other
structurings and other contexts of modernity, themselves also chaotic. This does
not in any way conceal tensions amongst and between numerous contexts and
practices, but makes them more evident as a flow of signification and meaning,
chaotic, more a multi-sensual kaleidoscope or patina than a scopic regime and
sight to gaze upon. Doing tourism is no 'trick,' although it includes many of them
along the way. Sights/sites followed from the sign and the map in a moment of
practice commingle with numerous other encounters and their dimensions.
People produce space, too, in a process of spatializing. In these ways we can come
closer to understanding the complexities of contemporary modernism and the
flow of practices that include doing tourism in the unsteady, temporal process of
identity.

Acknowledging the significance of mediated seductions of tourist
spaces/places needs to be borne through a closer attention to the negotiation of
such spaces by the tourist. In acknowledging the continuing influence of the
media, in its diverse forms, and the constitution of lay knowledge by human
subjects, there is potential to critique the role played by (other and pre-figured)
contextual significations in articulating places, which can engage and work with
the values and attitudes of tourists. Doing tourism is a moment semi-attached to
our lives and other identities, themselves fragmented and always in a process of
becoming. It becomes possible to consider the greater cultural and subjective
complexity of contemporary identities and practices that include doing tourism.
Of course, we are not 'tourists,' but human subjects doing tourism. Our tourism
geographies become 'live' (Thrift and Dewsbury 2000).

3   Tourism and the modern subject

Placing the encounter between tourist and other

Tim Oakes

Tea in the Sahara

> He did not think of himself as a tourist; he was a traveler. The difference is partly one of time, he would explain. Whereas the tourist generally hurries back home at the end of a few weeks or months, the traveler, belonging no more to one place than the next, moves slowly, over periods of years, from one part of the earth to another. Indeed, he would have found it difficult to tell, among the many places he had lived, precisely where it was he felt most at home.
>
> (Bowles 1949: 6)

This passage is from *The Sheltering Sky*, Paul Bowles' brilliant study of travel, exile, and loss. It describes Porter Moresby, an American traveling in French North Africa with his wife Kit and their companion Tunner, shortly after the end of World War II. Port embodies what could be called the subjectivity of the modern conscience: a restless search for authenticity that can only be fleetingly satisfied by a self-induced uprooting, by homelessness, displacement, and exile. Alienated from modern civilization—particularly its inclination toward mass destruction—Port is on an exotic and spiritual journey, a quest to reconnect the fragments of his being, and become an authentic person once again.

There are two important qualities of Port's journey that offer a useful departure point for our own journey into critically exploring the relationship between tourism and the modern subject. One is the paradoxical nature of Port's quest. The other is the way Bowles fashions a landscape of seduction upon which this paradox is revealed and played out.

The paradox is that Port knows there is no authenticity even as he continues to seek it out. The closer he gets to something pure, the more his own life dissipates. Arriving in Oran, on the Algerian coast, Port wants nothing more than to experience pure Africa, unsullied by the war and colonialism, and he pushes further and further beyond the coastal mountains and into the wilderness of the Sahara, growing increasingly weak from illness along the way. He loses his passport—his official identity—and does not wish to have it back. He puts all his effort into avoiding Tunner, the novel's symbolic link with the superficial life of the metropolitan West. Port is a traveler with an "infinite sadness" at the core of his consciousness, and he travels to be reconciled with it.

In Oran, Port hears a story that foreshadows his journey. It is the story of three girls from the mountains who, instead of seeking their fortunes in the coastal ports like most people, travel inland toward the desert, to "drink tea in the Sahara." They arrive in the M'Zab, where all the men are ugly. There they stay for a long time, dancing for the ugly men and dreaming of the desert and trying to save enough money to leave. One day, a handsome desert trader named Targui arrives from the south, sleeps with each of them, pays them in pure silver, and leaves. For months they dream of finding him, and finally realize they'll never make enough money in the M'Zab. So they leave for the Sahara anyway—using the last of their silver to buy a tea service—and follow a caravan into the desert, where they climb to the highest dune they can find in hopes of seeing Targui's city. But instead they die there, on top of the highest dune, with their tea cups full of sand.

When Port and Kit reach the fabled city of El Ga'a, beyond the mountains and surrounded by the sands of the Sahara, Kit notices that:

> Outside in the dust was the disorder of Africa, but for the first time without any visible signs of European influence, so that the scene had a purity which had been lacking in the other towns, an unexpected quality of being complete which dissipated the feeling of chaos. Even Port, as they helped him out [of the bus], noticed the unified aspect of the place. "It's wonderful here," he said, "what I can see of it, anyway."
>
> (Bowles 1949: 194)

Yet they are impelled to continue even farther, toward Sbâ and "the sharp edge of the earth," where Port will ultimately face what he knows is already there: just a teacup full of sand, complete emptiness.

One might call this the paradox of the modern subject. The purity that Port seeks is a freedom to unify the world ripped apart by the catastrophic mechanisms of modernity itself, accompanied by an awareness that such unity is forever out of reach. Port travels toward a utopia of authenticity that turns out, in the end, to be nothing more than a mirage in the desert. At the end of his journey the only truth he finds is that which drove him to escape civilization in the first place: the permanence of change. Authenticity, it turns out, is not to be found in a state of being (a teacup full of sand), but in the process of becoming, of contingency and transience (the *desire* for tea in the Sahara). Port intuitively knows this paradox before he even arrives in Africa; it is a deeply reflexive consciousness that fills him with an "infinite sadness."

But there is another element to *The Sheltering Sky* that necessarily complements and extends Port's journey toward the "paradox of truth." This is the cultural landscape which grounds Port's travels, and which, more importantly, conditions Kit's experiences following Port's death. While one might be tempted to argue that the desert is a purely symbolic landscape marking Port as a modern subject, it seems that Bowles' use of the desert goes far beyond its *symbolic* value in aiding the visualization of modernity's paradoxical subjectivity. Although it represents displacement and exile for Port, the desert is also *a real place*—a home to the other, the Arab. As a landscape, the desert is more than an exotic backdrop for Port's journey. It is also a *place*. The Arabs are at home in the desert; their lives display the ceaseless motion that the desert itself displays and demands for survival. It is they who make the desert a place. And as a place, the desert offers more than backdrop and stage; it offers an *encounter with the other*. Encounter in place, more than movement and journey through space, is what articulates the subjectivity of paradox in *The Sheltering Sky*. And more than anything, the encounters between traveler and other in Bowles' novel are experienced as *seductions*.

The other as temptress has an extended history that parallels imperialism, colonialism, and indeed modernity itself. Port's sexual encounter with Merhnia, an Arabian prostitute in Oran, certainly recapitulates this well-worn trope of orientalism. But when Port dies, and Kit is left there in Sbâ—alone beneath a sky ripped open to reveal the "absolute night" beyond—the desert becomes a place that turns the trope of oriental seduction on its head. Kit sees the emptiness that was Port's morose destination and continues into it, becoming the willing and captive 'wife' of an Arab trader who lives in Tessalit, across the Sahara at the edge of the Sudan. While she clearly surrenders to the desert's ultimate seduction, Kit surrenders to a *seduction of place*, more than one of movement or travel. She surrenders to encounter the home of the other in the most ultimate, meaningful way possible, through her body as a sexual subject.

It is not that Kit herself fulfills Port's desire to escape into emptiness. Instead she escapes into a place built and made meaningful by others—the Arab trader's home—*and does not travel through it*. Rather than maintaining the removed and objectifying distance of the traveler (seen, for example, in Port's sleeping with Merhnia), Kit closes the distance and actually *becomes the other herself*, experiencing seduction in reverse, as the Arab's hidden and captive lover. But Port's death shocks Kit into articulating an altogether different kind of subjectivity, one that is highly contingent, constructed through encounter in place, rather than travel across space. It is perhaps only through this radical departure from a journey in search of objective truth to a placed encounter with otherness that a more complete rendering of the modern subject can be achieved.

This chapter explores the theoretical arguments that would suggest Kit's experience might offer a corrective to the more commonly held view that the modern subject is a traveler. The modern subject, I will argue, cannot be limited by a narrow focus on the mobility of traveler or tourist. Rather, modern subjectivity is best conceived in the places of encounter, where traveler and other meet and are forced to negotiate the meaning of the place in which they find themselves. And because I am suggesting we direct our attention away from travel per se and toward places traveled to, the dichotomy between traveler and tourist—a dichotomy introduced at the opening of *The Sheltering Sky*—loses its power to articulate the modern subject.

This last observation would not have been possible without the pioneering work of Dean MacCannell, whose seminal study, *The Tourist* (1976), breaks down the traveler–tourist distinction by situating it in the context of modernity's broader contradictions. MacCannell provides the fundamental concepts with which to approach the tourist as a metaphor for the decentered modern subject. In the discussion that follows, I hope to extend MacCannell's approach to the tourist as modern subject, by focusing on his concept of "touristic space," and suggest that the modern subject is not a tourist acting out modernity's paradox in his search for "staged authenticity." Rather, the modern subject is reflexively constituted through *the placed encounter between tourist and other*. Subject formation occurs in the reflexive act of place-making that such encounters generate, rather than in the distanced and objectifying view of the world generated by travel. This approach, I will argue, is necessary if the idea of the modern subject is to withstand the challenges of postmodern critiques which have found a privileged yet melancholy and alienated white male 'traveler' (read Port Moresby) disguising himself as the 'universal' subject. While the tourist subverts the class privilege that the traveling subject generally assumes, it remains susceptible to the postmodern critique of the subject. By fashioning touristic space into *place*, however, I believe that we find a more meaningful and less problematic articulation of modern subjectivity. In tourism, place becomes a product of both tourist and other—those who are, in fact, living in the spaces of travel occupied by the tourist. They are the objects of the tourist's gaze, and yet are themselves agents in subject formation through their encounter with tourism, an encounter in which the tourist also becomes the other.

Modernity, postmodernity and reflexivity

It is necessary at the outset of this discussion to clarify the theory of modernity, which drives the following analysis. There are, in fact, two 'modernities' that haunt this chapter. One is modernity proper, which we understand to convey a rational and progressive outlook on the world, to pursue freedom and the improvement of life, and to feel a sense of faith in and anticipation for a better future. The other is perhaps more familiar as the 'postmodern,' and manifests itself as a critique of rationality and progress, and of historicism and the utopian future that legitimates it. This latter, more staunchly critical version of modernity is in fact no less 'modern' than the former; indeed the two come together to articulate the *paradox* of modernity (see Berman 1970, 1982, Oakes 1997). Postmodernity, then, remains a 'subset' of the modern conscience, rather than its confident successor. Agnes Heller makes this argument in *A Theory of Modernity* (1999: 4):

> Postmodernity is not a stage that comes after modernity, it is not the retrieval of modernity—it *is* modern. More precisely, the postmodern perspective could perhaps best be described as the self-reflexive conscience of modernity itself. It is a kind of modernity that knows itself in a Socratic way. For it (also) knows that it knows very little, if anything at all.

Heller's summary points to *reflexivity* as the significant point of differentiation between these two versions of modernity. As long as people have been able to critically reflect on the outcomes of their efforts to create a better future, 'postmodern' thought has shadowed modernity as its 'self-reflexive conscience.' For Heller, understanding reflexivity is crucial to recognizing that it is not progress or rationality that inhabit the soul of modernity, but *freedom*. It is freedom that drives *both* the self-reflexive conscience *and* the impulse for rationality and progress. The approach taken in this essay will be one in which modernity is conceived as an experience of dynamic *tension* between the modern project of rationality and progress and its 'postmodern' critique, which is in fact modernity's self-critique haunting its pursuit of freedom.

'The modern subject' necessarily assumes a specific version of modernity. This chapter seeks to establish a spatial approach to modernity in which reflexivity figures prominently as a constitutive feature of subject formation. Modernity is taken to contain, in any given instance, the paradoxical coupling of antithetical dualisms: for example, rationality and irrationality, progress and loss, modernity and tradition. Modernity is not just the leading side of these dualisms, but in fact envelops the dualisms themselves. This approach obviously implicates the way I have interpreted modern subjectivity. Drawing on a critical dialogue with MacCannell's work, I argue below that mobility-based conceptions of modern subjectivity have emerged in recognition of modernity's 'self-reflexive conscience,' but have failed to adequately contextualize the subject spatially and socially. The latter sections of the essay seek to establish that while the tourist offers important insights into modern subjectivity—as MacCannell argues—the tourist–subject needs to be recast as a place-based experience of encounter, rather than a displaced and authenticity-seeking traveler. This place-making subject is one which recognizes, as Port Moresby recognized, the paradox of freedom. As Heller puts it, "Reflected postmodern consciousness *thinks this paradox*; it does not lose it from sight, it lives with it" (1999: 15, emphasis in original). Yet whereas Port the traveler could *not* "live with it," Kit can. It is her experience of encounter in place that allows her to "live this paradox" and act upon it. That experience is not only one of encounter with the other, but an encounter in which she herself *becomes the other*.

The modern subject: exile, traveler, and tourist in a world of exhibitions

Why have exiles, travelers, and tourists been such compelling emblems for modern subjectivity? This section traces those particular theoretical roots of the modern subject in which the experiences of exile, travel, and tourism have been regarded as quintessentially 'modern.' While these theoretical roots derive significantly from a perspective on modernity in which reflexivity and paradox figure prominently, they ultimately render the subject in very problematic terms. Those problems—revealed particularly in feminist theory, poststructuralism, and cultural studies—will be taken up in the section following this one.

We begin at a point when the Cartesian subject of Enlightenment thought has been decentered and fragmented by social theory. The discussion begins, then, by accepting the critique of the Enlightenment subject initiated in the mid-nineteenth century, and approaches the modern subject as a contested field between those who have stressed its determined qualities and mistrust the autonomy of consciousness (for example, Marx and Freud), and those who have emphasized its reflexive qualities and emancipatory potential (and thus its *agency*) (Berman 1970, Dirlik 1994, Giddens 1979, Touraine 1995). Toward the end of the essay, I will address this gap between conceiving of the subject as socially determined versus the subject as reflexive agent, by turning to place and landscapes as terrains for subject formation. Conceiving of the subject as an exile, traveler, or tourist, however, has offered another—more familiar but also more problematic—means of addressing this gap. To understand, in other words, why Port Moresby must travel is to understand a modern subject that forever seeks to break away from the socially determined qualities of its constitution. Turning first to Foucault we find that duality lies at the heart of this ambivalently constituted modern subject.

For Foucault, the Enlightenment dualism of subject–object initiates a distinctly modern subjectivity. Referring to Foucault's *The Order of Things*, Paul Rabinow (1989: 18–19) writes:

> Modernity, the era of Man, began when representations ceased to provide a reliable grid for the knowledge of things. Modernity was not distinguished by the attempt to study man with objective methods—such projects had already a long history—nor by the attempt to achieve clear and distinct knowledge through analysis of the subject, but "rather [by] the constitution of an empirico-transcendental doublet called man. Man appears as an object of knowledge and as a subject that knows."

Foucault found that moderns are plagued by a schizophrenic sense of self, a subjectivity that depends on maintaining a clear distinction between self and other, between subject and object. There are at least two issues raised in Rabinow's passage that deserve further comment. One is the claim that modernity begins with a kind of reflexivity: the doubt that representations are reliable as conveyors of knowledge. The other is the role of objectification in the constitution of the subject.

In Foucault's rendering, the subject is unthinkable without subjection (*assujetissement*). This idea, explored also by Hegel in *Phenomenology of Spirit* and by Nietzsche, suggests that the subject is constituted through the experience of becoming subordinated to power. It is the experience of subjection, in other words, that conditions the subject to articulate itself and express agency. It is in this sense, I believe, that Foucault sees the duality of subject–object as fundamental to modern subjectivity. If subject formation depends upon an experience of subordination to power, it must be expressed in terms of objectification: the subject articulates itself in terms of its experience as an object of power. Subject formation, then, may be said to depend on the construction of an object world, a world of others.

Yet, if subject formation depends on the security of a distinction between subject and object, Rabinow's passage also tells us that such a distinction cannot be relied upon, for representations have "ceased to provide a reliable grid for the knowledge of things." Thus the paradox of modernity. Giddens (1990) has made a similar point about modernity, claiming that reflexivity always casts doubt on the truthfulness of knowledge. Modernity does not promise increasing degrees of scientific certitude, but rather promises unrelenting reflection, criticism, and reconstruction of knowledge. The 'objectivity' upon which scientific certitude and knowledge rest depends upon representations to convey the 'reality' of a thing. Being of the phenomenal world, we cannot claim a priori knowledge of objects, and so must rely on representations to *see* the truth of them. Being aware that such representations are constantly subject to readjustment and revision creates "anxieties which press in on everyone" (Heller 1999: 15), which Heller attributes to living the paradox of modernity. To be modern, then, is not to express a scientific certainty about the world, but to regard our knowledge of the world with some suspicion, and to strive for ever more reliable forms of representation.

This, perhaps, explains what John Berger (1992) calls our deeply ontological need to *see* things in order to truly experience them. It is the same visual predisposition that Benjamin explored in the Paris arcades project; that is, our collective infatuation with images and phantasmagoria (Buck-Morss 1989). Exhibiting the world for visual consumption objectifies it in a way that calms the demons of reflexivity. Related to this desire for visual refuge from modernity's paradox is a deep desire to establish a sense for what Baudelaire (1964) calls the "eternal and immutable" amid all the contingency of change. We dream of unifying the duality of the subject, of getting beyond our dependence on distinguishing ourselves from the object-world, and yet we remain ever frustrated in trying to do so. Of particular significance for our purposes, though, is the tendency to express this desire for transcendence by finding the "eternal and immutable" in other times, places, and peoples. It is here—in the exhibiting of the other as object—that we find another important refuge from the paradox of modernity.

Timothy Mitchell has called this refuge the "world as exhibition" (Mitchell 1988, Minca 2001). Vast exhibitions of objects, highly popular in Western Europe throughout the nineteenth and early twentieth centuries, reinforced the *objectivity* of the viewing subject, enabled a detached, even rational, perspective on things, so as to see them clearly. As Mitchell and others have argued (see Said 1978, Turner 1994), the idea of putting the world on exhibit explains much about the projects of orientalism and colonialism. The colonized Orient became both Europe's conquest and its exhibit, the foreign and exotic object to Europe's domestic subject. Thus, the great exhibitions of the nineteenth and early twentieth centuries were as much about European power as about reinforcing the subject–object dualism. Actual exhibitions, of course, made such detached viewing easy. But if the modern subject is a detached viewer, then the exhibition can be found everywhere one looks, not just under a glass case or up on a stage. This suggests a curious split in modern subjectivity, for while one must *experience* a phenomenal thing to 'know' it, one can only really 'see' it by maintaining some distance from it (that is, in order to *represent* it with some 'perspective').

It is precisely this schizophrenic split between subjective experience and objective representation that has inspired many to see the modern subject expressed most clearly in the form of the exile and traveler. Perhaps the earliest rendering of this was in Benjamin's characterization—inspired by Baudelaire's writings on Paris under Haussmann—of the *flâneur* (Benjamin 1983: 36–7; see also Tester 1994). The *flâneur* was an ambivalent consumer of images, a man of leisure who was at once alienated by and drawn to the urban maelstrom that defined nineteenth-century modernity. Benjamin wrote of the *flâneur* in the context of his work on the Paris arcades, "glass-covered, marble-paneled passageways," where "both sides of these passageways, which are lighted from above, are lined with the most elegant shops, so that such an arcade is a city, even a world, in miniature." The pedestrian *flâneur* strolled through these passageways, dismissive of the commodity fetishism that swirled around him, yet at the same time drawn to the images conveyed by the commodity form like a window-shopper:

> The *flâneur* still stood at the margin, of the great city as of the bourgeois class. Neither of them had yet overwhelmed him. In neither of them was he at home. He sought his asylum in the crowd. The crowd was the veil from behind which the familiar city as phantasmagoria beckoned to the *flâneur*.
>
> (Benjamin 1983: 170)

Both detached viewer and drawn by subjective engagement, the *flâneur* articulates the paradox of modern subjectivity: a reinforcing of the subject–object dualism even as one desires to *transcend* that dualism.

As John Urry (1995: 141) quips, "The modern subject is a subject on the move." The *flâneur* was only one sort of displaced person among many, outsiders trying to maintain some ever-shifting distance from the swirling maelstrom of the urban metropole. For Simmel, it was the exiled foreigner who most cogently expressed modernity's ambivalence toward the subject–object dualism (Touraine 1995: 202). Malcolm Bradbury notes (1976: 101), for example, that a great body of modern Western literature was written by writers—Joyce, Lawrence, Mann, Brecht, Auden, Nabokov, Conrad—experiencing some form of (forced or self-imposed) exile. The displacement of exile, and the chaos of uprooting, enforced an oxymoronic if not paradoxical *experience of detachment*. Exile made palpable the split between subject and object, while the ongoing tension of this dualism released tremendous amounts of modernist energy directed toward under-standing that which exile had relinquished: the "eternal and immutable," the home from which one had been uprooted. Thus, the exile metaphor also highlights another important outcome of maintaining the subject–object duality of moder-nity: "nostalgic melancholia" (Kaplan 1996: 34). In mimicking the exile, the traveler also expressed this melancholy.

Thus, while elite exile may be regarded as an experience of detachment, thus maintaining the subject–object dualism of modern subjectivity, travel has more generally been thought to represent efforts to *cultivate* such an experience, where one seeks and finds something authentic to view from an appropriately distanced perspective. Yet it is precisely at this point where we encounter MacCannell's tourist, inscribed with the same paradoxical subjectivity of modernity as Port's traveler. In *The Tourist* (1976), MacCannell proposes that we regard tourists as pilgrims of authenticity. This was a pioneering step not simply because MacCannell dares to attribute to tourists the solemn and melancholy journey of the modern traveling subject. MacCannell's analysis extends the above ideas on exile and travel to argue that tourism provided a means by which to understand modernity's underlying structural determinants. *The Tourist* was influenced by developments in 1970s structuralism. In linguistics, Saussure had argued that language—the articu-lation of our thoughts and of our consciousness—was a social, not individual, process. As individuals, humans were thus unable to control the signifiers they deployed in their speech. By stressing the socially determined qualities of consciousness, structural linguistics thus contributed to what had become a sustained attack by social theory—beginning with Marx and Freud—on the autonomy of the Cartesian subject. The progress and rationality of modernity's social institutions weighed upon modern subjectivity, determining the options available for the articulation of selfhood. The resulting 'decentered' subject of structuralism was both socially constrained and, as a result, freedom-seeking.

Travel, and the search for authenticity, was a way of acting upon the determined qualities of modern existence. Sightseeing becomes a "ritual performed to the differentiations of society" (MacCannell 1976: 18), a way of making sense of modernity's fragmentations, and a way of marking certain representations as 'authentic.' In this way, tourism serves to calm our anxieties over the unreliability of representations: it meets our impossible-to-satisfy need to experience *and* represent 'the real world' by producing 'staged authenticity.'

This is where, perhaps, MacCannell develops some of *The Tourist*'s most sophisticated insights. Drawing on Goffman's idea of front and back regions—another manifestation of modernity's subject–object dualism—MacCannell finds tourist displays to be similarly marked by front-stage regions (in which tourists are well-aware they are viewing a display) and backstage regions (marked as such so that tourists see them as 'real'). The most *meaningful* tourist displays, then, are those that allowed the tourist exposure to the 'real' activity going on backstage. The paradox, of course, is that there is no 'real' backstage, only a staged version of it, a representation of yet another kind of exhibit (MacCannell 1976: 101). The tourist's quest for authenticity, then, leads him or her into Mitchell's "world of exhibitions," where 'real' lives of others have been displayed *so realistically* that they are assumed to be authentic. Mitchell's "world of exhibitions" is termed "touristic space" by MacCannell (100), a labyrinth of representations concocted to shore up the subject–object dualism.

> It is only when people make an effort to penetrate into the real life of the areas they visit that they end up in places especially designed to generate feelings of intimacy and experiences that can be talked about as "participation." No one can "participate" in their own life; they can only participate in the lives of others. And once tourists have entered touristic space, there is no way out for them as long as they press their search for authenticity.
>
> (MacCannell 1976: 106)

Tourism, then, not only provides a refuge for the decentered subject, but also provides an arena for experience that does not require the shock of displacement, of exile, and escape. MacCannell's tourists get to have tea in the Sahara, but do not have to die for the experience. They do not have to die because touristic space is a mirage, a trick of the eye.

Poststructural interventions: the absent subject

The modern subject as tourist, in MacCannell's approach, is a subject radically decentered by the social determinants of modernity. The tourist is driven to seek authenticity, and yet remains constrained by modernity's inability to satisfy that search with anything more real than the mirage of touristic space. *The Tourist* offered a subject determined to act out the paradox of modernity, though perhaps without the freedom to *act upon* that paradox as an agent of change. Since *The Tourist* was written, however, the modern subject has come in for a great deal of poststructuralist criticism. Structuralism conceived a decentered subject that nevertheless offered the keys to understand the modern conscience. Yet structuralist theory also set in motion a postmodern discourse of deconstruction which would ultimately go so far as to claim the 'death' of the subject (Lyotard 1979, Baudrillard 1983). While claiming the 'death of the subject' is about as useful and tenable as claiming the 'death of language' or the 'end of history,' there have been a number of poststructural critiques of the subject that deserve careful attention. There are two related issues that concern us here: one is the question of contextualizing the subject in a field of power relations, and the other is the resulting question of difference and hybridity.

A feminist critique of the subject offers one approach to understanding the importance of power relations in subject formation. Part of the issue here is simply the obvious masculine gendering of the exiled–traveling-tourist subject. Thus, in *The Sheltering Sky* it is clearly Port who is the traveler; Kit accompanies him for love but harbors neither Port's innate sadness nor his need for a detached and objectifying view of the other. Nor did Benjamin, for example, imagine the *flâneur* as a woman. To experience detachment—wandering the streets of Paris or the towns of French West Africa—was a *privilege* of gender, and of class and empire. The feminist critique points not only to this gendering of the subject but to the broader question of power, and the ways specific histories and geographies of power condition the modern subject's ability to 'display' and objectify their world.

Along these lines, Caren Kaplan (1996) has argued that tourism cannot be separated from its colonial legacy and that, more to the point, the tourist, far from being a universal marker of the "totalizing idea" of modernity, is quite narrowly defined by class, gender, and general access to social power. Investing the tourist with the totalizing assumptions of modern subjectivity universalizes what are in fact "middle-class, Euro-American perspectives" (Kaplan 1996: 62). She goes on to argue that:

> The tourist in [MacCannell's]…formulation enables a critique of modernity but cannot subvert modernity's Eurocentrisms. The tourist, then, is not a postmodern cosmopolitan subject who articulates hybridity for anxious moderns but a specifically Euro-American construct who marks shifting peripheries through travel in a world of structured economic asymmetries.
>
> (63)

The issue of power, Kaplan argues, cannot be overlooked. This is the obvious question that has plagued structuralism from its inception: the difficulty of contextualizing structures in time and space (Giddens 1979).

Critiques such as Kaplan's have yielded a variety of responses devoted to reconstituting the subject in alternative ways (e.g. Kirby 1996, Pile and Thrift 1995, Touraine 1995). What many of these responses have in common (a significant exception being Touraine) is the conception of a fragmented, hybrid subject (see also Anzaldúa 1987, Bhabha 1996, Chakrabarty 1992, S. Hall 1991, Prakash 1992). This approach draws largely from the role that difference (as opposed to essence or transcendence) plays in distinguishing the different parts of a system, be they words in a language or meanings and ideas in a text. While difference in these terms was always the province of structuralism, Derrida's critique of the sign radically extended the implications of difference to suggest that the link between signifier and signified is in fact an arbitrary construction and, indeed, subject to manipulation. "The integral fusion of signifier and signified," initiated by Saussure and extended by Derrida, "entails that no philosophies which retain an attachment to 'transcendental signifieds' can be sustained; meaning is created only by the play of difference in the process of signification" (Giddens 1979: 30).

Extended to social theory, Derrida's departure from structuralism has inspired the idea that subjectivity is itself constituted solely through difference, meaning that the subject is in fact exploded into countless fragments or 'positions' that are each conditioned by specific histories and geographies of difference. Tourism, of course, thrives on difference, but MacCannell's tourist pursued difference as a way of acting out modernity's underlying structure, its "totalizing idea." The post-structural critique, instead, suggests that there is no underlying structure that determines difference, and that difference alone is what constitutes subjectivity. Constituted through difference, the subject is a hybrid collection of encounters with 'otherness.' Thus, Stuart Hall has proposed that we adopt a radically new conception of identity that recognizes the process of differentiating the other within. This he terms "living through difference," which entails "recognizing that all of us are composed of multiple social identities, not one" (Hall 1991: 57). The experience of diaspora in identity-construction, Hall (1995: 206–7) argues, offers a way of understanding the role of difference. While it is an idea similar to exile, diaspora emphasizes not displacement but multiple homes (contemporary and ancestral); it suggests a permanence instead of exile's transience. This is important in that our attention shifts to places and landscapes turned into homes, rather than homes from which one is uprooted and away from which one travels.

In his 1999 introduction to the second edition of *The Tourist*, MacCannell acknowledges feminist critiques, and the hidden qualities of power that naturalize the tourists' ability to travel and to view their world. He further sees the increased mobility and diaspora of others (refugees, guest-workers, etc.) as an important development that would require adjusting the tourist-as-modern-subject thesis. But about the claim that the subject is constituted solely through difference, MacCannell is unequivocal: the tourist remains determined by the structuring forces of modernity. The tourist's poststructural critics, MacCannell argues, are driven by the same impulses, and look for the same thing as the tourist: a resolution to modernity's contradictions, and some relief from the anxieties of "living with" paradox. In resolving these contradictions through mobility, and through encounters on the move, the tourist still reveals to us the "totalizing idea" of modernity. Urry (1990: 82) seems to make a similar point, when he suggests that in today's 'postmodern' society, "people are much of the time tourists whether they like it or not."

Thus, for MacCannell, modernity continues to be transformed by the imperatives of mobility, by tourists increasingly drawn to the peripheries, and by the increased mobility of formerly marginal peoples to the centers, such that a feeling of displacement is increasingly the norm for everyone. Modernity marches along unabated, demanding even greater degrees of mobility, radically displacing the cultures of the world. Travel, exile, displacement, and nostalgic melancholy still define the subjectivity of modernity, only now the "human community" has come to realize that "it can no longer contain everything that it does contain" (MacCannell 1992: 2). In response, poststructuralism has led "the community to conceive itself as empty and impossible while it awaits realization" (3). That is, the subject has become absent, no longer able to shore up the subject–object dualism even as it tries to transcend it. This is apparently because the tourists, upon returning home, have found those increasingly mobile others right in their own back yard. "Critical theory," MacCannell (*ibid.*) writes, "has prepared us for the absence of the subject, for an empty meeting ground, but it does not help us to get beyond this historical moment."

Here, again, the tourism experience suggests a way forward. The continuing historical development of modernity's 'conquering spirit' has led to a situation in which the objects of the tourist's quest, the 'authentic others,' have themselves developed fragmented subjectivities. In MacCannell's (1992: 3) terms, what the tourism encounter reveals is the formation of "new cultural subjects" that are necessarily hybrid, and yet remain "stubbornly determined" by social power. Suggesting a response to the poststructuralist challenge along the lines advocated by Giddens (1979), MacCannell (1992: 4) seeks to move beyond a myopic focus on difference and hybridity per se as the emancipating destination of subject formation. Instead, his concern is to see difference and hybridity as a point of departure in the modern subject's continuous pursuit of freedom. Difference is a "precondition for the inventiveness and creativity which will be demanded from all of us if we are to survive the epoch of globalization of culture currently dominated by advanced capitalism." Yet the tourist remains incapable of such "inventiveness and creativity" unless a broader context for tourist-subjectivity is conceived. In the following section I seek to establish an approach to the tourist-subject that meets these concerns.

Placing the subject: the spaces of tourism

What MacCannell seems to be suggesting is a greater role for conscious, discursive reflexivity in conceiving modern subjectivity. Such reflexivity, however, seems to involve more than simply "living with" the paradox of modernity, but also a freedom-seeking attempt to *act* upon the structuring forces of modernity. While the detached experience of travel may be enough to convince the exile–traveler–tourist that he or she is a "miniature clone of the old Western philosophical Subject" that imagines itself transcendent and free (MacCannell 1989: 15), the reflexive subject must supersede the experience of travel itself. The reflexive subject is conceived not so much through mobility and displacement, as through the encounters with otherness that such mobility yields. How else can reflexivity be introduced to the tourist if not through the encounter? Encounter, as I am using it here, is something quite different from the experience of detachment afforded by travel. Port Moresby 'encounters' Merhnia, but this is an encounter that only reinforces his subjectivity through her sexual objectification. Rather, I am thinking of Kit's encounter with the desert as a place, an encounter that expresses a reflexivity capable of challenging the subject–object dualism. Encounter of this kind can only be conceived as a product of *place*, as opposed to travel.

The tourist can no longer trust that sightseeing will reinforce the subject–object dualism, because his or her gaze is often returned by the others who increasingly assert their own subjectivity in the encounter. The burlesquing of tourists at a Zuñi Pueblo described by Sweet (1989)—where masked caricatures of visitors from 'New York' or 'Los Angeles' were inserted into dance performances—is echoed in tourism encounters throughout the world (see Abram 1997, Linneken 1997, Oakes 1998: 1–10, Wood 1997). The tourist is increasingly confronted with the disquieting realization that he or she is not transparent, but has become an object in the subject formation of others. Here, reflexivity has become inescapable, leading MacCannell to suggest that the tourism encounter in such situations reveals a "new cultural subject" that cannot help but recognize its own constitution of difference in the fragmented image reflected back upon itself.

The issue of reflexivity has been raised in contexts other than those where tourists finds their gaze returned. Maxine Feifer (1985), for example, suggests the term "post-tourist" for those who consciously "play" with the assumptions of authenticity in the tourist's solemn quest. Post-tourists are reflexive, in that they understand the whole touristic encounter to be staged, that one cannot avoid "being a tourist," no matter how hard one tries. There is a kind of pleasure in travel here that falls beyond the purview of the nostalgic and even melancholy tourist-as-subject. Thus, the solemn modernist asks, perhaps in the curmudgeon-voice of Guy de Maupassant: why visit the Eiffel Tower? What truths can be gained? Feifer's (1985: 267) post-tourist answers with a wry smile: it is as obvious as Andy Warhol's soup can—it is the tourist thing to do.

If reflexivity is to play a major role in reconstituting modern subjectivity in relation to tourism, then we seem to be left with the need to contextualize the tourism encounter. This cannot be achieved by maintaining a narrow emphasis on the mobility of the modern subject. If the subject is to reflexively derive "inventiveness and creativity" from its encounter with difference, then that encounter must be clearly situated in time and space. This was, after all, the goal of Giddens' theory of structuration. The idea of "practical consciousness," for instance, represents an effort to understand how one's maneuverings through the temporal and spatial fabric of everyday life can be conceived in terms of the agency of the subject. Yet while his work is highly valued for its contribution to a theory of action in the social sciences, Giddens has been criticized for a lack of *social space* in his theory (Cloke, Philo and Sadler 1991: 129). As Kathleen Kirby (1996: 7) remarks, "space helps us recognize that 'subjects' are determined by their anchoring within particular bodies or countries." More to the point, perhaps, is her argument that it is necessary "to view subjectivity as a place where we live, a space we are, on the one hand, compelled to occupy, and, on the other hand, as a space whose interiority affords a place for reaction and response" (35). Rather than beginning with mobility, as Urry does, I propose we begin with place, and work toward finding a role for mobility within the history and geography from which placed subjects are constituted.

This suggests an explicit call for a spatially reconstituted subject in response to the need to address reflexivity, agency, and action without losing sight of social determination. Kirby (1996: 150) echoes this call in the following way:

> Space provides precisely the substance we have been looking for to provide a multidimensional analysis of subjectivity, one that can be truly material without losing sight of the vitality of the inner life of individual subjects, that can incorporate "experience" into broader categories such as global economic relations, while maintaining the flexibility and the fluidity for imagining ways of transforming future subjects.

While Kirby's focus is more on the psychic mode of subject formation, her insights help inform a project of constituting the place-making subject. The work of Henri Lefebvre (1991) also offers some conceptual direction in focusing on the spaces of tourism (as opposed to the mobility of tourists and previously peripheral others) in an effort to reconstitute the modern subject in these terms.

Like Foucault, Lefebvre finds in modernity a world where representations presuppose one's experience of the phenomenal world. Indeed, "representations of space" have become so central to modernity that they are capable of *producing* space. In Mitchell's terms, the space thus produced is the "world of exhibitions." Representations of space are necessarily ideological, and are mobilized in the service of power, for they conceive an idealized space in which the needs of capital, of the state, and other forms of social power, are met. In the service of power, representations of space produce what Lefebvre (1991) terms "abstract

space." "*Abstract space* is characterized by both the fragmentation and homogenization of space, and both processes are the result of the commodification of space"
(Stewart 1995: 614). But Lefebvre leaves a space for subjects to produce "differential space" by resisting and appropriating the contradictions inherent in abstract
space. And yet, the "differential space" of Lefebvre's subject is a space continually
being reinvested with place-based meaning. It seems possible that the subject may
reject (discursively, or even non-discursively) the experience of "being-out-of-
place" and seek ways to invest space with symbolic meaning, collective memory,
and meaningful agency. More to the point, the tourist encounter may offer a
framework for understanding how subjects draw on encounters with others to
produce "differential space." Place, as I am conceiving it, attempts to draw on
Lefebvre's ideas of spatial production by suggesting a stage on which the tourist
encounter occurs (Pratt 1992). That stage is built upon many "layers of historical
accumulation" (Massey 1988), the savings bank of collective memory, from which
the subject derives meaning and the will to act.

The concept of place, as I am using it here, is not a bounded entity, nor is it a
spatial equivalent of 'community' (Agnew 1989, Anderson and Gale 1992, Pred
1986). Place is not a locality, nor is it simply a more local version of region or
nation. These terms may assume a distinct territorial quality, defined by a
bounded unity of some sort (these being qualities of Lefebvre's "abstract space").
Place is not a 'local scale,' but rather *transcends scale*. Place is more about action
than about scale or region. This emphasis on agency is suggested by what Doreen
Massey (1992, 1993) has called a "progressive sense of place," and what Dirlik
(1994: 108) terms "critical localism." This is not, in other words, place associated
with Heidegger's "being-in-the-world," an idea which 'progressive' modernists of
all stripes (e.g. Harvey 1989, Touraine 1995) find anathema to modernity's critical and emancipating potential. Rather, place is a point of intersection among "a
particular constellation of relations" in space and time (Massey 1993: 66). More
than this, place is where subject formation occurs in encountering this constellation of relations. Such a subject, Arif Dirlik (1994: 112) suggests, is not limited to
a particular scale of action, but is 'translocal.'

While this notion of place depends extensively on mobility and on other link-
ages that 'jump scale' and traverse great distances, the power of place is found in
its insistence on grounding the flows of people, capital, information, and other
media in a precise location—a point of intersection—where such flows meet the
deposits of collective memory, the sediment of past encounters between flows
and subjects struggling to build some security in an ever-changing world. These
deposits of collective memory, Massey's "historical layers of accumulation," are
invoked again and again as people negotiate their encounters with flows.
However, place is also conceived as a site of subject formation that is not bound
by scale, that derives from, but is not completely encompassed within, mobility.
Place is an ideal stage for MacCannell's "new cultural subjects," for it provides a

space for the reflexive subject while at the same time maintaining a critical focus on the forces of social determination within which subject formation occurs. What is being suggested, then, is that the modern subject be reconstituted as a place-making subject. Such a subject remains, to use MacCannell's words, "stubbornly determined" by the broader forces of political economy, information, and media; and yet subject formation does not come about simply through one's mobility, but more significantly through the ways mobility and other practices are called upon in an effort to inscribe space with meaning. For too many people around the world, mobility remains a difficult, if not impossible, proposition, yet these people nevertheless claim a modern subjectivity.

The placed subject articulates a sense of loss in the face of modernity, but also sees countless opportunities for new forms of action, pleasure, inventiveness, and creativity in this experience. This subject does not leave home for a respite from the fragmentations of modernity, but struggles to create places of difference from those fragmentations, stitching them together through its mobility-induced encounters. Tourism offers a rich set of resources for this project, and so it is in tourist encounters that we still see this modern subject so clearly articulated.

Conclusion: rereading *Cannibal Tours*

To conclude, I want to offer one final reading of MacCannell, this time focusing on his essay "Cannibalism Today" (1992: 17–73). I hope to draw on this essay both to illustrate the abstract ideas offered above and to suggest that MacCannell's case actually supports the focus on place-making that I am proposing.

In "Cannibalism Today," MacCannell offers a reading of Dennis O'Rourke's documentary film *Cannibal Tours*, about a group of Euro-American tourists traveling by luxury boat up Papua New Guinea's Sepik River to see authentic, head-hunter natives. The tour is, clearly, a packaged journey into the "heart of darkness," and the film is an attempt to demystify the touristic search for authenticity and detached experience of the other. Throughout the film, the audience is offered interview footage, with translation, revealing the attitudes of the New Guineans in juxtaposition to those of the tourists. Not only does this reveal the typical gap in understanding between two groups of vastly different people, but more importantly, we learn that the New Guineans understand much more about the tourists than the other way around. MacCannell calls the New Guineans in this encounter "ex-primitives," for the film tends to reveal how "there is no *real* difference between moderns and those who act the part of the primitives in the universal drama of modernity" (34). The New Guineans are reflexive about their encounter with the tourists—much more reflexive, in fact, than the tourists themselves turn out to be. The tourists and the ex-primitives, MacCannell argues, do not represent absolute differences but the differentiations inherent in the new cultural subject. The only difference that MacCannell can see between moderns and ex-primitives is in the language that they speak. Moderns cannot detach themselves from their myths, their denial, their values. Ex-primitives are *necessarily* detached; they are more conscious of it than the tourists. Thus, the New Guineans are clearly aware of the *role* they play in the touristic encounter—they understand their value as a 'primitive' attraction. The tourists, however, are less aware of the role they play, the role of performing a ritual aimed at absolving the modern guilt at having "killed off the childhood of humanity" (24). The "heart of darkness," as in Conrad's novel, is a guilty conscience.

For MacCannell, tourism mediates and reveals differentiations in the new cultural subject, differentiations that develop in response to the spread of the global capitalist economy. MacCannell wants to see in O'Rourke's tourists all the guilt and repressed desires (homoerotic and sadistic) of the modern subconscious. His reading turns our attention to the more psychic aspects of modern subject formation, but it still shares many similarities with *The Tourist*. While he is clearly aware that the New Guineans, too, need to be regarded as "new cultural subjects" in their own right, his primary interest lies with the tourists themselves, in the way that they are compelled (determined, even) to play out the tragedy of modern life.

> Modern civilization was built on the graves of our savage ancestors, and repression of the pleasure they took from one another, from the animals and the earth. I suspect our collective guilt and denial of responsibility for the destruction of savagery and pleasure can be found infused in every distinctively modern cultural form.
>
> (MacCannell 1992: 25)

Ultimately, then, we are left with a structurally determined tourist that remains stubbornly unreflexive. I want to make two points here. First, that the "ex-primitives," not the tourists, are the ones who articulate the kind of reflexivity that MacCannell's "new cultural subject" seems to require. Second, that it is the touristic encounter, and the spatial nature of that encounter, that yields such reflexivity. The reflexive subject, in other words, is a product of encounter, but more precisely of an effort to derive meaning and a sense of place from that encounter. As the New Guineans attempt to derive meaning from their encounter with tourists, they engage in an act of place-making, by articulating their differences from the world of the tourists. For the New Guineans, those differences, as revealed in *Cannibal Tours*, are all about money—who has it, and who does not. The tourists' money is the only thing that can explain their ability to travel such far distances, and their ability to possess such fantastic machines. Most importantly, money explains the *behavior* of the tourists, their maddening stinginess when purchasing Sepik crafts, their inability to *understand* the lives of those they travel so far to see— "they exhibit an unimaginable combination of qualities: specifically, they are rich tightwads, boorish, obsessed by consumerism, and suffering from collectomania" (27). The New Guineans position themselves as an emplaced people relative to the tourists and their money. The New Guineans have no money, and this fact is used to frame their articulations of identity in spatial terms: the tourist's money is not *earned*, it is simply a fact of the place where the tourist lives, just as a lack of money is a fact of the place of the New Guineans. Indeed, their *performance* as primitives for tourist consumption enables a sort of 'anti-money' discourse to emerge, where place-based identity asserts a non-monetary moral superiority over the outsiders:

> [The New Guineans] deny the economic importance of their economic exchanges [with tourists]. They will explain that they are exploited absolutely in their merely economic dealings with tourists, but also as far as they are concerned, at the level of symbolic values, these exchanges count for nothing. By the ex-primitives' own account, their economic dealings with tourists are spiritually vacuous and economically trivial, producing little more exchange than what is needed to buy trousers.
>
> (MacCannell 1992: 29)

For MacCannell, this suggests mutual complicity in the tourist encounter, between the tourist's need to absolve modern guilt, and the ex-primitive's need to perform primitiveness. But it also suggests a reflexive process of differentiation, a process to which both the tourist and the New Guinean contribute. The space of tourism—inhabited by reflexive modern subjects—becomes a place of difference. While it is clear that the New Guineans deserve the appellation "new cultural subject" as much, if not more, than the tourists, the point being made here is not that one group is more qualified to express reflexive modern subjectivity than another, but rather that the modern subject emerges out of the encounter *between* these groups, and cannot be constituted in any other way. It is not the *guilt* of the tourists that informs subject formation (particularly since they seem so unreflexive about it). Instead, the subject is constituted in the particular time and place where that guilt (or whatever else may motivate the tourist to journey up the Sepik) encounters its other. This subjectivity belongs as much to the New Guineans as the tourists, for both occupy the tourist space in which it has been constituted. In addition, the need to invest that space with meaning, to make it a place, is what drives this process of subject formation.

Constituted in these terms, the modern subject is necessarily contingent upon the particular historical and geographical 'instances' in which such encounters occur. I do not mean to suggest that there is necessarily a large socio-economic gap that must be filled by tourism for such instances to occur. In the case of *Cannibal Tours*, the gap is indeed vast in socio-economic terms. But differentiation occurs in a virtually infinite variety of situations, as infinite as the possibilities afforded by any given intersection between history and geography, between flows across space and memories through time. Place defines the site of this intersection. By focusing on tourist space, and place-making more specifically, we move beyond the modern subject as simply a mobile subject, a traveling or exiled or sightseeing subject, to a historically and geographically situated subject, constituted in part through mobility, but more importantly through encounter and the differentiation that encounter yields.

To return, then, to Kit in the desert, one might argue that her terrifying experience following Port's death is in fact *more* alienating and displacing than Port's travels ever were. Indeed, she is seduced by the desert as a thoroughly determined and highly vulnerable subject. The quote from Kafka that initiates Kit's departure into the Sahara suggests that she moves beyond the point at which her subjectivity can be reclaimed: "From a certain point onward there is no longer any turning back. That is the point that must be reached." And yet, Kafka's desire to never return reminds us that as long as subjectivity is conceived as *travel beyond certain points* it will remain as empty as Port's final destination. Ultimately, *The Sheltering Sky* does not itself escape the displacement and alienation that inspired the modern traveling subject. Kit's actions subvert the traveling subject through her encounter with place, but she realizes her freedom at a tremendous cost, returning to Oran a mere shell of her former self. Yet her experience also points toward the possibility for a different approach to modern subjectivity, one which reveals the crucial *processes* of differentiation conditioned by place in subject formation. We need not be seduced, like Kit, beyond the sheltering sky to appreciate how tourism may be helpful in conceiving the modern subject in these terms. Kit experiences a seduction of place, not travel, and that is what takes her into a full experience of modernity's paradox. The tourist, then, does not play out modernity's paradox by traveling in search of authenticity, as MacCannell has suggested, but by encountering the other and, indeed, *becoming* the other, in a landscape of places.

4 The souvenir and sacrifice in the tourist mode of consumption

Jon Goss

Allegory and the economy of salvation

> Within the development of culture under an exchange economy, the search for authentic experience and, correlatively, the search for the authentic object become critical. As experience is increasingly mediated and abstracted, the lived relation of the body to the phenomenological world is replaced by a nostalgic myth of contact and presence. "Authentic" experience becomes both elusive and allusive as it is placed beyond the horizon of present lived experience, the beyond in which the antique, the pastoral, the exotic, and other fictive domains are articulated.
>
> (Stewart 1993: 133)

This chapter explores the intersection of consumption and tourism, or what might be called the "tourist mode of consumption," in which material objects are represented as the embodiment of an original essence that is estranged from its historical and geographical origins. The exemplary object of tourist consumption is the souvenir, a thing that refers metonymically to a temporally and spatially distant origin, and metaphorically evokes collective narratives of displacement and personal stories of acquisition. It is, like the fetish and the gift, a peculiar object in that it possesses material traces of human intentions and actions, or the social relations of its 'production,' that lend it an (un)certain interiority, which is taken to be its subjectivity.

Marx famously argued that capitalist society is the first in history to have eliminated religion, but that it returns in repressed form in the fetishism of commodity production; late capitalism, under conditions of globalization, might similarly be said to be the first society to have more or less eliminated the 'obligational economics' and 'reciprocal economies' of simple societies, and transcended temporal–spatial distance, but affection and displacement return in 'spirit' through the consumption of the commodity as gift and souvenir. The commodity is simultaneously fetish, gift, and souvenir, an object of animation, affection, and a collector's item (Benjamin 1985: 55).

Each of the commodity forms—fetish, gift, and souvenir—interiorizes the Other, arrests time, and appropriates space: the fetish, the love and labor embodied in its human maker; the gift, the affection of the giver and the time between its giving and its return; and the souvenir, the hospitality of the host, memory of the past, and spirit of place. A certain sacredness inevitably inheres in each of these forms as they intermediate between transcendent essence and mundane materiality, but the souvenir is the modern commodity *par excellence*.

Walter Benjamin (1985) argued, for example, that the souvenir is what modernity makes of the commodity: it is emblematic and symptomatic of the subject's felt estrangement from the world of Objects/Others under conditions of Enlightenment rationality and commodity capitalism. The souvenir exemplifies the capacity of the commodity to evoke nostalgic desire for an authentic anterior and exterior reality that is felt to have been lost to us, and it attaches equally to the religious, antique, and exotic, as well as to the collective and personal past. It potentially realizes the functions of both fetish and gift, and so represents the capacity of the commodity to substitute for the entire field of object relations, for a lost world of immanence that is re-stored in the souvenir shops of tourist consumption, and so brought into private possession. It is a model of contemporary consumption, of the object consumed from a position of temporal–spatial displacement and it is the model of culture itself as commodity, of culture as 'cultural production': that is, the culture of tourism (Stewart 1993: 145).

Benjamin argues that the souvenir makes an exemplary commodity, as it 'interiorizes allegory,' telling the story of temporal–spatial displacement but also embodying the trace of its lost origins and so promising the possibility of a restoration (Buck-Morss 1989: 189). That is, the souvenir speaks in the "language of longing" (Stewart 1993: 135), simultaneously proclaiming its material displacement (its metonymic meaning) and promising return or restoration in the personal memory or story of acquisition. Like the figure of allegory, it signifies the non-presence or non-being of what it represents, a "longing for the origin whose loss is the necessary condition of that longing" (Greenblatt 1981: viii).

Allegory is figurative, a way of looking at the world for concealed truths, for correspondences between the characters and objects of the narrative world, as if they were personifications and materializations of 'higher' meanings or essences. It is a general effect of language itself, which opens up a space of desire between sign and meaning, object and narrative, but is intensified by the printed word and media of mechanical reproduction (Stewart 1993: 23), hence its particular purchase in scriptural religion and the poetry and drama of early modernity. Allegory is a figure that occupies the void that accompanies the remotion of God, the banishing of the dead from life, and the frustration of proper mourning (see Caygill 1998: 54). Like tourism and commodity consumption, in fact, it gestures longingly to a world of immanence and anticipates return, resurrection, and reconciliation to the loss of an imagined unity of meaning and materiality (de Man 1983).

The discourse of tourist consumption displays an "allegorical intention" (Buck-Morss 1989: 186–7), relentlessly reducing the world of things to signs of something

Figure 4.1 A garden of Eden fed, clothed and sheltered the children of Hawai'i

else, transforming the landscape itself into a text, a "chaotically scattered heap of metaphors" (Benjamin in McCole 1993: 143) that work through their multiple correspondences to evoke a melancholic remembrance of lost value or spirit, and longing for immanence, a world in which meaning inheres in objects themselves. In 'Western' culture, a generalized nostalgia has its theological expression in stories of the Fall, with banishment from a state of prelapsarian innocence and the loss of the originary "Adamic language of Names," and secular expression in the loss of authentic relations with Nature and the world of objects that is the corollary of modernity and progress. Tourism exploits both forms of nostalgia, promising a glimpse of a world before the originary estrangement and modern disenchantment, into the Past and of the Other. This conflation of alterities is perfectly represented, for example, by an advertisement inaugurating a new marketing campaign for Hawai'i in 1972. In Figure 4.1, a doubled nostalgia articulates prelapsarian innocence to authentic everyday life prior to modernity, thus conceiving history as "the Passion of the world" (Benjamin in McCole 1993: 136), and the tourist consumer is promised the prospects of transport not merely to another and an Other's place, but projection from the profane time of the historical present into the sacred time of revelation and mythical origins: that of Absolute Past and Absolute Other (Eliade 1954: 35).

Allegory often tells of journeys that function as sustained metaphors for the progress of life towards its inevitable destination in death. Surprisingly perhaps, tourist landscapes, usually associated with the liminal and ludic, are often similarly morbid, and thus sites of consumption in the original, dreadful sense of the word: that is, wasting away, toward death. Most obviously, 'dark tourism' involves travel to sights or 'black spots' that are literally shrines to martyred heroes, murder, and mass killing, but the museum shares etymological origins with the mausoleum, and many 'heritage' and 'nature' tourism sites mark figurative deaths of civilizations, cultures, or local ways of life (Lennon and Foley 1999, Rojek 1993b: 137), and historical or ongoing degradation of natural environments and extinction of species. Similarly, some contemporary retail concepts are positively sepulchral, suffused with a profound sense of loss so that their images and objects seem to be literally relics of the death of the natural, historical, and cultural Other. The most obvious examples include, respectively: 'natural nostalgia' stores such as Nature Company, Endangered Species, and others, where dying ecosystems, fugitive fauna and flora, and natural objects like fossils are brought (back) to life in souvenirs of restored naturalness (see Goss 1999); the so-called heritage industry (Hewison 1987), which in its museum stores and 'olde shoppes' restores dying forms of the commodity, labor, and traditional value(s) in antiques and arts and crafts; and in the exotic handicraft stores where extinct forms of non-commodities, religiosity, and aspects of a "primitive ontology" (Eliade 1954: 35) are captured in 'ethnic products.'

The discourses of consumption and tourism are thus informed by the "melancholic metaphysics" of allegory (Benjamin 1977): surrounded by overwhelming evidence of physical dissolution and temporal and spatial estrangement, and saturated in the pathos of death, decay, and departure, they nevertheless perform a "stunning reversal" (McCole 1993: 147), or "dialectical trick" (Hanssen 1998: 100), that celebrates the resurrection of life, restoration of value, return of the departed, and recovery of meaning. They deploy what Beatrice Hanssen (1998: 96) calls the "economy of salvation": representing temporal process as relentless 'natural history,' even as they confront the inevitability of death and the awful abyss of eternity, they nevertheless invite an imaginative "leap of faith" in(to) eternal life (Benjamin in Buck-Morss 1989: 175). The souvenir, like the relic of holy pilgrimage, attests to the persistence of our faith, or rather the intensity of our collective desire, for the material presence of meaning, for immanence of the world.

Landscapes of tourist consumption are in effect sacred spaces that respond to our impossible desire to experience the absolute authenticity of death, "the jagged line of demarcation between physical nature and significance" (Benjamin cited in Hanssen 1998: 68), the singular event that marks the end of life and meaning, the destination from which there is no 'real' possibility of return. Tourist landscapes simultaneously accept and refuse death, presenting the tragically inevitable obsolescence of things and mortification of life, but staging the resurrection of value and spirit, as the terrible moment of material loss and nullification nevertheless realizes 'higher' meaning. The landscapes of tourist consumption therefore pose questions on the nature of the relationship between matter and spirit, lifeworld and eternity, and realize in ritual practices and souvenir objects an ancient desire for the possibility of traffic between them.

Tourist consumption is thoroughly modern, perhaps in the felt intensity of the spatial and temporal displacement that motivates it, and certainly in the social and material investment in the narratives of restoration, but it nevertheless persistently evokes primitive religiosity and righteous journeys such as pilgrimage, with the rhetorical effect of re-sacralizing discretionary travel and re-sanctifying the souvenir. By evoking primitive faith in the animation of objects and authentic forms of travel, discourses of tourist consumption establish the presence of the "subject supposed to believe" (Zizek 1997: 106), of the primitive and historical Other who believes for us, who believes what we want to believe and what we really *do* believe, but would otherwise find hard to *really* believe: that material world of tourist consumption is enchanted, place and objects possessed of spirits of the dead. If this was once sacrilege, in the modern literary tradition, allegory stands accused, with other traditional narratives such as fairy tales, myths, and 'natural histories,' of 'bad faith' for betraying the real by assimilating the particular to instances of the general (as opposed to the figure of the symbol), and by locating meaning in the thing itself rather than the sovereign subject confronting it (Nägele 1991: 84). A similar attack is made upon the "preformed complexes" (Percy 1975: 47) and "pseudo-events" (Boorstin 1964) of tourist consumption, where meaning is foreclosed as it is 'circulated' in promotional texts and contrived experiences, and the particularity of geographically and culturally distinct sites is reduced to signs of general history of progress-decline and loss-restoration.

'*Going* shopping'

> The consumer is a person on the move and bound to remain so.
>
> (Bauman 1998: 85)

Consumption and tourism are not reducible to the same phenomenon, of course, and numerous typological studies attest to culturally specific variations in forms and motivations for both practices. Nevertheless, their semiotic structure and social structuration overlap, in part because of the universalization of commodity consumption and globalization of the means of discretionary travel, and in part also because of the 'transparency' of their dominant discourses (see Olson 1999: 18–20), which exploit emblematic images and objects that can be read and appropriated by diverse audiences who share a generalized "subjective sense of insufficiency" (Seabrook 1988: 15). The assumption made here then is that even if there is no universal modern subject, worldly consumers daily devour signs of exotic and historical Otherness, and regularly tour the world in search of experiences of spirituality and authenticity (Lash and Urry 1994: 271–3), such that, for better or worse, "we are all (more or less), consumer-tourists now."

Tourism sights would not be complete without souvenir stands, of course, and shopping is a major component of tourist experiences in many destinations. The built environments of travel, from hotel lobbies to airports, increasingly resemble shopping centers, while shopping centers are reconstructed as idealized tourist destinations, from the hillside villages of Mexico and the Mediterranean, to endless variations on tropical island paradise and urban festival marketplaces. Megamalls such as Mall of America and West Edmonton Mall, with their travel services, hotel accommodations, gift stores, and themed shopping environments, are major tourist destinations in their own right (Goss 1999).

There is always already a sense of touristic displacement built into the concept of 'going shopping' (in contrast, at least in British English, to the everyday 'doing the shopping'), and even mundane mass-produced commodities are often marketed and displayed in exotic and/or historic *mises-en-scene* associated with the world of tourism (see Sack 1992, Goss 1999). What Celia Lury (1997: 79) calls "tourist-objects" trade directly upon their spatio-temporal displacement from idealized origins, and their embodiment of exotic nature and culture—their status as souvenirs from lost worlds. Nelson Graburn (1976: 5) explains: "there is a cachet connected with international travel, exploration, multiculturalism, etc. that these arts symbolize; at the same time, there is the nostalgic input of the handmade in a 'plastic world'." The concept of an "ethnic revival" (D. Hall 1991) implies, however, not only the doubling of desire in the spatial and temporal Other, but also that desire for the exotic Other is not new. There are elements of primitivism in Virgilian poetry, for example, but the aesthetic was most intensely applied in the panoramas, arcades, department stores, world's fairs, museums, winter gardens, and public places of the nineteenth century, when also the modern subject particularly felt the "vertiginous" effects of time–space compression (Harvey 1989: 277).

In nineteenth-century Euro-America, shopping was already part of the emerging mass tourist economy in large cities, where these sites were the main attractions (see Harris 1991: 66), and the souvenir was already the form the commodity took in such "phantasmagoria" of industrial capitalism (Benjamin 1985). Allegory provided the narrative means to make sense out of all this material, for according to Benjamin (1977: 224), "allegory establishes itself most permanently where transitoriness and eternity confront each other most closely": it is an aesthetic response to the experience of accelerated social change and the accompanying crisis of representation in the object world.

As Benjamin, the historical materialist, understood, the allegorical mode of representation is peculiarly suited to the world of the commodity (see Buck-Morss 1989: 188), most obviously because technological change so rapidly renders objects obsolete, but also because of the peculiar "metaphysical subtleties" of the commodity system through which things relate to each other in the marketplace: hollowed out of their intrinsic material and essential human qualities, their use-value, and invested with arbitrary signs of collective desire and anxiety, they take on a spectral, "supersensible" quality, that is their exchange-value, which the modern prophet Karl Marx could only *represent* in the language of allegory and religion. Drawing upon metaphors of theatricality, magic and necromancy, Karl Marx (1977: 83) sought to expose the fiction of the "very queer thing" called the commodity, and so to free the political economy from ideology. Marx exposes the fetishism of the commodity that, like the Christian relic or fetish of 'primitive' religion, comes to life through anthropomorphic projection in the phantasmagoric commerce between things, as it speaks the universal commodity-language of exchange, and so "mimes the living" (Derrida 1994: 153). In so doing, however, he reduces the ineffable 'thingness' of the object world to the human condition, so that the "forgotten residue in things" (Bataille 1991: 35) and source of their sociality is reduced to the mortified life/time of labor they embody, ghost of the inalienable form of the human species.

For Benjamin, the messianic materialist, however, the specters seen by Marx were manifestations of aura, the fleeting experience of temporal–spatial proximity with "the material origin—and finality—that human beings share with non-human nature, the physical aspect of creation" (Hanssen 1998: 212). Like personifications of allegory, commodities seem to speak of or upon some other plane, for or from some original subjectivity, and thus like the objects we call art, antiques, primitive artifacts, and even nature, they evince an uncanny ability to look at us, to return our look of desire (Benjamin in Wollen 1982: 237). There is thus a certain "empathy of the soul with the commodity," particularly for the modern urban *flâneur*, the prototype of the contemporary global consumer–tourist (see Buci-Glucksmann 1994: 76), who pursues transitory images of the commodity world in search of signs of authentic meaning and spiritual presence, for a glimpse of the sacred through modernity's natural, historical, or cultural Other.

My argument, from here on, however, will focus on the cultural Other, because, as the souvenir effectively incorporates the commodity as fetish and gift, 'the primitive' embraces both nature and history, and the possibility of their reconciliation (see Goss 1999: 63–5). As Marx and Freud both demonstrated in their respective appropriations of the anthropological fetish, primitive religiosity provides a model for modern experience of the object world: representing the persistence of the desire for an animated world with the historical consciousness of the tragic loss of the 'innocent' conditions of that desire. If the souvenir is the exemplary form of the commodity under capitalism, the primitive artifact is surely, therefore, the exemplary form of the souvenir.

Allegorical topographies and tourist consumption

According to Benjamin (1977: 92), allegory spatializes time or sets history to stage—the flow of time as natural history is "petrified" and "scattered" in images of ruin and narrative history triumphs over natural temporality (Stewart 1993: 31). This "spatialization of time" memorializes time in space and privileges the aesthetic of Being, with its nostalgia for spatial fixity and temporal stability, over Becoming, or "annihilation of space by time," and the celebration of temporal flux (Harvey 1989: 273). In this sense, landscapes of tourist consumption function as "heterochronies," that is forms of what Michel Foucault (1986) calls hetero-topia—or "real places...which are something like counter-sites, a kind of effectively enacted utopia"—that make an "absolute break" with traditional temporality, death and progress, in order to renew faith in mythological time. Again, landscapes of tourist consumption are structured by allegory: saturated in the pathos of material loss, represented in verbal narratives, visual images, and objects in various stages of decay and obsolescence, and in the spatial displace-ment evoked in themes of departure and arrival, they are also sanctuaries from the effects of time and modern life, denying the effects of disenchantment, distance, and ultimately of death.

The topography of the allegorical landscape includes archetypal natural, archi-tectonic, and vehicular metaphors for the various passages and ultimate passing of human life, that play productively on linguistic meaning, in the form of such "emblematic props" as islands, forests, caves, tunnels, labyrinths, ocean depths, mountain heights, volcanoes, waterfalls, and rivers; in towers, bridges, paths, and shrines; and in caravans, carriages, trains, boats, and ships. The meaning of such metaphors is complex, but the function of images and objects of antiquated vehi-cles and vessels, for example, might be to provide for our imaginative transport to past and distant places, to embody the qualities and values of traditional labor and material, and to represent the technical operations of salvage and restoration to (often priceless) value (see Goss 1996). Nautical metaphors are particularly prevalent in landscapes of tourist consumption, which favor waterfront locations and images of historic marine transportation, and it is significant that Foucault (1986: 27) identifies the boat as "the heterotopia *par excellence*...a place without a place, that exists by itself, that is closed in on itself and at the same time is given over to the infinity of the sea." The boat is also the archetypal vessel in the alle-gorical tradition, symbolizing the journey of life subject to currents, tides, and winds of fortune—the drift of meaning—and of death, the transport of the bodily vessel and transportation of the soul beyond the human horizon into the afterlife. Finally, and most importantly, it is a modern metaphor for the authen-ticity of oceanic voyages, for a mode of travel made obsolete by the development of aeronautical technologies, and for an original and originary settlement of exotic lands, made impossible by the condition of postcoloniality.

Obsolete forms of transportation evoke possibilities of genuine travel as an arduous journey, the archetypal form of which is a religious or sacred quest. Many other observers have been tempted to characterize tourism as a modern pilgrimage, the tourist undergoing similar, if diminished, ritual experiences, thence returning to society personally transformed (Graburn 1983, Leed 1991; see Turner and Turner 1978: 1–39). In pilgrims' tales, righteous travelers face physical and moral dangers—tantamount to what Anthony Leed (1991: 8) calls a "fictional death"—in order that they are figuratively reborn. Rituals conducted at the sacred site and incorporating its sacred objects were the means by which pilgrims projected individual biographies and journeys into mythical time and space (Eliade 1959). Significantly, this capacity for temporal projection is exactly the function that Stewart (1993) ascribes to the modern form of the souvenir, and like the ancient relic its value derives from an anterior death, whether literally of human life, nature, and/or culture, or symbolically of a previous self.

The contemporary consumer has also been described as a "spiritual tourist" (M. Brown 1998), driven by "a hunger to find spiritual truth" (Roof 1999). Reference here is to the "postmodern" turn to New Age products and services, but Marx long ago identified the religiosity of consumption, appropriating Hegel's concept of fetishism as "the religion of sensuous desire" (cited in Pietz 1993: 133). Marketers and Marxists are equally convinced that in commodity aesthetics traits of 'primitive' societies are conserved and exploited: things are invested with 'soul' (Dichter 1960) and, advertising, for example, is "a highly organized and professional system of magical inducements and satisfactions, functionally very similar to magical systems in simpler societies, but rather strangely coexistent with a highly developed scientific technology" (Williams 1980: 185). Pursuing this analogy, this magic is compensatory for the felt loss of an authentic way of being in the world that is attendant upon the "disenchantment of the world" (Weber 1946: 138–40), the "withering of experience" (Adorno 1987: 40), and the loss of the "thingness of things" (Heidegger 1971: 171) under the relentless materiality of modernity and the commodity system. The sign of our alienation, the commodity is nevertheless the focus of our desire for transcendence (Belk 1991: 47). It is a religious object: consumerism is the form that faith takes under capitalism, shopping is one form of its ritual, and tourism is its primary form of pilgrimage.

Sites of tourist consumption are also what Erving Goffman (1967) calls "action places," institutionalized settings for licensed revelry, places where customary norms of behavior are relaxed, and a social drama of *communitas* is performed as we imagine acting out fantasies of self-transformation. As in the sites of traditional pilgrimage, with their markets and fairs, the structure of liminality in consumption and tourism gives license to a libidinal economy, based on the "erotics of display," where objects and the Other are invested with sexual potential (Haug 1986: 55)—it hardly needs stating that discourses of consumption and tourism evoke desire for sexual as well as commercial exchange with the exotic object/Other. This is, of course, the sense in which Freud conceived of the human experience of the subjectivity of objects—as sexual fetish, where an impossible (sexual) desire for the unattainable whole object is displaced onto an arbitrary, partial (asexual) substitute object.

While the figure of the pilgrim evokes authentic travel, artisan labor figures the possibility of authentic forms of material production, prior to or beyond contemporary capitalism. Whether there is a radical difference in object relations between capitalist and pre-capitalist societies (Gregory 1992), let alone between historic and modern society, or it is simply that things are more or less hopelessly 'entangled' in multiple forms of object relations in all social formations (Carrier 1994, Frow 1997, Thomas 1991), discourses of consumption and tourism are heavily invested in the notion of an authenticity located in modernity's Other in the form of 'pristine' nature, 'primitive' cultures, and pre-modern 'heritage,' like the first travelers searching for "revelation of something ineradicably present" (Leed 1991: 9). Similarly, contemporary consumers quest for signs of 'authenticity,' the new measure of marketing (Lewis and Bridger 2000), as antique and other forms of the "anteriority of objects" lend meaning to the abstract system of commodities (Baudrillard 1996: 106).

Landscapes of tourist consumption thus re-present in images and texts, and re-store in commodified form an originary use-value in the form of 'unique,' 'one-of a kind,' 'traditional,' 'handcrafted,' and 'primitive' objects. However, if the boat is particularly evocative of both religious experience and authentic transport, a similarly special role is played in the topographies of tourist consumption by the sacred objects of 'primitive' religions, potent images, and objects of authentic religiosity. Cultural performances reproduce elements of traditional rituals and employ their power objects—authentic fetishes ontologically and temporally anterior to the fetishization of commodities. In the tourist landscapes of Hawai'i, for example, visitors witness live enactments of ancient myths and legends of Polynesia, and ritualized worship, reproduced in souvenir programs, videotapes and audio-recordings, at the Polynesian Cultural Center; in a shopping mall they can see magnificent full-scale reproductions of *tikis*, statues of Hawaiian war gods, and purchase miniatures in the form of key rings, refrigerator magnets, and T-shirts; and even at the Arizona Memorial they can learn that Pearl Harbor (*Wai Momi*) was a site of sacrificial temples (*heiau*) to Ka'ahapahau, the guardian shark goddess, who had to be appeased by a native Hawaiian priest retained by the US military during construction in the early part of the twentieth century. If, as noted by Derrida (1994: 147), "survival and return of the living dead belong to the essence of the idol," the performance of rituals and deployment of the objects of a primitive religiosity lends the mundane material exchanges of contemporary tourist consumption a 'higher' meaning.

Death, departure, and sacrifice: the meaning of 'aloha'

In landscapes of tourist consumption, images of genuine travel and the authenticity of objects, located before and beyond the temporal and spatial horizons of modernity, are articulated to narratives of life, death, and potential resurrection. The basic model is mythical: the 'hero' departs on a sacred journey never to return in body, but he is memorialized and his spirit materialized in the landscape. The end of the tourist–consumers' journey is the souvenir, an object image of displacement similarly invested with the aura of the ever-departed.

Images and texts in the landscapes of tourist consumption, for example, represent historical and mythical moments of departure and arrival that evoke nostalgia for 'real nostalgia' only experienced by the historic and cultural Other; that is, for the lost poignancy of partings and greetings whose emotional intensity derives from uncertainty that the parties separated by travel will ever meet again. There are several excellent examples in the tourist landscapes of Hawai'i: in the IMAX movie at the Polynesian Cultural Center, called "Polynesian Odyssey," we experience the poignancy of leave-taking and gift-giving between a young (male) Polynesian canoe voyager and his (female) lover forever left on the shore; while at the Arizona Memorial there are images of the battleship leaving on its 'maiden' voyage and of 'innocent' sailors posing for photographs and shopping for souvenirs, shortly before their last 'voyage' to the bottom of Pearl Harbor. Such images and objects of authentic journeys are vital to the experience and social structuring of modern travel, for the conventional (and elitist) definition of the tourist is one who travels with a clear intention of returning home within a short period of time (Urry 1990: 3). Genuine travel, then, is undertaken as if without the certain expectation of return.

The allegorical landscapes of tourist consumption in Hawai'i, for example, saturated in signs of death and departure, are nevertheless full of figures of prosopoeia, animation, monument, and other means of material–spiritual exchange, that promise its transcendence. Narratives at each sight mourn historical loss of life and celebrate contemporary recovery of its spirit: at the Arizona Memorial it is the bodily deaths of the battleship's crew, the loss of pre-war innocence, and the revival of American fighting spirit; at the Polynesian Cultural Center it is the demographic and cultural decline of oceanic peoples, the loss of ancient tradition, and the contemporary regeneration of *mana*, the spirit of Polynesia; and at the Aloha Tower it is the decay of the waterfront, the obsolescence of maritime passenger travel, and the loss of genuine urbanity and exchange, which are restored in the new spirit of Honolulu and the waterfront marketplace. Some typical forms of souvenirs of the death and resurrection of the Other include: a videotape of the spectacular evening revue at the Polynesian Cultural Center, called "Mana!: The Spirit of Our People" and ending with a performance fittingly called *Ho'omana'o* (Remembering) that is described as "a song of horizons past, present and future unfolding before us as the miracle of life" (PCC 1995); and the souvenir flags at the Arizona Memorial which have been raised and flown momentarily over the wreck, for which one receives in the mail the all-important authentication in the form of a 'Certificate of Flag Presentation.'

Tragically, modernity seems to deny the possibility of authentic death, so that it has become the object of nostalgia, for according to Derrida (1993: 58), "the dominant feeling for everyone is that death, you see, is no longer what it used to be." Hence, narratives of tourist consumption tell the possibility of the authentic death of the Other: the one who dies readily and easily, who gives itself absolutely in order that the Self might live, but whose spirit is remembered, embodied, or re-membered in memorial objects. Sacrifice is the ancient means by which exchange is performed between the profane and sacred, the material and spiritual, as biological life and earthly material are ritually consumed and assimilated to Life, or the Absolute Spirit, in the Hegelian sense (see Jameson 1999: 57). It seems strange to find it symbolically reenacted in sights of tourist consumption.

The paradigmatic case of the consumption of the material trace of the spirit of the Other who has been sacrificed in order that we might realize our relationship to an originary authenticity, and renew our faith in the continuity of the world and the divine, is of course the Eucharist. At the risk of sacrilege, I suggest that the souvenir is exactly such an object: mark of mourning for life originally given as gift, sign of its spiritual resurrection, token of its original affection and compassion, and coin of the continued currency of exchange between materiality and spirituality. Condemning tourist consumption for its sacrileges against spirituality and modernity, however, for its corrupted religiosity and spurious authenticity, would only repeat the archaic gesture that vainly attempts to drive merchants from the temple and echoes virtuous denunciations of religious kitsch by the self-consciously 'better class' of pilgrim (see Harris 1999).

Sacred sites were always already places of commerce, where pilgrims purchased provisions, absolution, and reproductions of holy relics, and were entertained during secular interludes in devotional activity (Leed 1991: 25, Rowling 1971). In fact, shopping and sightseeing were integral parts of the pilgrim experience, always already socially structured in a hierarchy of relative righteousness—indeed, the internationally organized pilgrimage 'trade' eventually provided the model for tourism. As Leed (1991: 37) points out, "pilgrimage devotion, the market, and the fair are all connected with voluntary, contractual activities (the religious promise, the striking of a bargain, the penny ride on the merry-go-round), and with a measure of joyful, 'ludic' communitas." Originally, the church was the market, as divine presence was evoked to sanctify secular dealings, and secular dealings could in turn evoke divine presence, while in ancient sacred centers, traders in sacrificial animals and currency simply facilitated the ongoing ritual exchange of objects, between those of profane and sacred value (Williams 1991).

Landscapes of tourist consumption are similarly sacred, with the souvenir as its material–symbolic center, as the objective means to bring the unique and unrepeatable event of death into the economy of repetition and exchange. In the tourist consumption landscape, however, this sacrifice is articulated to authentic travel or the originary authentic voyage. In this context, there are two gifts of life: that of the traveler, the mythical figure who dies upon an authentic journey in the act of founding new life, and of the one who greets him, and gives herself to him in absolute hospitality. I use the gendered terms advisedly, for as Leed (1991: 221) points out, "travel is a gendering activity," in the sense that myth and ideology constructs male mobility as a "spermatic journey," so that despite, or rather because of, the hero's death and dismemberment, he projects identity into future and distant territory. This is evident in the founding acts of heroic masculinity celebrated at tourist sights in Hawai'i, for example: the settlement of the islands of Polynesia by oceanic canoe voyagers and the establishment of the Church of the Latter Day Saints in Polynesia evoked at the Polynesian Cultural Center, the annexation of Hawai'i, construction of the harbor and naval base, and protection of US interests in the Pacific at the Arizona Memorial. Interpretive texts tell of Polynesian martial traditions and rituals of royalty, the heroism and 'fighting spirit' of ordinary seamen, and the civilized dignity of early travelers to Hawai'i.

The territory is, of course, female, and not surprisingly, given the conflation of sacrificial and sexual economy, her worthiness defined by feminine innocence. At these sights we witness the warm welcome, and physical and symbolic embrace of the heroic male traveler by the sessile and sensuous female represented by beautiful Polynesian 'maidens' dancing the hula and offering leis with 'aloha.' The 'aloha spirit,' commonly described as "the world's loveliest greeting or farewell," and widely acknowledged as the most important resource to tourism and commodity production in Hawai'i, is exemplary of the "emotional surplus" upon which the "hospitality industry" depends (Urry 1990). It has been used to define the experience of tourism and consumption in Hawai'i since the moment that founds the possibility of all other founding moments in narratives of tourist consumption of the islands: the arrival of Captain Cook at Kealakokua Bay in 1778 and his 'discovery' of the natives' innocent gift of hospitality as they purportedly freely gave of their possessions and sexual services to his men (see Goss 1993b: 684). By the 1860s, the word 'aloha' was already imprinted on souvenirs (Kanahele 1986: 485), and it has since lent its name to literally thousands of commodified products and services. The so-called Aloha Spirit (patented in 1936) is marketed mercilessly by the state and private industry in Hawai'i, and material objects and commodified services are invested with the essence of the Other that is available to the tourist consumer for purchase and possession.

Aloha is perhaps the most complex, and certainly the most contested, concept attributed to the Hawaiian people (see Trask 1993). For the visitor, it is typically glossed as simply greeting and leave-taking, or more generally 'love,' although many tourist texts and performances hint at the 'deeper' meanings uncovered by anthropologists. For Sahlins (1985: 3), for example, it means an empathy for all human beings—"a kinship substance with the other and a giving without thought of immediate returns"—that applies equally to land (*aloha 'aina*) where ancestors are buried, and a reciprocal though unequal 'love' between rulers (*ali 'i*) and their people, a mode of the commoner's submission. For Kanahele (1986: 483), on the other hand, *aloha* was the basis of meaning in the gift-giving "primal economics" of native Hawai'i, and it has connotations of empathy, compassion for suffering beings (see also

Sahlins 1985: 3), and the genealogical relationship between people and their ancestors or *'aumakua* (see also Trask 1993: 187). *Aloha* thus connotes sacrifice, the
absolute gift of love and life made by one's forbears, whose death is not only the
condition and means of continued life. The Aloha Spirit of tourist consumption is a
'holy' spirit, a ghostly trace of the Absolute Other whose eternal life and infinite
capacity for compassionate giving—whether as God, Nature, or 'land' (*'aina*)—is
one side of the transaction at the heart of all economic exchange, the Other of
which, is "the *an*economy of the gift of death" (Derrida 1995: 97). The once innocent
primitive who haunts landscapes of tourist consumption is the ghostly host who
provides the means for our continued faith in the operation of *the* economy, and the
possibility of exchange between materiality and meaning. The primitive, and concepts
such as *aloha*, attests to the possibility of faith in the continuation of spiritual life in
the face of bodily death, the conservation of value despite the obsolescence of
objects, and the persistence of memory despite separation in time and space.

In the contemporary consumer tourist economy, the Aloha Spirit is thus exemplary of the operation of fetishism, gift, and souvenir, offering the possibility of a
spiritual surplus to commodity exchange: it is in the words of one particularly
crass advertisement, "the smile of the waitress who really cares if you enjoy your
meal" (see Goss 1993b). In the 'forgotten history' behind this smile, we can see
more easily the connection of tourist consumption to sacrifice. The absolving
compassion and love of the Other in death makes the sacrifices ritually reenacted
in these landscapes into sacred gifts of life, rather than profane acts of violence: in
the context of Hawai'i, this would surely include the horrors of population
decline through European disease, the terror of world war and weapons of mass
destruction, and the systematic dispossession of land, denial of native rights, and
environmental degradation that have occurred with urban land development.
Images of heroic sacrifice, compassionate spirit, and loving smile absolve tourist
consumers of terrible historic and contemporary acts made in their name in the
founding and maintenance of the 'Aloha State.'

Judaeo-Christian mythology is effectively what Robert Bellah calls the "civil
religion" (Williams 1991: 3) of the United States, and religious imagery pervades
promotional advertising for the commodity 'Hawai'i' so that in some cases, as we
have seen, a trip to the islands might seem almost literally like a pilgrimage. And
we can see at work in the economy of tourist consumption what Nietzsche refers
to as "the stroke of genius called Christianity," whereby the creditor is called upon
to sacrifice life to pay the debts of the debtor and the sinned against are made to
take upon themselves the sin of the sinner, so absolving them (see Derrida 1995:
114)—and it is thus that I have argued that *aloha* might be interpreted as forgive-
ness for a partially forgotten history of imperialism and colonialism in Hawai'i
(Goss 1993b). In the hypernarrated landscapes of tourist consumption the ancient
economy of sacrifice is articulated to Western (Hegelian) historicism, resulting in
the myth of regeneration through violence, such that the physical and cultural body

of the historic or exotic Other is "dismembered," to be reconstituted and appropriated as an innocent abstraction (see La Capra 1999: 703). The ultimate "dialectical trick" of the allegory of tourist consumption is the displacement of death onto the Other, who sacrifices its life for us, repeating symbolically the original means the ritual murder through which the spirit of the divinity was conjured (Girard 1977).

Conclusions

> Imperialist nostalgia revolves around a paradox: A person kills somebody, then mourns the victim. In more attenuated form, somebody deliberately alters a form of life, and then regrets that things have not remained as they were prior to the intervention.... In any of its versions, imperialist nostalgia uses a pose of "innocent yearning" both to capture people's imaginations and to conceal its complicity with often brutal domination.
>
> (Rosaldo 1989: 70–1)

The discourses of tourist consumption ascribe the general felt absence of "soul" in the modern world to a particular historic loss that occurs through the separation of religion and economy, or the spiritual and material, with the development of industrial society. While this liberates a powerful desire for spiritual and authentic experience of the Object/Other, at the same time the tourist consumer is subject to the dominant "proteophobia" of Enlightenment discourse (Bauman 1993: 164), which, suspicious of fetishism and reification, denies the possibility of material mediation between self and Other. The result is a contradictory desire, and hence the oxymoronic conceits such as 'authentic reproductions,' 'live recordings,' 'original copies,' 'real nostalgia,' and, in Hawai'i, 'genuine Aloha.' Consciousness of the impossibility of this desire, and of the paradox whereby progressive expansion of the commodity field necessary to sustain tourist consumption also necessarily undermines the possibility of existence of its ideal Object/Other—which transform tourists into "belated travelers" who turn up after the exotic referent has been transformed into a commodity sign of Otherness (Behdad 1994: 13)—only lends it increasing urgency and intensity. At the same time, differential distribution of knowledge and the material means to realize the marginal possibilities of tourism leads to a complex social and spatial structuring of experience (Bourdieu 1984). Tourist consumption is structured by the dialectic of desire and disappointment, snobbery and shame, as consumer tourists attempt a "fantasized dissociation" from dominant practices and symbolically distance themselves from others in the "positional economy" (Frow 1991: 146, Urry 1990: 43).

An alternative to this contradictory experience is to accept that critical distancing and adoption of the Enlightenment distinction between rational and mystified object relations is not viable, and with the post-tourist and ironic consumer, treat the object playfully 'as if' it were authentic (Cohen 1988: 383). But playfulness belies the seriousness of fetishism and its effects, and plays into the social hierarchy of tourist consumption. Instead, we might recognize the reality of enchantment, and, with Benjamin, treat modernity's faith in rationality, and in functional relations with objects and others, as itself a form of mythical thinking, for magic, status, and religion in the form of the fetish, gift, and souvenir are fundamental to all forms of exchange (Belk 1991: 17–18, Gregory 1994: 233). The modern experience is the repressive sublimation of collective desire for more 'natural' and 'original' relations with objects and others upon the past and the other, so that a felt absence in social life as we live it now is experienced as a loss of life as it was once lived, and progressive desires are transformed into regressive nostalgia.

Benjamin recognized the power of images from the past, not as vehicles of nostalgia in memorial narratives, but as "wish images," manifesting the collective dream on the possibility of utopia, a dream whose future orientation has always been 'forgotten' so that it is experienced as the restoration of timeless paradise, rather than the ever potential future (Missac 1995: 108). Benjamin believed in harnessing the progressive potential of allegory to redeem meaning in the material world, but refused the regressive tendencies both of the Baroque, in its passive melancholy and dream of spiritual transcendence, and of modern allegory, in its angered but helpless attachment to "petrified unrest" (in Buck-Morss 1989: 196). He believed in the potential of the "dialectical image" to draw the "dream image" into an awakened state, to reactivate consciousness of the original dream of modernity to fulfill the human potential that remains always asleep in the once-upon-a-time of historical discourse. Historical discourse imaginatively invests present meanings in past objects and others, rather than allowing images from the past to 'flash forward' into the present.

What might this mean in the landscapes of tourist consumption? In Hawai'i, it might mean reading images of 'ancient' Hawaiian villages and the pomp of nineteenth-century royalty with images of contemporary native Hawaiian communities, of juxtaposing the romance of Waikiki's beaches with the reality of homeless Hawaiian encampments on the windward and Waianae coasts of O'ahu; or images of bombs destroying the USS *Arizona* with images of the bombed island of Kaho'olawe or live ammunition training in Makua Valley; or images of the majestic passenger liners arriving on 'Boat Day' with those of the crowded steamships bringing labor from Asia to the sugar plantations in the islands; or of aircraft disgorging tourists at Honolulu International Airport. It would mean seeing past images of sacrifice in the present context of political–economic struggle of native Hawaiian people over representation of their culture, and control over land and natural resources. What is at stake is the possibility of an alterity in Objects and Others that resists assimilation into the history of individuation and secularization (Nägele 1991: 177) precisely because it refuses the "gifts of civilization" (Bushnell 1993). It is thus to accept the irreducibility of the Other as nothing less than the presence of the Absolute in the Other, without the desire to sacrifice its body so as to possess its spirit.

5   Tourism economy

The global landscape

Ginger Smith

By most measures, tourism is the world's largest industry and, until September 2001, it was also one of the world's fastest growing industries. Tourism is an umbrella industry comprised of diverse services sectors, which has grown substantially in the final quarter of the twentieth century. Between 1950 and 2000, the overall global tourism industry experienced an average annual growth rate of 11 percent (WTO 2001a). The growth of the industry owes to the rise of 'mass tourism,' which has been made physically possible especially in the final quarter of the twentieth century by revolutions in transportation and communications technologies, especially the jet aircraft and, later, computer reservations systems. Based on its industry definition, tourism encompasses

> the activities of persons travelling to and staying in places outside their usual environment for not more than one consecutive year for leisure, business and other purposes not related to the exercise of an activity remunerated from within the place visited.
>
> (WTO 2000)

Most 'travel,' therefore, is 'tourism,' and almost all travel uses tourism infrastructure and facilities. Financial services have also played a large part in the economic growth of tourism, just as travel has been a major component in the phenomenal growth of the financial industry in the same period.

In terms of national macroeconomic accounts, tourism is measured as trade: international tourism receipts are classified as export earnings and international tourism expenditures are classified as imports. For example, in 1999, comparative worldwide export earnings show that tourism exceeded all other international trade categories, at an estimated $555 billion, edging out the next largest category, automotive products, by approximately 5 billion (WTO 2001a). International tourism is the world's largest export earner, and the largest non-agricultural employer. By 2000, trade in tourism and its associated industries amounted to $3.5 trillion worldwide, which was equivalent to 10.7 percent of global GDP and 207 million jobs (Miller 2002). As a major source of foreign exchange, tourism accounts for over 100 million jobs in developing countries. The growth potential for tourism in developing countries is considerable, given that in 2001, they only accounted for 147 million international tourist arrivals, or about 21 percent of the global total of about 700 million arrivals (WTTC 2001, in Wayne 2002b).

The growing need and desire for people worldwide to travel, for business, pleasure, personal reasons, and often for all three at the same time, will only increase in the future. Assuming no further unexpected barriers to growth, such as another 'Asian economic crisis,' major terrorist attacks, or other unforeseen social upheaval, by 2020 world tourism arrivals are projected to grow and reach 1.6 billion persons (at an average growth rate of 4.3 percent per year) and world-wide tourist spending is expected to reach $2 trillion (growing at 6.7 percent per year) (Miller 2002).

How can we make economic, business, and educational sense out of this growth? What can these statistics tell us about the world economy and the impor-tance of travel and tourism in it as a subset of services industries? These questions require a 'big picture' approach to tourism trends, ranging from an assessment of the industry's structure, to its global, regional, and local geographies, and issues concerning specific types of tourism, including ecotourism and medical tourism. These factors are important drivers of change in the tourism economy, and, given the industry's size, are significant and influential in the world economy.

This chapter views the tourism economy as a microcosm and reflection of the global economy in the information age in the late twentieth and early twenty-first centuries, as technology, information, and capital have converged to produce new and increased flows of information and knowledge. The conditions of this conver-gence are spatially distributed in the international system and produce geographies of economic power with global reach and locally specific dimensions. As a microcosm of the contemporary world economy, international travel and tourism is, in its core industries and infrastructure, a complex and interrelated set of functions whose expansion has depended on advances in communications tech-nologies and economies of information (Smith 1991a: 1–6).

This essay uses tourism as a laboratory to explore the global economy from three major perspectives. The first perspective is an assessment of the global tourism industry from the framework of an "integrative model of international relations," which highlights both the major role of tourism in the world economy and how the industry's conditions reflect global trends in economic and financial activity and organization. A second perspective demonstrates how the tourism industry mirrors downturns and disruptions in the world economy, through an assessment of impacts of war and terrorism using examples of the 1990–1 Persian Gulf War and the September 11, 2001 terrorist attacks. The third perspective introduces world regional trends in tourism activity and explores tourism from arenas of social transformation with significant future impacts, including global environmentalism, demographics, and health care. These arenas are already reflected in new interregional tourism trends, especially ecotourism and, increasingly, medical tourism. The discussion focuses on aspects of the growth of medical or health tourism, which especially reflects economic globalization as neoliberal policy promotes private health care and higher costs in industrialized countries and people seek alternative lower cost health care options. New global patterns of health tourism also appear to be on the rise in the post-September 11 era. In these regards, this chapter both establishes the significance of tourism in the world economy and poses fundamental questions about its expanding roles.

Tourism as information economy

In its purest geographical form, international tourism is the movement and management of capital and information embodied in people crossing the global landscape. These material processes and conditions are embodied in the international business and leisure traveler. Tourism service industries institutionalize the management of this capital and demographic profiling in databases for the information-hungry frequent traveler and other frequent traveler industries. The American Express Company is an early leader of this organizational form:

> When a customer walks out of our travel office, ready for a business trip or a vacation, he or she probably thinks that they have just bought "travel," as though "travel" was a commodity. But we don't sell "travel." What we do sell is the means to manage travel; in other words, information.
>
> (Cupp in Smith 1991a: 292)

The buying and selling of such information, through communication and information technologies supporting global trade in service industries, are major forces of the global economy in the information age. The neoliberal restructuring of the world economy in the 1990s has decreased barriers to trade and financial flows, and has broadened world markets and opened geographical and political borders. These complex events have had significant impacts for tourism, including "visa- and passport-free excursions, extended accessibility to formerly restricted areas, unification of divided states, and the growth of international economic and trade alliances" (Timothy 2001: 120). In the ways that changes in the world economy concern cross-border integration of markets within and among national and regional economies, these trends also describe the role and growth of international tourism as a trade-in-services export for countries. They reflect the growth of tourism and tourism-related companies, from small-scale local tour companies to transnational corporate interests, such as Mytravel (previously Airtours; covering Europe and North America, headquartered in the UK), First Choice

(European, based in the UK), Kuoni (a Swiss company with holdings in Europe, Asia and the United States), Club Med (a French company with properties around the world), and Expedia (an Internet company based in the United States). As Knowles, Diamantis, and El-Mourhabi (2001: 76–7) have observed:

> Technology, information and the reduction of boundaries have created new forms of service companies, not only the large transnational corporations such as the Disney Corporation, but also the small niche specialist that can take advantage of the Internet, international communications, and market positioning or targeting.

Strategically managed smaller companies now can compete with larger companies for international visibility and positioning via the World Wide Web.

Terrorism and tourism restructuring

In the tourism industry, economic restructuring manifests itself in aspects of tourism supply and demand, technological innovation, organizational consolidation, and product diversification. With the spate of terrorist activities in 2001 and 2002, and the economic downturn experienced by most economies, supply and demand have been revised to reflect a growing trend in home country travel and to countries perceived as 'safe.' Hotels, reacting to lower occupancy rates, are concentrating on technological upgrades in room services as part of standard equipment, targeting both business and leisure travelers, and an increased emphasis on personalized attention to visitors as special clients.

Unsettling events in 2002 included upheaval in US and international airport security systems, bankruptcy filings as a result of deep losses by two major US airlines (United and US Airways), and drastic changes in airfares and frequent flyer programs. Airlines now view information-rich frequent flyer programs as a means for product diversification to increase the value of brand loyalty for premier gold and platinum status members. After losing more than $7 billion in 2001, "airline companies probably lost about $8 billion in 2002" (Alexander 2003). These losses have led airlines to yet another product diversification—new fees attached to paper tickets ($25) versus electronic ones. An attempt to charge passengers $100 to fly standby on flights on the same day as their originally scheduled flights was quashed by irate air travelers. What did stick was a restriction on discounted tickets. Whereas in the past, passengers could apply the price of unused discounted tickets to new purchases, in 2003 if not used on a scheduled flight or rebooked prior to take-off of the original itinerary, the ticket value is lost (Alexander 2003).

These changes are designed to influence the behavior of tourists as consumers, whether leisure tourists or the high-revenue-generating frequent business traveler, in order to salvage airlines buffeted by changing global circumstances. Behind the scenes, and coordinating many of these changes, were tourism 'facilitators' that control powerful telecommunications technologies, specifically airline computerized reservation systems, and tourism financiers and global-reach bank and non-bank credit card corporations, such as The American Express Company, Mastercard, and VISA. In the arena of policy, law, and regulatory institutions these facilitators also include the World Tourism Organization (WTO), the World Travel and Tourism Council (WTTC), and the 'other' more commonly known 'WTO', the World Trade Organization, and its General Agreement on Trade in Services (GATS). Important regional alliances and coalitions are also influential in the industry, such as the Pacific Asian Travel Association (PATA) and Caribbean Tourism Organization (CTO) (Smith 1991a: 60–3).

In previous research on the organization of tourism in the global economy, I introduced these groups of agents as four key nodes of activity in an "integrative model of tourism as international relations" (Smith 1991a: 61). This model seeks to explain the international tourism infrastructure as a central hub in the technological interaction of economic sectors comprising the tourism industry (Figure 5.1). This infrastructure includes transportation (airlines, car rental, rail companies), lodging (hotels, bed and breakfasts, vacation rentals, and timeshare units), food and beverage, travel suppliers (travel agents, tour operators, cruise lines), and financial organizations (banks, credit card companies)—all of whom depend on centralized information systems to conduct tourism services. What facilitates interaction among these agents or groups are communication technologies— from global Internet-based computer reservation 'call center' services for airlines, hotels, and rental car companies, to fax machines and wireless PDAs. These communications technologies combine the information-bearing capabilities of the tourism industry with the means of the international financial system to form new industry products and means of business operation. Travel products have evolved from travelers' checks to new products and modes of payment and types of business organizations, such as E Bookers (in the UK and Hong Kong), which is a European Internet booking company, Galileo (computer reservation system) owned by Cendant Corp, Sabre (owner of the online Travelocity, serving business travel), and the Pegasus computer reservation system.

The impacts of e-commerce, the buying of goods and services through the Internet, which is growing at phenomenal rates, are fundamentally changing the travel and tourism industry. Growth in Internet usage increased in Western Europe from 6 million users in 1999 to 29 million projected by 2002 (Cook 2000a), while 72 percent of Americans (203 million people) were online in 2001, up from 67 percent in 2000 (UCEA 2001/2002: 3). In 2000, airline website ticket sales comprised 73 percent of all online travel sales dollars, and online travel agencies and hotel, car rental, and cruise sites were growing at a projected 200 percent annually (WorldRes.com 2000).

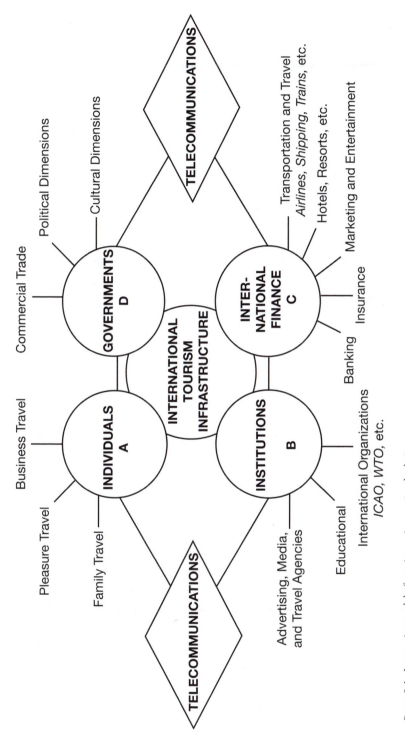

Figure 5.1 Integrative model of tourism as international relations

Source: Smith (1991b)

The power of information accessibility through the Internet allowed airlines to take cost-cutting measures with dramatic impacts on travel agency business. For decades, 10 percent of the ticket price was standard commission for travel agents. Since 1995, the major US air carriers have implemented five reductions or 'caps,' ultimately limiting agent commissions to $10 one-way and $20 round-trip. In March 2002, even these small commissions came to an end when Delta Air Lines, third largest in the United States, eliminated many of the commissions it was paying to over 26,000 US travel agents, retaining commissions only for those agencies guaranteeing a certain percentage of bookings (Alexander 2002). Northwest, American, and Continental Airlines quickly followed the new policy. Distribution costs, including paying commissions and printing and mailing tickets, were the third-largest cost for airlines, representing nearly 8 percent of their expenses. As a consequence, booking fees have risen and travel agents have had to diversify their products and services to survive (Alexander 2002).

Links between telecommunications technologies, transnational banking, and tourism infrastructure especially demonstrate the relationship between tourism and the larger financial services industry. By assessing links with airline computer-ized reservations systems it is possible to see some interesting aspects of these relationships. Companies such as American Express serve as facilitators for finan-cial and tourism transactions through American Express-branded travel agents. American Express also promotes its own travelers checks and credit cards. Airlines routinely co-brand with MasterCard and VISA credit cards, rewarding card users with airline frequent flyer miles. These miles can be used to buy a variety of travel and consumer products. The airline computerized reservations systems are also linked with those of rental car companies. This is apparent when making airline reservations, which reliably conclude, whether over the telephone with an airlines' agent or on an airline reservation website, with the question: "do you also need a rental car?" In turn, it is impossible to rent a car from any major car rental agency without using a personal, corporate, or government credit card.

Credit cards, especially as debt instruments, are the finance of international travel and tourism. Between 1980 and 1990 in the United States alone, the credit card market grew from 128 to 267 million cards in circulation, with charges on these cards growing from $67.4 billion to $300 billion (Smith 1991a: 308–9). The top two credit card companies today, Visa International Inc. and MasterCard International Inc., have 6,000 member banks and together control 75 percent of the $1.3 trillion credit card market (Glovin 2001: E7). American Express has 18 percent of the market, followed by Discover, which is owned by Morgan Stanley. With their ease of use, credit cards have facilitated the purchasing process in travel, allowing buyers to make large purchases conveniently and in their home currency.

Since negotiations began on the GATS at the end of the Uruguay Round of Trade negotiations in 1994, trade in tourism services has played a significant role in the progressive liberalization of markets through the reduction and removal of barriers to international trade. Over 120 countries have made commitments to the World Trade Organization to liberalize trade in tourism services, more countries than for any other trade sector—"a symbolic significance of agreement in one area that might facilitate agreement in another area" (Wayne 2002a: 1).

With the World Trade Organization overseeing nearly 98 percent of global trade, and with a large percentage of this committed to the liberalization of trade in tourism services, we should expect tourism to increase in economic stature and financial power in the twenty-first century for developing economies. Should we, therefore, also expect tourism as a set of service industries to increase in stature as a symbol of free trade economies, a potential focus of criticism and protest by the anti-neoliberal platform, or even attack by terrorist groups? These growth and anti-growth tensions pervade the landscape of contemporary tourism as global economy.

Tourism disrupted: war and terrorism

Diminished travel and tourism activity in world and regional economies, which is usually concurrent with diminished economic activity in general, often results from political, economic, environmental, and social instability. This is especially pronounced for countries and regions reliant on international tourism and its set of services industries. The United States had been relatively shielded from such impacts on its tourism industry until September 11, 2001. The impacts of the Persian Gulf War, a decade earlier in 1990–1, foreshadowed some of the consequences of such an event.

Tourism's vulnerability to the negative conditions of political instability reflects dramatically and rapidly in the services landscape and various substrata of travel-related industries (Smith 1991b: 5). Such ripple effects were never more apparent in recent history than during the Persian Gulf War, from July 1990 to March 1991, and, even more so, following the attacks against the World Trade Center and the Pentagon on September 11, 2001. In the immediate aftermath of these crises, the complex interdependence and integration of world politics with world markets manifested in 'collateral damage' of global proportions to international travel and tourism. Accompanying multiplier effects heightened these impacts, through rapid information flows and their near-immediate effects on prices, currencies, and markets, and secondary effects on consumption and labor demand. During the Gulf War, even though the geographical threats of war were perceived to be far from the shores of the United States, they impacted international travel plans for both industries and consumers. In 2001, the United States experienced a major domestic attack, with the likelihood of longer-lasting reverberations in the national and global economies.

During periods of upheaval, the tangible evidence of tourism's connectivity to global economic markets can take diverse forms. In 1990, for example, wildly fluctuating jet fuel prices were foremost among factors affecting the international transportation industry—rising from 57.5 cents a gallon (3.8 liters) in July 1990 to $1.28 by October, and back to $1.05 by December. During this period, the cost of fuel for the American Airlines corporation alone, on an annualized basis, jumped more than $1.5 billion (Smith 1991b: 6). Other evidence of the integrative nature of tourism in economic affairs was the marked decline in international visitor receipts by one-half of earlier projections for 1990, from an estimated $527 billion to $225 billion. There was also a 1.5 percent fall in global air traffic for that year. This decline was unique in that it appeared to be triggered by a new pervasive and relentless fear for personal safety relative to travel, one of the most paralyzing and demoralizing of human motivations (Smith 1991b: 11).

Multiple sectors of the US travel and tourism industry organized an unprecedented coalition to bolster US domestic tourism in the face of severe declines in international travel both into and out of the country. Key US industry leaders, including American Airlines, Marriott, and American Express, mounted the Go USA: Travel Moves People campaign, promoting "America – It's time to get back to business" and "Call your travel agent." It was the first coalition ever formed among the disparate and competitive sectors making up the US travel and tourism industry. The campaign drew together industry elites in a unique, heterogeneous constituency whose influence accrued from broad-based, cross-sector diversity and effective mass representation. These industry sector leaders put aside their competitive differences and pooled vast resources to do something concrete about the industry's plight in the face of the Persian Gulf War. Avis and Hertz, perennial competitors who have gone as far as to base their marketing campaigns on their status against each other, both supported the Go USA campaign—said to be the first time the presidents of Hertz and Avis ever agreed on anything (Smith 1991a: 340–54).

Effects of September 11, 2001

Links between tourism and declines in economic activity during the Gulf War now appear as early warning signals for the potential of even larger scale impacts of terrorism on tourism and the world economy. Following the terrorist attacks on the World Trade Center and the Pentagon, September 2001 arrivals to the United States were down almost 30 percent from September 2000, which was the largest single monthly decline in US history (TIA 2001). While business travel was expected to recover more quickly from a 50 percent reduction in the period September–November, 2001, the US Federal Aviation Administration (FAA) reported that passenger traffic was not expected to recover from the terrorists' attacks until 2003, at the earliest—barring, or course, another war or continued terrorist threats. Already in the context of an economic slowdown in 2001, the US tourism industry-wide decline accelerated after September 11. In what remained of 2001 alone, US tourism job losses of 453,500 in the 2001 US tourism sector produced a net revenue loss for 2001 of $43 billion (TIA 2001).

The effects of September 11 also deeply impacted local tourism industries. The Washington, DC Convention and Visitor Corporation suffered crippling cancellations in its hotel and conference center registrations. In the first few weeks after September 11, in Washington, DC, visitors to the Smithsonian Institution museums decreased by 75 percent, while hotel room occupancy plummeted from 80 to 1 percent (Hanbury 2002). One-half of the 2 million student and youth travelers who come to the Capitol on school programs, and generate $1 million in revenues, were cancelled for the entire year within a few days after September 11.

As a result of September 11, the tourism industry has come to recognize both its strength and vulnerabilities in the production, distribution, and management of information critical to the well-being of global tourism. In reaction to the terrorist events, industry leaders organized themselves to promote travel and tourist activity in several key ways. Similar to the industry's response to the travel slowdown during the Gulf War, the Go USA: Travel Moves People tourism industry coalition, headed again by major industry organizations, reemerged in a new form to encourage new management partnerships with travel-related firms and organizations. Through networked interfirm relationships, the coalition sought ways to achieve rapid responses to management issues after such catastrophes. In October 2002, the presidents of leading travel and tourism associations formed The Global Coalition to demonstrate unity and leadership and encourage industry partnering to promote effective security measures globally and restore consumer confidence. CEO Link, an initiative led by senior executive volunteers from the Business Roundtable, whose members include some of the nation's largest corporations, formed to establish a communication network that would maintain industry and government communication in the event of a national emergency. This network is being designed and developed by AT&T at its own expense and will include a wireless telephone network and secure website (Miller 2002).

In response to September 11, and in light of the international tourism as international relations model, we should not be surprised to learn that in 2002, US federal aviation authorities and technology companies started "testing a vast air security screening system designed to instantly pull together every passenger travel history and living arrangements, plus a wealth of other personal and demographic information" (O'Harrow 2002). To prevent future terrorist attacks, the government is working to link travel reservation systems to private and government databases. Because of the globalization of airline computer reservations systems through partnerships and alliances, the reach of such a plan will circle the globe in search of traveler information. The increasing sophistication of information technology could, for example, link a debit card to its owner if it is used to buy tickets for four people sitting on different sections of the same plane or sharing the same addresses in the past. Refined searches could reveal an array of unusual links and travel habits among passengers on different flights. Such profiling contributes to a "threat index or score for every passenger," with passengers with higher scores singled out for additional screening. As controversial as it is powerful, proponents claim it as an effective new tool in an unobtrusive network enabling authorities to target potential threats far more effectively; critics fear it as "one of the largest monitoring systems ever created by the government and a huge intrusion on privacy" (*ibid.*). Despite enormous sums spent on security, the government response has not been able to quickly reassure the traveling public in the United States.

Tourism futures: trends toward regionalism

The World Tourism Organization maintains tourism statistics by countries and world regions. Despite ideas about the globalization of mass tourism, general trends of the data, discussed below, indicate just how uneven the travel mobility and tourism economy continues to be. Still, tourism demand has been growing rapidly, and is also growing regionally, as international tourists seek new places and new countries to experience, and, if post-September 11 trends continue, domestic and international tourists seek vacations closer to home. Similarly, tourism supply by region reflects increased national marketing efforts, the organization of services at regional levels, and recognition of increasing tourist interest in travel oriented toward 'well-being' and lifestyle values. On the world scale during the 1990s, tourism has been growing fastest in the Asian region; the following discussion of niche tourism focuses on one of these new niches, health and medical tourism, and its growth in Southeast Asia.

Changes in tourism arrivals by world region provide some indications of international tourism trends. The leading tourism region by the number of international arrivals continues to be Western Europe. In 2000, 58 percent of an estimated 699 million worldwide international tourists traveled to and between the countries of Europe, and France has been the leading destination for at least a decade. The Americas are the second most popular region for international tourists, with 18 percent of international arrivals, followed by East Asia and the Pacific at 16 percent of the world total. Africa drew 4 percent, followed by the Middle East at 3 percent and 1 percent for South Asia. Within the Americas region, North America dominates substantially: in 2000, nearly 40 percent of all international tourist arrivals in the Americas went to the United States, followed by Mexico at 16 percent and Canada at just under 16 percent. Not surprisingly, most of these international visitors were cross-border travelers between the United States and its neighbors. The leading international destination in South America has been Brazil, followed by Puerto Rico, Argentina, and the Dominican Republic. In a country-based comparison, the world's top ten destinations for 2000 were France, the United States, Spain, Italy, China, the UK, the Russian Federation, Mexico, Canada, and Germany (WTO 2001b).

Regional tourism rankings are somewhat different when growth rates are considered. The fastest growing tourism region for much of the 1990s has been East Asia (which includes Southeast Asia) and the Pacific, which grew at nearly 15 percent from 1999 to 2000. Within the region, China drew over a quarter of international tourists in 2000 and has contributed substantially to regional growth. Although part of China since 1997, Hong Kong is still treated as separate in most international statistics, and within the region it holds second place, receiving nearly 12 percent of international arrivals in 2000 (a good portion of which were from mainland China). Although based on a very small share of international arrivals, South Asia has experienced higher than average growth, with 11 percent for 1999–2000 (WTO 2001a).

Tourism spending by region and country generates yet again a somewhat different picture of regional trends. For example, while Japan ranked ninth within the East Asian and Pacific region in 2000 for international tourism arrivals, the Japanese ranked fourth in the world for total international tourism expenditures, outranked only by the United States, Germany, and the UK. The United States leads all other countries as the world's number one tourism earner in terms of international tourism receipts, due to its size and comparatively high costs; it received nearly 18 percent of the world total for 2000. Among the top ten countries in 2002 in terms of tourism receipts, the only non-Western country was China (WTO 2001a).

The size and volatility of the tourism industry makes industry organization a key target for regional economic restructuring. Regional reorganization especially characterizes the airline industry. One aspect of organizational regionalism, especially notable in the United States after September 11, is a significant trend for some airlines to pare costs by shifting routes to smaller regional airlines. The US Federal Aviation Administration predicted that the number of regional jets would rise from 732 in 2001 to 2,970 by 2013 (Associated Press 2002). One reason for this is that sub-nationally both regional tourism firms and destinations in the United States experienced the least amount of loss post-September 11. These included mid-sized companies (50–99 employees), rural areas, the Western US region, and businesses serving primarily domestic markets (Cook 2000b).

Travel demand for regional tourism is being fueled by increased fears of travel-related terrorist activities in urban areas and major travel routes, and may be changing the global tourism landscape permanently, thereby contributing to growth of regional and rural tourism worldwide. According to Cook (2000b), possible post-September 11 effects may include enhanced markets within residential, 'closer-in', and 'drive-to' categories; domestic travel doing better than outbound; more interest in 'visiting friends and relatives'; and a shift in activity preferences with increased popularity for national heritage sites and more 'serious' educational and self-improvement endeavors. In light of these changes in travel preferences and trends, what characteristics will future travelers embody?

We can approach what the future of travel might look like from trends and forecasts in a variety of arenas. For the purposes of this essay, these include the graying of tourism, increasing world environmentalism, and rising interest to travel for personal improvement and health services.

The graying of tourism

One critical phenomenon affecting travelers and their behaviors is the 'Floridization' of the developed world. In the United States, 18.5 percent of the population of Florida currently is aged 65 and over, and populations of the developed world generally are aging fast. In the following countries, 18.5 percent of the population will be over 65 within the first quarter of this century: Italy in 2003, Japan in 2005, Germany by 2006, the UK and France by 2016, Canada by 2021, and the United States by 2023 (WTTC 2001). An over-65-year-old traveling population presents a significantly different profile for travel marketing and management. An even larger group of travelers, 'junior matures' (ages 55–64) will drive tourism growth during the early decades of the new century; they travel more often, farther from home, stay longer, include two or more destinations in one trip, engage in more activities, and spend more per trip (Cook 2000a). Older women, too, are becoming a strong travel category requiring accommodations, services, and activities appropriate to their needs, preferences, and budgets.

In tandem with the aging of the population and perceptions of a less secure world is evidence of an interesting trend toward people assuming greater responsibility for their own quality of life and societal values regarding environment and landscape conservation. At senior and middle executive and management levels, the effect of this is a growing interest in personal services. Such services have become increasingly visible in larger airports (including a quick rise in massage therapy salons in the United States post-September 11), and in upscale health resorts with personal trainers, coaches, nutritionists, career counselors, concierges, and health care providers (The Herman Group 2002: 1). An example is the rapid growth rate of the spa industry in the United States, which experienced aggregate growth of 114 percent between 1999 and 2001 (ISPA 2002). The following discussion focuses on these trends, what the industry terms 'niche tourism.'

Niche tourism

New travel markets are increasingly emerging with trends toward more personalized travel experiences. Key travel segments now include timeshare, cruises, adventure travel, ecotourism, cultural heritage tourism, 'volunteer' tourism, and health and medical tourism. Four million households worldwide in 190 countries own vacation interval timeshares, and cruising is on the rise, especially short cruises within single regions (Cook 2000a). While short cruises have increased in popularity, trends in the size of new cruise ships are moving in the other direction, especially for high-end, long cruises. One ship, described as a "floating city," has been proposed for launch in 2006; it is effectively a small town.

> The size of four aircraft carriers, the 'Freedom Ship' is a mile-long vessel with commercial enterprises including restaurants, banks, grocery stores, and schools. It will also include several airstrips and a 25-story steel building onboard. The ship will follow the sun and circle the world every two years, stopping in all the major ports along the way. The cruise ship will accommodate up to 30,000 day visitors and with another 20,000 able to stay in the ship's hotel rooms.
>
> (Rach 2002: 1–2)

Overall, the cruise industry has grown steadily at approximately 8 percent per year since 1998.

Adventure tourism and ecotourism continue to be a high-growth sector, with projections of a 10 to 15 percent annual growth rate, especially in existing established destinations such as Costa Rica, parts of Mexico, and Peru. In addition, more seniors are selecting this sector of travel, with more women than men opting for activities like sailing and backpacking in recent years (Cook 2000a). The industry is beginning to distinguish between 'soft' and 'hard' ecotourism, the latter referring more to adventure tourism and extreme sports. The popularity of ecotourism will likely lead to this niche becoming increasingly mainstreamed. International recognition of the World Earth Summit (Rio + 10) in Johannesburg, South Africa, in 2002, coinciding with the International Year of Ecotourism 2002, served to strengthen the links between mass and sustainable tourism. In the United States, the National Geographic Society established a Sustainable Tourism Resource Center information site to foster this trend. The principal locus of the center, under the direction of Jonathan Tourtellot, an editor of *National Geographic Traveler*, is an interactive website dedicated to ecotourism, culture and heritage, transportation, and destination appeal. Topics covered include wildlife habitats, historic districts, unique music and local cuisine, and more, where "The goal is to encourage everyone, whether resident, visitor, or professional, to conduct tourism in a way that supports the geographical character of the place being visited—its environment, culture, heritage, aesthetics, and the well-being of its citizens" (Tourtellot 2002: 1).

Thus travel organizations and other industry institutions are interrelating trends of 'Floridization,' ecotourism, and also cultural heritage tourism. Major associations such as the AARP (formerly the American Association of Retired Persons) have formed travel policies and programs to address the mounting interests of its members worldwide in participating in ecotourism and heritage tourism. A long-standing member of the United Nations, the AARP promotes international cultural awareness and respect for the traditions of indigenous populations. As an example of its participation, in 2001 the AARP established a website marking the UN International Year of Ecotourism and to support its "Ecotourism and the 50+ Traveler" colloquium.

An increasing tendency among contemporary travel consumers is to view travel as a means for enhancing the quality of life, through seeking personal and health services, or by building on a philosophy of doing well while doing something good for society. In the former case, such services include everything from spa, meditation, and other non-essential health services, to elective and sometimes medically necessary surgeries, oftentimes in higher quality or lower cost locations. In the latter case, these activities range from volunteerism to education, training, and career development as well as scrutiny of the travel and tourism industry's cultural and environmental impacts. "One of the reasons why people are traveling in record numbers relates to consumer demand for hybrid types of leisure experiences where a more complex array of 'human needs' require being met, particularly physical fitness and mental well being" (Spivack and Chernish 1999: 1). Health and tourism also have strong economic multiplier effects, generating activity that reverberates through the service sectors. For example, in the 1990s an average of 6,000 people travelled to Cuba yearly for such specialty treatments as night-blindness surgery and orthopedic therapy, with the result being a major tourism boon to Cuba's state and local economies. In general, the effects of the growing elderly population combined with the neoliberal economic policy turn worldwide, promoting privatization of health services, especially high-quality elective treatment, are placing intense pressures on pension and health care services providers and giving rise to increasing demands for health tourism.

Health tourism in Southeast Asia

Diverse countries have been destinations for spa and health tourism for years, especially in Europe. Tourist guides of Germany, for example, describe the spa tradition as "a health craze phenomenon that is rarely surpassed in any other culture. Taking the 'cure' or the 'waters'...is...also covered by German health insurance" (Flippo 2002: 59). In other regions of the health tourism landscape, on the Caribbean island of St. Lucia, guests at the Hilton can make appointments with the island's resident bush doctor, who "dispenses hands-on healing and natural herbal medicine from his garden" (*Bangkok Post* 4 February 1999). South Africa has attracted 'scalpel tourism' or plastic surgery beauty holidays (Kruger 2001). But the rise of global tourism in more basic or non-elective health and medical services is more recent, reflecting simultaneously the globalization of medical science and culture in rapidly developing countries, and the 'graying of tourism' combined with high costs and sometimes long waits for health care in the industrialized world. Southeast Asia is one region where this niche has distinctively evolved. After a quarter century of rapid development in Southeast Asia, technological standards in the region are high, and many of the countries have well-developed tourism infrastructure, while costs overall remain relatively low. Singapore, with its developed country status, became a destination for advanced medical procedures in Southeast Asia in the 1980s. But Singapore's place in this niche has yielded to its lower cost regional neighbors (Wee 2001), Malaysia and Thailand, which, combined with their stylish and aggressive marketing, have created new standards in medical tourism.

At Bumrungrad Hospital in Bangkok (whose majority shareholder is Bangkok Bank Ltd), a US patient described her visit this way:

> Someone dressed in a beautiful Armani suit with little high-heeled shoes simply took me around from appointment to appointment and they immediately did all these tests, one after another. I went down and had lunch at the Starbucks in the lobby of the hospital, came back up and the doctor had on his desk the most beautiful file, all bound with tabs and everything, with all the results of the tests.... Something like that...you know, is impossible in America. I mean it's inconceivable.
>
> (quoted in Mydans 2002)

In Thailand, the low costs combined with world-class service with style confound first-time Western patients, or medical tourists. Bumrungrad Hospital's CEO is a former administrator of the University of Southern California Hospital in Los Angeles, who has become a booster of medical tourism, even having looked into how to award frequent flyer miles for services rendered. The hospital maintains a travel concierge to assist with airline reservations, bank transactions, and visas; and the food service boasts a monthly guest chef. Bumrungrad "has almost single-handedly shifted the regional hub of medical care from Singapore" (*ibid.*), and in 2002 one-third of its patents were international (Intel Corporation 2004). Costs for medical service are several times lower than private health care facilities abroad: an outpatient visit costs less than $10 and the average hospital bed is $50 a night. Its top five categories of international outpatients are Americans, Japanese, British, Bangladeshis, and Chinese, and after September 11 it has increasingly served "a flood of Middle Eastern patients who now avoid the United States for fear of discrimination. In response, the hospital...hired extra Arabic interpreters, stocked up on Muslim prayer rugs and opened a kitchen serving religiously acceptable halal food" (Mydans 2002).

In Malaysia, medical tourism has become a focus of state planning. The Eighth Malaysia Plan (2001–2005) targeted 44 of the country's 224 private hospitals for medical tourism development. The state economic rationale is service industry expansion, after a generation of rapid development based largely on manufacturing. Again, lower costs are the draw: in Malaysia the average cardiac bypass surgery cost $6,315 in 2002, compared to $10,417 in Singapore and up to $90,000 in the United States (Ahmad 2002). Malaysia's private hospitals tend to serve medical tourists from Indonesia, Brunei, India, and the Middle East, since the country's Islamic Malay culture provides comfortable surroundings for Islamic travelers, who have also increased in numbers in Malaysia since September 11 (Abdullah 2002). Like medical tourism in Bangkok, Malaysian health tourism is being located in established tourist destinations, especially Malacca, Penang, and Kuala Lumpur.

Conclusion

The new tourism trends targeting life-enhancing experiences reflect both needs and desires, and the search for meaning in the new global landscapes. In sum, these experiences increasingly will be anchored in visiting friends and relatives, the perennial favorite of travel, including visits to expanding networks of otherwise virtual friends. We should expect and welcome big increases in travel that broaden and enrich our lives—educationally oriented, elder hostel-type programs, heritage and cultural tourism that sustain natural and human environments. We will see strong growth in health and life-enhancement related travel that includes special diagnostic and treatment centers as tourism destinations. Health spas will increase in reputation as places for physical and spiritual revitalization. Continued interest in the outdoors and being active will further stimulate ecotourism and combine 'soft' and 'hard' adventure travel with volunteering and performing community service, becoming part of the shared tourism legacy. International tourism, embodied in living outside of one's country of birth, will become a living lifestyle (Cook 2000a).

The rapidly changing world economy is forcing expansion of national, regional, and international boundaries to include new and interdependent interactions—human heritage and technological, political economic, socio-cultural, and sustainable environmental. The central role of travel information capital, its movement and management, is now viewed as an increasingly powerful national "good" and international resource. In other words, the globally distributed nature of the convergence of technology and tourism conveys new forms of power in the international system—power of economics, trade, and investments, of control and access to information, money, and of individual safety and national security—of which international tourism is a microcosm. This chapter offered a glimpse into the vitality and dynamics of the set of service sector industries comprising international tourism as one way to understand this complex global landscape.

Part II

The city

There are no clear rules about how one is supposed to manage one's body, dress, talk, or think. Though there are elaborate protocols and etiquettes among particular cults and groups within the city, they subscribe to no common standard.

For the new arrival, this disordered abundance is the city's most evident and alarming quality...there are so many things...such a quantity of different kinds of status, that the choice and its attendant anxieties have created a new pornography of taste.

Jonathan Raban, *Soft City* (1974: 65–6)

At the end of three days, moving southward, you come upon Anastasia, a city with concentric canals watering it and kites flying over it. I should now list the wares that can profitably be bought here: agate, onyx, chrysoprase, and other varieties of chalcedony; I should praise the flesh of the golden pheasant cooked here over fires of seasoned cherry wood and sprinkled with much sweet marjoram; and tell of the women I have seen bathing in the pool of a garden and who sometimes—it is said—invite the stranger to disrobe with them and chase them in the water. But with all this, I would not be telling you the city's true essence; for while the description of Anastasia awakens desires one at a time only to force you to stifle them, when you are in the heart of Anastasia one morning your desires waken all at once and surround you. The city appears to you as a whole where no desire is lost and of which you are a part, and since it enjoys everything you do not enjoy, you can do nothing but inhabit this desire and be content. Such is the power, sometimes called malignant, sometimes benign, that Anastasia, the treacherous city, possesses; if for eight hours a day you work as a cutter of agate, onyx, chrysoprase, your labor which gives form to desire takes from desire its form, and you believe you are enjoying Anastasia wholly when you are only its slave.

Italo Calvino, *Invisible Cities* (1974: 12)

6 Silicon values

Miniaturization, speed, and money

Dean MacCannell

It is a paradox, perhaps, to speak of Silicon Valley in a geographic context. This is in spite of the unambiguously geographic marker in its name. Why? Because the inventions, discoveries, products, and values which produced the enormous wealth in the valley, and attracted so many bright entrepreneurs and innovators, over the past 25 years, are technically *anti-geographic*. I will argue that this is true at all levels from landscape, to products and performances, to behavior ideas and beliefs.

First, consider products that substitute programs for geography. Primary among these are the virtual geographies and virtual landscapes programmed for gaming, training, military, and other simulations. These digital environments are proffered by their creators as substitutes for actual terrain and places for life-forms:

> Artificial life scientists consider computers to be worlds in a variety of senses. Often they mean that computer simulations can be seen as self-contained, self-consistent symbolic models that generate objects and behaviors corresponding to things in the real world. Increasingly, they mean that computers are literally alternative worlds.... Making sense of this contention is important, for it makes possible the claim that real life-forms can exist in the computer.
>
> (Helmreich 2000: 65)

> Computers are figured as places to begin again.
>
> (93)

> The word for world is computer.
>
> (65)

It does not take a dialectician to extrapolate from these statements that the world outside the computer (the domain of 'geography' conventionally so called) will be devalued and neglected by this kind of thinking.

We can draw a straight line from the "word for world is computer" to the actual condition of the landscape of Silicon Valley. The Valley is the teleotype of the 'geography of nowhere.' It is generic exopolis, homogenized, undifferentiated; a vast internodal matrix of tract housing, apartment complexes, strip malls, and industrial parks. In his latest book, Joel Kotkin (2000: 39) comments:

> Although rarely noted amid the hype over Silicon Valley, the fact [is] that the Santa Clara Valley now suffers from many of the undesirable traits of traditional urban areas—from increasing smog and traffic to juvenile crime and soaring home prices.

One could argue that the degradation of Silicon Valley is even greater than 'traditional urban areas.' Older urban areas that have been overlaid with suburb-like infill may retain some amenities such as transportation systems, museums, parks, libraries, etc. Silicon Valley has more than its share of the 'undesirable traits' of over-development with precious few of the amenities.

A visit confirms that the Valley does not reliably have even the most elementary amenities of 'traditional urban areas,' where people ordinarily live and work. Unreliable presence of sidewalks is one of the issues:

> As a fairly recent resident of the Silicon Valley, I...am absolutely stupefied by the random placement of sidewalks. I cannot even walk from the Sunnyvale Caltrain station to my apartment at Wolfe Road without encountering "disappearing sidewalks." I am forced to walk on the road or negotiate a slalom through the trees placed where the sidewalk ends. Silicon Valley is the worst planned and executed urban environment I have ever seen.... The valley [is] the world's biggest strip mall.
>
> (Wozniak 2001)

We know that the degraded condition of the Valley is a product of a certain civic indifference (evidently not shared by the author of the last quote). The motive for moving to the Valley has not been because of any distinction or quality of place and space that it offers. The motive is the one that can blind us to every negative aspect of our immediate surroundings, i.e., to get rich quick.

Silicon Valley today resembles other areas where rapid development has surged ahead of regional planning. It is a patchwork of manufactured landscapes, strip malls, master-planned developments, industrial 'campuses,' and some older traditional communities, all newly merged in an undifferentiated matrix of international beige suburbia. It is the current high-water mark of 'flexible accumulation' strewn across what was once the world's richest prune-growing region. The business parks of Silicon Valley, if less attractive, are as visually undifferentiated as the orchards they replaced.

The Valley has a very strong identity as the locus of the business culture and values that are the engine of new computer industry wealth. But it is otherwise devoid of place distinction, so much so that it repulses some industry entrepreneurs. Commenting on the reason why he did not locate his business in Silicon Valley, George Morrow, an early computer pioneer, remarks that he wanted to avoid the "Mickey Mouse environment down there where you can't tell one company from another unless there is a sign out front, and where there is no sense of job identification or loyalty" (Mahon 1985: 168). The landscape of Silicon Valley is refractory to any geographic sense of place distinction or loyalty; it is fully evolved into a negative geography or an anti-geography.

It is noteworthy that the condition of the Valley, i.e., as anti-place, is seen as problematic by some residents; so place-making becomes a project. In an attempt to halt the complete rewriting of the landscape with new wealth, local community groups and governing bodies have tried to prevent the construction of 'monster houses' (tearing down or remodeling average homes and building to the full extent of the lot). And a local preservation movement has focused on the Eichler house, an inexpensive (when it was built), modernist tract home found in the older towns in the Valley. It is symptomatic that the nostalgic impulse has made as its cause saving a symbol of 1950s middle-class suburbia. The great wealth of the Valley, and the innovation on which it has drawn, might have been invested in the creation of experimental new forms of quality public space or creative housing alternatives, but it has not been.

Silicon Valley's geography of nowhere is spreading. The same international beige, suburban sprawl with palm trees is being squeezed into redevelopment projects in San Francisco South of Market: in the Mission District, Noe Valley, and Bernal Heights. At the height of the Silicon Valley boom, these San Francisco neighborhoods were sought out by commuting high-tech workers as the 'in' places to live. International sites are also in the nimbus of Silicon Valley residential development. In a study of the 'mirror home' phenomenon, researchers in Environmental Design at the University of California, Berkeley found that transnational migration of high-tech workers is leading to the global spread of 'silicon landscapes.' Designers of new suburbs in Taiwan measure and photograph streetscapes and residential facades and interiors in Silicon Valley and faithfully reproduce them in Asia and elsewhere. Interestingly, what supposedly makes these transnational clones of Silicon Valley desirable is often their retrograde design features. The Berkeley researchers commented:

> Regarding the home…from high-tech policy makers to local real estate developers, agents, and architects all try to duplicate the sense of suburban Silicon Valley in the hillsides of the Hsinchu region. In terms of the home dwellers, from those who have American life experiences to those who only imagine the American dream, most believe that the prototype of their homes

is an American single-family detached house standing on a piece of lawn fenced in and protected by a guarded gate.

(Chang and Liu 2000: 6)

Commenting on his residence in the Hsinchu region, the heart of Taiwan's high-technology industry, one of the respondents in the Berkeley study stated:

I have lived in the United States more than ten years, mostly in San Jose area…but we finally decided to come back here. Though I have lived here for seven years, I don't quite feel I am in Taiwan. I seldom leave home, and the interior decorations of my home are the same as my previous home in the States…. My father-in-law always told me, "Though it is very chaotic outside, as long as I enter your home, I feel I am in America."

(1)

Taiwanese respondents migrating in the other direction stated that as long as they stayed close to home in Silicon Valley they felt as if they had not left Taipei: for example, "I can enjoy my perfect Taiwanese life in Silicon Valley these days" (3). The drive among these transnational professionals is to erase any aesthetic or evident cultural differences between high-technology places, to reduce the psychological impact of geographic movement to a minimum.

The landscape that emerges from these design decisions closely corresponds to the one mapped in hand-held GPS devices: the status of any given point on the face of the earth is equivalent to every other point. Ideally, all points are connected by a program that provides clear and simple instructions on how to move in the most efficient way from anywhere to anywhere. There is nothing left over, nothing left out, nothing left to discover, nothing hidden from view. It is the digital ideal that fails to include and therefore authorizes sprawling abject reality.

Geography, miniaturization, and speed

We might question the utility of geographic conventions in the context of this abjec-tification of space. Here, I want to explore the usefulness of conventional place analysis in increasingly de-spatialized, de-contextualized landscapes. (I believe that classic conceptions of space and place continue to be useful.) Specifically, I want to examine the particular inflexion of concepts of place and space in Silicon Valley that leads to an anti-place ethos at the level of the landscape and the human body.

Begin with the paramount values in Silicon Valley, the values on which its wealth and empires have been built: *miniaturization* and *speed*. I will argue that there have been specific shifts in the meanings of miniaturization and speed before and after the computer, which, once explicated, will help us understand other deformations of the time–space continuum in the silicon landscape.

Miniaturization is inherently a spatial concept—perhaps the one most stubbornly resisted by the subject matter of geography except in the sub-field of cartography. Before the computer, miniaturization involved making reduced scale models and other kinds of smaller images of people, places, and things. This took place mainly in the realms of arts and crafts. Artisans have played freely with the concept of miniaturization. In the arts, smaller-than-life models are found across a wide range of scales. At one end is the three-dimensional silicon carving of Frank Lloyd Wright's architectural masterpiece, 'Falling Water,' made recently by a collaborative team of artists and engineers. This model can only be seen with an electron microscope. At the other end is the not-much-smaller-than-life copy of the Eiffel Tower on the Las Vegas strip. Both are technically miniatures, and participate in the aesthetics of miniaturization, but the scale of the miniature is also suggestive.

Claude Lévi-Strauss goes so far as to suggest the possibility (he does not say this directly) that a larger-than-life equestrian statue of a fallen hero might qualify as a 'miniature' if it somehow succeeds in implying, however larger than life it may be, that it is nevertheless not completely up to the task of representing the heroic status of the personage or the events it tries to depict. Lévi-Strauss explains that miniatures are seductive of human understanding. They seem less formidable, simpler, easier to grasp than the actual object and events they depict (Lévi-Strauss 1966: 23). They turn the process of understanding inside out. Rather than paralleling the movements of the reductive sciences from discrete elements to the totality, in the miniature, the totality is grasped all at once. The miniature in art always gives up some quality or qualities of its object. The object loses volume, three-dimensionality, or color, or details are left out. Still, we are able to grasp what is being represented. The miniature makes us feel powerful— in control. "A child's doll is no longer an enemy, a rival, or even an interlocutor. In it and through it a person is made into a subject," says Lévi-Strauss (*ibid.*). Aesthetic miniaturization thus has its own distinctive ethical structure. The miniature is always understood not to have the status of the original even if, as in the case of the doll, it creates its own original. Miniatures are not idealizations. They take the full-sized original as their ideal. Therefore they represent the idea of idealization.

The goals of computer hardware miniaturization are very different from those outlined above. The first difference is the object of miniaturization is not an external person, place, or thing. The miniature is not a metaphor. It is a smaller-scale reproduction of an existing component of the machine itself. The difference between the previous iteration and its miniaturized copy is always the same: increased efficiency, reduced size, less weight, and eventually lower cost of manufacture. Thus, in the computer industry we do not find 'miniatures' at all scales. Components just keeps getting smaller and smaller. The standards for computer miniaturization are also different from art miniatures. In art there is an instructive aesthetic difference between the original and its miniature, a difference recogniz-able by a subjectivity that can properly be called 'human.' As we come to understand what has been left out of the miniature, we also come to understand what is essential in the original.

When computer components are scaled down there is an opposite kind of requirement: namely, that the data stream they handle must not be degraded. There can be no performative difference between the two components unless the smaller device can actually handle more data than the larger one, as occurs, for example, in the latest generations of data storage media. As scale decreases, reproduction of detail must stay the same or increase. Retention of detail has proven to be an embarrassment for certain bit-mapped type fonts that take on a fuzzy appearance as they are scaled down. In the interest of legibility, more instructions have to be added as the typeface gets smaller.

I will argue that the differences between the two versions of miniaturization, the orientation toward space that they imply, can account for the abject quality of silicon geographies at the landscape level, and at the level of the human body and its various enjoyments.

But first, speed

Just as there are two distinct types of miniaturization, there are several distinct kinds of speed. Jacques Derrida, writing on the apocalypse, states, "*At the beginning there will have been speed*.... [A] race...a *competition*, a rivalry between two rates of speed. It is what we call in French a *course de vitesse*, a speed race" (Derrida 1984: 20, italics in original text). Derrida is well aware that to call the arms race a 'race' is a metaphor, because the contestants are literally going nowhere. In order to speak of the 'speed' of the arms race it is first necessary to drain all geographic referentiality out of the idea of a destination. The destination here is the end of time, not the end of a measured mile on the Bonneville Salt Flats or Black Rock Desert.

It is possible, then, to distinguish at least three types of speed. First is the 'speed' in the speed contest properly so called. It is the speed of the machine age that can be experienced by the human body as wind in the face and gravity force in acceleration and turns. This kind of speed always involves a specific traverse of the landscape from point A to point B, as occurred in 1997 on the Black Rock Desert in Nevada when English RAF pilot Andy Green drove a 105,000 horse-power car between the measured mile markers at 764 miles an hour (1,222 kilometers an hour), faster than the speed of sound (Zukovic 2000).

The second kind of speed has only incidental geographic coordinates. It is the speed of the nuclear age. The contest does not involve geographic movement or changing places. It refers to the time it takes to achieve an organizational or production goal; to achieve market dominance; an arms race; the time it takes to guarantee the destruction of a competitor or opponent.

The third kind of speed has no geographic specificity. This is the speed of the digital age. It refers to the time it takes to move information between machines, and the computation time it takes to cycle tasks that are required for the manipulation of digitized information, especially tasks that are frequently repeated. This version of speed has its roots in Fordist manufacturing procedures, codified in the 1920s by Frank Bunker Gilbreth using chronocyclegraphic, or stop motion, photography. But it assumes its most elegant forms in Silicon Valley. For example, before bit-mapping technologies, only the last line of type on a computer screen was available for editing. The discovery of an error meant ripping out everything that came after it. Bit-mapping allowed instant access for editing purposes to information anywhere on a computer screen. The first Ethernet cut the time of transmission of one page of data from computer to printer from 15 minutes to 12 seconds.

The spatial coordinates of digital speed are between different machines in an office, between offices in a firm, and between computers connected by the World Wide Web. Nanotechnologies permit the implantation of digital communication devices inside the human body. Anything that retains geographic specificity or the hint of embodiment, such as stumbling into a Japanese language website, constitutes an impediment to the instantaneous transfer of digitized information. As *Los Angeles Times* writer Bill Sharpsteen (in Zukovic 2000: 128) comments, "A car equipped with a surplus jet engine seems crude compared to the infinitely swift, silent power of the Pentium computer chip."

The fact that there is no conflict between miniaturization and speed in the creation of new information technologies is well known. In fact, reducing the current flow in circuits and the distance that electrons must travel are keys to higher speed computing. From an engineering perspective a way to make computers faster is to make them smaller and vice versa. What I want to examine here are some of the consequences of extending this engineering ethos to the level of the human body, socio-cultural relations, and the landscape. According to Michael L. Dertouzos, who was head of MIT's Computer Science Laboratory in the mid-1990s, miniaturization has brought us to the threshold of digital immortality. He believes that within 30 years nanorecorders will be able to transfer every experience of a person onto a microchip (Robischon 1996). Advertisers have jumped the gun, showing images of a schematic of the brain, one with a caption reading "now available on disk."

Paul Virilio was among the first to suggest the consequences of silicon values (i.e., the convergence of speed and miniaturization) for geography and the person. In a 1996 interview in *Wired* magazine, Virilio commented:

[T]he infosphere—the sphere of information—is going to impose itself on the geosphere. We are going to be living in a reduced world. The capacity of inter-activity is going to reduce the world to nearly nothing. In fact, there is already a speed pollution, which reduces the world to nothing. In the near future, people will feel enclosed in a small environment. They will have the feeling of confinement in the world, which will certainly be at the limit of tolerability, by virtue of the speed of information. If I were to offer you a last thought— interactivity is to real space what radioactivity is to the atmosphere.

(quoted in Der Derian 1996)

Are we to assume from Virilio that the high-tech workers of Silicon Valley feel geographically confined, restricted, inhibited? They certainly do not. Humans who are caught firmly in the grip of the 'Next Big Thing' are capable of incredible feats of compensatory self-mythification to see them through.

This is not to suggest that Virilio's analysis is wrong. Only that the denizens of Silicon Valley do not see their situation as he sees it. In fact, much of the culture of Silicon Valley can be read as an elaborate scheme of denial of any suggestion that their lives are 'confined,' 'small,' and 'limited'. The most persistent cultural thematic in the Valley is 'NO LIMITS.' Consider the following from Derrick de Kerckhove, a reigning pop-philosopher of the Web:

As you eliminate your body on the Web, you recuperate it in your physical loca-tion. Sometimes you have a body, sometimes you don't. If you don't have a body, you're not there. If you have a body, you are *so* there that your relation-ship to the world is what I call proprioceptive. It's tactile. It's not visual as it was during the Renaissance. In the Renaissance, what was your identity? It was the outer limit of skin, a head that processed information, a dumb universe shown as spectacle. Identity became a point of view. Today, identity is a point of being. We add new possibilities of mixed identities, collective identities, just-in-time identities, fabricated identities. There's great flexibility, but the core business of self remains, just…all over the planet by electronic extension.

(quoted in Kelly 1996)

The same geo-spatial anxiety is expressed more succinctly in the following adver-tisement for a laptop computer:

It's pretty clear what the rules are now. 1. If you stand still, you lose. 2. Your office isn't some set place. It's wherever you have to be to do your job. 3. Being away is no excuse for not being on top of it. A daunting state of affairs. Unless you are properly equipped—with Hitachi Mobilized Computing Technology.

(*Wired* 1996)

The only human condition which would encourage thoughtless discourse about having 'no limits' or 'no boundaries' would be a life that is bounded absolutely.

People who move easily through varying natural or social landscapes are highly aware of boundaries and respect them as a condition of their freedom. Interestingly, the rhetoric of having 'no limits' is often found conjoined with the phrase 'Shopping—the final frontier.'

Rhetoric notwithstanding, the most meaningful space in Silicon Valley is the triangle, approximately 18 inches (45 cm) on each side, that connects keyboard, eyes, and monitor. This is the space in which the most action takes place and most of life is lived. Even during 'wet' meetings, where actual human bodies huddle together, the participants often connect to cyberspace via laptops and hand-held devices. The computer screen is touted as the window onto the world, and even the universe. But it is not the whole universe or even the whole world. At best it is an encyclopedia version of totalization—it purports to be everything, but nothing stands out—the more that gets added to it, the more impoverished it becomes.

Can we say that the landscape is really in the computer? No. The computer is a fantasy keyhole to a morally and aesthetically undifferentiated universe. Its famous interactivity is not different from that of the telephone or the mail, only more and faster. When the computer itself does look back at you, it sees you only as a 'hit' or a source of error. It puts all of its users in the subject position of the pervert (MacCannell 2000: 55). For true perverts, everything is bathed in the same bright light. There is no possibility of a return of the gaze, because there is no *other*. "The core business of self remains, just all over the planet by electronic extension." Encountering no resistance, the ego swells to fill the entire landscape. A full-page ad in *Wired* (1995) magazine showed a racing motorcycle on a computer screen with a scrawl taking up the entire page asking, "Ever get the urge to take out a cow at 160 mph?"

Miniaturization and speed infect all other areas of life creating specific exigencies for the non-perverts. The Valley holds out the prospect of becoming rich quick, in no time, and without the need for any space beyond the 18-inch triangle bounded by eye, hand, keyboard, and screen. In the boom days in the Valley, there was a near perfect correlation between increasing the speed of information processing and the speed of wealth acquisition. We now know that the prospect has attracted thousands of workers to the Valley who would never become wealthy. And some of those who did become wealthy, gained and lost fortunes overnight.

Silicon leisure: love and sport

The 'get rich quick' ethos extends to other areas of life. Not just wealth, but love can be instantaneous. The personal ads of Silicon Valley newspapers and the San Francisco alternative press are filled with notices of intense affairs of the heart that last perhaps as long as 5 seconds:

You: stepping off a southbound train. Me on the yellow safety strip. I projected onto your one moment's glance a lifetime of unrealized expectations. If only we.... Sushi?

Friday 9/24 at Candlestick [Park].... When our eyes met I felt you were looking straight into my heart. I can't stop thinking of you.... Let's share some Merlot in a cozy place.

Red sports car @ 16th and Mission, 9/03. You: driving with backwards baseball cap yelling 'Hey baby you sure do look fine!' Me: walking. I pretended not to notice.... [I] wanted to chase your car.... Another chance?

(Guardian Connections 1999)

These cries are emblematic of erotic servitude to miniaturization and speed. There is no time and no space for the trajectory of a sexual relationship. The ad writers elaborate fantasies around purportedly lost enjoyments as a way of getting back something they never possessed. And it is only here, embedded in erotic anguish, that we find concrete geographic reference points: the train station, the ball park, a well-known street corner.

Instantaneous romance and passion have also evolved into cyber heartthrobs. A Japanese pop star named Kyoko Date, whose hit single, "Summer of Love," topped the charts, is described as being "5-feet-four and a fashionable anorexic 95 pounds" (Parry 1996), and a soccer buff. She has adoring fans and a line of commercial endorsements, and she is 100 percent digital. Her programmer–manager explains, "Most teenage pop idols have their songs written for them, their voices electronically altered and their images pre-packaged...so there is no reason why a computer animation can't do the job" (*ibid.*). If Kyoko Date should seem to be too remote or 'stuck-up,' for the price of a computer game boys can spend the night raiding tombs with Lara Croft, controlling her every move.

Something that has escaped the notice of commentators on Silicon Valley is a large complex of recreational sports activities strongly linked to Valley culture. These include mountain biking, said to have been invented in nearby Marin County, skate boarding, inline skating, street luge, indoor rock climbing, river running and other paddle sports, bungee jumping, and whatever else that is a recognized competitive event at the annual X-games held in San Francisco. Perhaps the infamously jerky driving style of high-tech workers should also be included here. The kinetics and aesthetics of these activities bleed back and forth between their real and virtual versions in video games, called 'twitch sports' after the spastic movements of people engaged in playing them in virtual space arcades. It is also the case that these sports (in their actual, not virtual, versions) require high-technology gear, which is often designed by the same engineers who make contributions to the digital revolution.

What interests me here is the possibility that this group of recreational and competitive activities is not merely historically associated with the digital revolution in Silicon Valley. Are jump and twitch sports, and the X-games, also products of the high value placed on miniaturization and speed? An argument can be made that the single-minded concentration that these kinds of sports require is the physical analogue of the mental concentration required to write error-free strings of code. They also tend to be *solitary* activities, engaged in by a single skater on a U-shaped ramp, or a single climber on a plywood wall. The actual space of the game or sport is a tight envelope around the body. Even if the body is moving through the landscape, as in mountain biking and running rapids, the macro-landscape has little to do with the event. It is what is immediately above and below the body at any particular moment that is all important.

Technically, the X-game player does not move through the landscape. The same perspective is always deployed around the subject no matter where he or she is in the landscape: it is always *this* edge, *this* drop, *this* wave, *this* overhang, *this* mogul that is all important—nothing like 'field position,' or bases occupied. The landscape is there simply to supply a sequence of opportunities for virtuoso displays of twitch, jump, or 'radical air,' and impossible-seeming recovery from the moment. The primary skill set and core activities of these games are the exquisitely executed jump, twitch, jerk, twist, spin, squirm, and wiggle. This is the logical effect on sport of adding the constraint of miniaturization to the existing ethos of speed.

In the effusive and breathless rhetoric surrounding these sports we find a reprise of the theme of 'NO LIMITS.' An advertisement for a 4 × 4 recreational vehicle asks, "What's it like living life out of bounds?" A sports watch advertiser quotes someone who has gained fame riding down waterfalls: "This is no romp in the rapids. It is nature on my terms. Without limits."

The geography of the X-game is the same as that of the yuppie geeks in everyday and work-a-day life who render every position they might occupy in the larger landscape equivalent by relating to that space and to others via web-connected cell phones, beepers, and laptops with faxes, GPSs, wireless networking, and the like. A common set of values is foundational to both the work and play activities. Silicon culture resolves the totemic division endemic in suburban American high schools between jocks on the one side and nerds and geeks on the other. With all the apology that is due to James Joyce for appropriating his famous maxim ("Greek–Jew is Jew–Greek"), let me suggest that in Silicon Valley Geek–Jock is Jock–Geek.

Conclusion

What is lost when the landscape disappears behind the seduction of quick money? A primary function of the landscape is (or used to be) to situate possible narratives. The landscape is structured like a language. What is beyond the landscape is the realm of the imaginary. As the landscape shrinks, the imaginary expands. This is the basis of the drive to 'get rich quick.' This drive, like all others, effectively motives behavior only if it is not fulfilled or realized. So the people of Silicon Valley, for the most part, did not get rich quick. But the drive, the ethos, had its impact on the landscape nevertheless. The result is a self-propelling fallacy. As the landscape and the stories associated with it disappear, there is nothing to do, nothing to see, nothing to know. The only desire that survives in a world without landscapes, without stories, is to get rich quick—and certain perversions.

It is commonplace to assume that 'get rich quick' is fully supported by a collective desire. Who has not experienced that desire? Can we call it innate? Is it heresy to ask what is the foundation of that desire? Is there something behind or beneath it? Here at the end, let me propose a hypothesis: the desire to get rich quick is always found in a miniaturized landscape that promises instantaneous transport from any point to any other point. It makes no difference whether it is 'get rich quick' because the horizons are so narrowed that the subject must fantasize escape; or they must narrow their horizon because 'get rich quick' is their only goal. 'Get rich quick' is the fantasy that we have woven around the loss of landscape, the body, their stories, their memories. The degraded landscape of Silicon Valley is an historical product of that fantasy. The fantasy of instant wealth is the only thing we have been told might compensate us for the loss of decent common ground. Hopefully, we will soon learn that nothing, especially not a baseless fantasy, compensates for a loss of decent common ground.

7   Bellagio and beyond

Claudio Minca

There are places in the world so exceptional they don't require superlatives.
They are best described with the simplest of the words.
Bellagio is such a place.

(www.bellagiovegas.com)

On a flight from New York to Milan a few years back, my attention turned to the aeroplane's hanging TV screens that update passengers on the duration of their flight, their current position in the virtual airspace. What I saw materialize before my Italian eyes, however, was an entirely new geography of Italy. The blue–green representation splashed across the shaking screens reminded me, in fact, of the blank map of the 'boot' that my elementary school teacher used to present us with in order to test our knowledge of the 'important' places of the country. Only on the screens before me, the order of the 'important' places that had sedimented within my mental map (as well as within those of most other Italians, I presume) was largely unrecognizable—subverted by the handful of signs dotting the virtual map, offering a concise and facile summary of Italy.

Besides the plane's Italian destination (Milan) the computerized map, in fact, marked the position of only four places: Rome, Florence, Venice and, surprisingly enough, Bellagio. Bellagio? Why should a small town on Lake Como, largely unknown to most Italians, be part of such a select and selective representation? Who was that map speaking to? Was it an error, a superficial tourist cartography of Italy? Bellagio is so popular among a certain stratum of Americans, I would find out later, because it (apparently) ideally incarnates the typical nineteenth-century Northern Italian lifestyle represented by a variety of novels and movies. Bellagio is where two centuries ago leading Lombard aristocrats built their villas—the most famous of these being the Villa Giulia, the Villa Melzi, the Villa Trotti and the Villa Serbelloni, which now hosts the Rockefeller Foundation's Centre and is rented out to a variety of conferences and aptly termed 'educational holidays.'

But Bellagio represents very little to most Italians (especially those who do not live in Lombardy); it is but one of the quaint and quiet tourist destinations on Lake Como. I remember, actually, teaching a course in Rome several years ago, where I met a student from Bellagio, who was incredibly frustrated by the fact that his claims of living in a world-famous destination were taken as a joke by all the other students on the course who had never heard of Bellagio before. Bellagio is a wholly Anglo-American tourist icon, a spot on the map that does not entirely belong 'here' (that is, in Italy); it is a globalized place subject to multiple interpretations: a floating signifier within the networks that criss-cross above Lake Como (such as the path of the Delta Airlines plane) but that never touch it, if not in semantic fashion; it is a commodified lifestyle that its multi-million-dollar Las Vegas incarnation represents at its best. It is there, in fact, that I would like to begin the story (and geography) of a globalized place.

A place called Bellagio

Bellagio is a place. Or at least it is supposed to be. Perhaps we should reflect briefly upon what we mean by this designation. If we adopt Tim Oakes's (this volume) conceptualization of place, the place called Bellagio can be thought of as a 'stage' upon which the tourist encounter occurs: a stage "built upon many 'layers of historical accumulation' (Massey 1988), the savings bank of collective memory, from which the subject derives meaning and the will to act." As the new place scholarship reminds us, 'place' should be considered neither a 'bounded entity', nor the spatial equivalent of 'community'; not merely the local scale but, rather, transcending scale: for place "is more about action than about scale or region" (Oakes, this volume). In fact, it is only by granting such emphasis to agency, suggests Oakes, that we can valorize that which Massey (1992, 1993) has defined "a progressive sense of place." This idea of place, then, is an arena of intersection among "a particular constellation of relations" in space and time (Massey 1993: 66), the point at which subject formation occurs.

I find this reading of place not only particularly appealing, but also quite useful for any interrogation of the meaning(s) of place in the information age; even more so, perhaps, for any geographical analysis of the paradoxes and tensions which characterize the relationship between the modern subject and certain highly significant—and signified—spaces. If the modern subject might, indeed, be "reconstituted as a place-making subject," as Oakes's (this volume) reasoning seems to suggest, then the Bellagini (that is, the inhabitants of Bellagio) could/should be conceptualized as actors who, in/through the place called Bellagio, 'encounter' Modernity, its flows and its contradictions. If Bellagio is the site—the point of intersection—within which flows of people, capital and information meet the deposits of collective memory (the memory of Bellagio, that is), then we should consider Bellagio a 'place.' Again, this is not to say that it is bordered 'through description.' Rather, it is a specific location within which diverse modern subjects (including the Bellagini) meet within "a particular constellation of relations" in space and time.

Bellagio is thus both the physical/material as well as the metaphorical space within which this negotiation of meaning between the inhabitants and other 'interpreters' of 'Bellagio-ness' takes place. It is, however, a rather odd place— not due to some strange habits of its inhabitants or anomalies in its physical features. It is an odd place since, as I would find out, it is located both in Lombardy—and in Nevada at the same time.

On 15 October 1998, the most expensive hotel in the world, the Bellagio, announced its glorious opening along the Strip in Las Vegas. Yet the opening of the Bellagio did not simply signal the birth of yet another spectacular development geared to attract the crowds of tourists and gamblers of the Strip: the Bellagio was nothing more than another hotel inspired by faraway places in the gamblers' mecca. What was being inaugurated that fateful day, rather, was an entirely novel way of conceiving what a place might be—or, more precisely, what a touris*ted* place might be.

As I will argue, the Bellagio not only recalls in anaesthetized fashion some of the most easily recognizable features of (the 'original') Bellagio's landscape, but, rather, has succeeded in creating an entirely new 'place'—with its very own atmosphere, ensemble of meanings, 'tradition' and, of course, landscape. After all, the idea of landscape is nothing other than a way of granting meaning to place through the adoption of a particular (physical and conceptual) perspective (Guarrasi 2001); thus, the idea of the landscape as a 'text' (Duncan 1990) to be interpreted, as the embodiment of a distinct value system (Cosgrove 1984b). The hotel complex rising up along the Vegas Strip, however, does not merely reproduce some of the traits of this specific perspective of/on Bellagio's material/metaphorical space. Here, rather, the representation of Bellagio's landscape has been 'grounded' through the construction of an *alternative* place called Bellagio—by inverting the ways in which we usually conceive of the relation landscape/place in our 'reading' of space.

Bellagio thus becomes a place whose meaning emerges in a number of different 'sites'; an ambiguous but extremely rich 'point of intersection'; a window upon the meaning of place in Modernity. The Bellagio (Las Vegas) has, in fact, become a place of its own—a place whose relationship to its Italian counterpart deserves closer investigation. What I hope to propose within the pages of this chapter is an analysis of the relationship between modern subjectivity and tourism *through* place, adopting the notion of place as a codification of the spatio-cultural coordinates of the above relationship.

As the reader will note, the bulk of the sources dealing with the 'American' Bellagio are Internet-based; this choice of references is not incidental. Just a few months after the hotel opened, a local Italian newspaper would, in fact, proclaim in its headline "A war against the fake Bellagio" (*La Provincia*, 23, 28 December 1999), noting that a heated "battle" was raging between the "authentic Bellagio" and the "artificial Bellagio"; a "battle waged in the spaces of the Internet but

Figure 7.1 The Bellagio Hotel, Las Vegas (photographed by the author)

Cezanne, Gauguin, van Gogh, Matisse and Picasso and which has become a *must see* in Las Vegas. The hotel's shopping arcade, 'Via Bellagio,' is apparently almost equally edifying: as Muriel Stevens (1998), a columnist for the *Las Vegas Sun*, would characterize it, Via Bellagio is "an outstanding form of entertainment [that] allows Las Vegas to join the best-known shopping cities in the world—New York, Paris, London and Milan…bringing a bit of Italy to Las Vegas." In fact, Italian opera music plays in the background, its sounds diffusing from the elevator to the slot machines. Fourteen distinct dining establishments and a state-of-the-art aquatic amphitheatre built specifically for the synchronized swimmers and acrobats of the Cirque du Soleil complete the Bellagio universe, something much, much more than a simple hotel or even a mega-resort. What is on offer here, in fact, is a lifestyle, an aesthetic and cultural model held up for admiration, emulation and consumption; a philosophical statement on hospitality and the contemporary urban context.

As an analyst for Bear Stearns (www.businesswire.com, December 1999) would affirm:

> the *Bellagio* property takes Las Vegas into a new era of attracting highest-end visitors, those who can meet or vacation anywhere in the world they choose…. There has never before been a property like the Bellagio…. It is a visionary achievement.

And it is not about gambling. The Bellagio is about opulence, extravagance and romance. It is about the crafting and selling of a unique aesthetic experience, with every feature of its development speaking to the visitor/spectator of such an enchanted experience. Let us try to decipher just what this experience is all about.

Experiencing Bellagio(s)

The Bellagio offers the ideal context for the satisfaction of the new global(ized) petit bourgeoisie's desires for luxurious commodified landscapes (see e.g. Knox 1991, Urry 1995). The hotel's Botanical Conservatory—an explicit sensory experience filled with seasonal flowers and plants, changed every month at the cost of approximately $8 million a year—seems to represent an ideal backdrop for aspiring middle-class families' self-portraits. The visitors' interpretation of the Conservatory and its gazebo is particularly revealing in this sense. Most seem, in fact, to be irresistibly drawn towards a specific location within the garden, positioning themselves carefully so that their souvenir snapshots frame their presence here against the backdrop of the summery white gazebo, as though in the shadows of a famous monument. The Conservatory's gazebo is thus transformed into a tourist icon, a virtual memorabilia, a stage for the reconstruction of a luxurious Italian lifestyle; a dream-like context for a middle-class clientele willing to buy the image first—to be completed with a variety of products recalling that same image later (readily available in the gift shops just around the corner from the gardens).

The crowds moving through the gardens, waiting for their turn at the photographic altar, seem moved by some mysterious force, some invisible hand guiding group after group through the very same dance of gestures to be immortalized with a photo or video-camera shot, as though recalling an ancient ritual, some sort of pilgrimage deeply rooted within their culture—a pilgrimage, in fact, paying tribute to the aesthetic model celebrated by the Bellagio, to the very idea of Bellagio as a middle-class referent for the *dolce vita*. Watching the queue of tourists waiting for their turn to step up to the gazebo in order to snap that prized photograph actually leaves quite an impression, recalling in strangely familiar fashion other similar queues, in other places—a variety of diverse places, to be sure, though all characterized by the very same reverence for the sacred, the devotion of the pilgrims for the icon housed within (Figure 7.2).

This pilgrimage also closely resembles the strange 'commute-in-reverse' that characterizes the weekends in many large European urban centres; the 'myth of the city' that moves the peripheral dwellers, the inhabitants of the suburbs and high rises stretching out around cities such as Paris and Rome, into the city:

> The *coatti* and the *borgatari* in Rome or the *banlieusards* in Paris congregate in the centre each Saturday afternoon, attracted by the city and its myth, its

Figure 7.2 In the Bellagio Hotel gardens, visitors wait in line for their turn to reach the photographic altar (photograph by the author)

lights and endless opportunities. Immersing themselves in the phantasmagoria of goods and images, they greedily try to swallow up the city, though without success. They move through the streets like an occupying force—eager, over-bearing, but unable to perceive what lies around them. They are witness to a representation, to a dream to which they have no access; a dream that both attracts and excludes them. Their ranks strolling through the streets remind one of lines of prisoners, of a march of the defeated.

(Amendola 1997: 27)

Visitors to the Bellagio also recall these ranks: fully aware that they have been admitted, albeit temporarily, into the sanctuary of a lifestyle that they will never be able to achieve—but that still remains a model to be emulated, an ideal to be looked up to, an atmosphere to revel in—and from which to bring home some memorable object or image. A lifestyle (or, better yet, an idea of a lifestyle) that feeds upon many other Bellagios, all loyal to the subliminal message of the 'Bellagio-sign,' to the realm of meaning incarnated by that distinct brand name, that *griffe*, that ambiguous allusion to the 'place-called-Bellagio.' We can note several from among the varied Bellagios around the world: a gated community in Thousand Oaks, California (www.pardee-homes.com, December 1999); 'Bellagio Gourmet' hot chocolate, that "lush, creamy cocoa that has graced Caffe d'Amore for over twenty-nine years," inspired (or so the label proclaims) by the "Bellagio Classic European Tradition…that conjures up images of ancient courtyard cafes and thousands of distant cathedral bells" (www.caffedamore.com, December 2001); the 'Ciabatta di Bellagio,' an entirely American invention passed off as an authentic Italian specialty and sold at a ridiculous price in up-scale natural foods supermarkets in the Western states; and, finally, the 'Stile Bellagio Design Collection,' a series of table lamps, wall sconces and chandeliers inspired by the "Bellagio stile" (www.visionsdesigngroup.com, December 2001).

As a variety of observers have noted, the function of art has long been that of sustaining and legitimating the image of the state, of the local administration and/or of private and public benefactors. Public art, in particular, often expresses the intentions of power by communicating through the symbolic language and myths of the past in order to celebrate the present (Amendola 1997: 96). What better icon than a 'Gallery of Fine Art,' a prestigious collection of signature pieces, to celebrate the most expensive hotel in the world? What better form of legitimation for a casino that does not want to appear to be a casino but a place, the embodiment of a lifestyle, an aesthetic experience?

Turning to the *Las Vegas Sun* once again, we find facile confirmation of the role of this 'noble presence' in legitimizing and celebrating the 'grandeur' of the Bellagio. The *Sun*'s columnist Gary Thompson (1998) describes the Gallery, in fact, as "something you simply *must see*"; quoting the Village Voice's art critic Peter Schjeldahl, he suggests that a visit to the Gallery "requires…a *suspension of intellect* [emphasis added] in sheerly beholding. If we manage to silence our chatty brains in the art's presence, the art will explain itself with wordless eloquence straight to our hearts." The Bellagio thus manages to speak directly to our hearts through its magnificent works of art.

Thompson's (1998) commentary insists, moreover, on the fact that

> neither the Mirage Resorts Chairman Steve Wynn, nor *Bellagio* (understood as the hotel) will directly make a dime's worth of profit from displaying the art. …After deducting about 37.5 percent of the proceeds from $10-a-person ticket sales to cover the expenses of displaying the art, they'll contribute the remainder to local charities. Personally, I'd rather see the money go to charities than politicians.

In other words, the Gallery is something akin to a charitable operation, a gift to the world and, in particular, to the state of Nevada. In fact, as Thompson (*ibid.*) is fast to stress,

> At least 40,000 Southern Nevada school children will get free access to the collection each year, and it's bound to move at least some of them. Might one or two, inspired by the masters and blessed with talents of their own, aspire

to become an artist? And if Las Vegas can give birth to the world's greatest hotel, why not some future great artists?

Incredible but true, the Bellagio could inspire a new Van Gogh: Vincent from Nevada.... However much it may be in the public interest, the Bellagio Art Gallery is subject to corporate directive. In 2000, in a corporate takeover, MGM Grand Inc. bought Steve Wynn's Mirage Resorts Inc., and Wynn left the Bellagio with much of his art collection (Macy 2000). MGM Grand managed to retain ownership of a few Picassos and Renoirs, which it subsequently installed in its two restaurants bearing the artists' names: now diners at the Bellagio's Picasso restaurant can feel surrounded by the authentic oil paintings and ceramic works by Picasso; while paintings by Renoir 'decorate' the Renoir restaurant at the Mirage Hotel. With its original signature art collection removed and dispersed, the Bellagio Gallery of Fine Art closed; but "the idea of a permanent gallery had become a marquee-sized calling card for the hotel" and so it reopened "with an exhibition of holdings from the prestigious Phillips Collection in Washington, DC, followed by a showing of Steve Martin's personal art collection" (Coast Staff 2002). It has gained some independent legitimacy with a series of exhibits including works by Alexander Calder, the Fabergé collection from the Kremlin (matched by a Russian-themed dining experience) and works by Andy Warhol.

The new essence of dream-like architecture fused with the spatialities of hyper-reality is perhaps best embodied precisely within hotels—by definition, root-less worlds, created with the express purpose of fulfilling desires. But the desires and the stereotypes upon which they are built are never abstracted from a cultural context; they emerge, rather, from a distinct set of models and tastes, a distinct middle-class consumer aesthetic (Amendola 1997). And the whole of the Bellagio beckons you to abandon yourself to its opulent forms, to let yourself be led through its luxurious atmospheres, lulled by the myth that guides you and that, for once, seems within reach. To materialize a dream, such environments no longer need to reproduce reality but rather, as many planners and architects have suggested, the *idea* that people have of reality (Amendola 1997: 118). The latest generation of grand hotels are, in fact, monuments to hyper-reality and to dreams. The Bellagio's theme song, "Con te partirò" ("I'll go with you"), by Andrea Bocelli, which was a major European hit in 1995, only confirms this hypothesis: it is moving, touching; it provides the perfect romantic backdrop for the dream-like play of lights and water-jets that beckon the Strip's passers-by and hotel guests alike. As a sort of appropriated corporate national anthem, it holds the promise of a world of the good, the true and the beautiful; the magic of a landscape that the music continually transforms and models before our eyes. The Bellagio thus becomes a cultural ideal, a subliminal message, a sensual invitation to become part of a grand project, of a way of thinking, of a sense of place traced by entirely new spatio-temporal coordinates, of an aesthetic experience of the urban context that the city no longer seems able (or perhaps never was able) to offer.

To be successful, however, any such commercialization of place must render this latter immediately recognizable. Its so-called *genius loci* must be emphasized and communicated to the point of making it become the emblem, the mark of that place and of the experience that it should embody (Amendola 1997). What, then, is the *genius loci* (or *genre de vie*) that the Bellagio draws upon, which it seeks to communicate, in turn? What value system is represented within its landscape perspective? If any and all place marketing strategies necessarily rely upon a 'bank of collective memory,' upon layers of historical accumulation at a specific 'point of intersection,' what, then, is the meaning assigned to the place called Bellagio within this American/globalized imaginary?

Tourism, place and subjectivity

Having gotten to this point, we cannot but ask what all of this has to do with the Bellagio that lies in Lombardy. Is the Bellagio but another replica of a real place, just a more sophisticated one? Or it is something entirely different? And what do the Bellagini have to do with this reproduction of a place/landscape that supposedly emerges from the sedimentation of *their* local culture, from the deposits of collective memory about the place where *they* live? Do they have any authority left to speak about 'Bellagio'—and thus to determine the accuracy of other 'Bellagios' with respect to the 'Original'? In other words, is the Bellagio, Las Vegas, the simulacrum of a place—or a place itself? If so, what is its relationship to the northern Italian lakeside town? Again, who has the authority to decide what the place called 'Bellagio' is—and should be? Is it simply a matter of power?

The issue at stake here is nothing less than the very meaning of place: or, better yet, the uses of the meanings of *others'* places. In an attempt to understand this interplay, we can begin by turning again to Oakes's (1998) work on the relationship between tourism and Modernity, an analytical framework which can allow us to begin to conceptualize the relations between tourism and place within Modernity—and thus within the Bellagio.

In conceptualizing the tourist as the "emblematic figure of modernity," as Oakes (1998) maintains, two accordant issues emerge. First, if we accept the idea that the modern subject is a traveller, we must also ask what can we say about

> the places that serve as the traveller's destination; are the inhabitants of these places incapable of a modern subjectivity independent of that experienced by their fleeting visitor? Surely they cannot be reduced to an exotic backdrop for the modern exile.
>
> (Oakes 1998: 19)

We must also be sensitive, however, to the fact that, as Kaplan (1996: 57) suggests, "the tourist-traveller-exile cannot be universalized to stand for every subject position in modernity"; rather, following MacCannell (1989), we should conceive the tourist as

> striving for meaning and continuity in a fragmented world; she seeks the exotic and authentic in an effort to displace the instrumental rationalism that has come to dominate her life. But her very search is also presaged upon the false modernist belief that tradition and authenticity are modernity's antithesis, not its own creation.
>
> (Oakes 1998: 19)

In other words, "equating the tourist with the modern Subject" holds the danger of only revealing "what Kaplan calls a middle-class Euro-American cultural myopia" (Oakes 1998: 19).

With this line of reasoning, Oakes stresses the need to recognize a subjectivity that is often denied to the 'locals' (in Oakes's empirical study, the inhabitants of villages in the Chinese province of Guizhou)—inhabitants struggling with the variety of modernizing pressures induced by the growing presence of tourism. We should be careful to distinguish, however, between such subjectivity and that which Oakes (1998: 7) terms the "false modern": that is, "the utopian, teleological modernity of nineteenth- and twentieth-century historicism, of the nation state, and of the institutions of rationalism and scientific objectivity."

What Oakes (1998: 7) suggests, rather, is that the local inhabitants participate instead in a sort of "authentic modern": not authentic in terms of some discernible modern "essence" but rather, "in terms of a process-oriented approach to modernity, one in which human subjectivity is ambivalent but irrevocably engaged in a struggle over the trajectory of socio-economic change." The villagers, according to Oakes (1998: 10), assert a "modern subjectivity" precisely in their ability to stage the tourist experience, to play self-consciously a series of described roles for tourists, even to turn irreverently those tales into acts of humour and subversion, succeeding as a tourist village not because they were more traditional, but because they were good at "*playing* tradition."

Although the choice of terms separating the two conceptions of modern subjectivity ('false' vs. 'authentic') may be debatable, I find the conceptualization particularly useful for a reflection on the relationship between tourism and places. The reader may be wondering what all this has to do with the Bellagio and its opulent clientele. What does the most expensive hotel in the world have in common with the inhabitants of a Chinese village? The connection emerges from Oakes's reflection on the relationship between the *observers* and the *observed* within a tourist*ed* place: a place within which these two (or perhaps even more) subjectivities/projects come together; within which they manifest their disparate strategies, their visions and experiences of modernity. The exhibition of tradition (or, more generally still, the exhibition of placeness, of the identity of place) is thus the result of an incessant and unfailing contradiction between *the tourists' desire for authenticity* and *the local community's desire for modernity* (a local community which, all the while, considers such placeness as an essential referent in the definition of their identity). A contradiction negotiated within the elaborate staging of tradition, consumption and commerce that, in Oakes's (1998) case, welcomes the tourists into the village. In this sense, then, tourism provides a "particularly meaningful metaphor for the experience of modernity in that it displays both modernity's relentlessly objectifying processes as well as its promise of new and liberating subjectivities for those participating in either side of the tourism encounter" (Oakes 1998: 11).

If we accept the above hypothesis, we must necessarily abandon any (even marginally) mechanistic/deterministic interpretation of the relationship between tourists/observers/spectators and the accordant visited/observed/exhibited/staged places. In particular, however, we must abandon the idea of place as essence, as a detached and definable entity delimited by spatial and historical boundary lines, as a crystallization of tradition, a trove of values, the incarnation of local culture. We must move on, rather, to a conceptualization of place as an (id)entity which cannot be 'fixed' in space by our descriptions, which cannot be captured within photographs or cartographies which claim to capture its essence. Identity, place and thus also the modernity of that place can only be conceptualized as, rather, a *dialectical process*, as a relentless negotiation of meanings between diverse social actors—ongoing negotiations and struggles which define the history, the memory of that determinate place; negotiations and struggles which are never 'accomplished,' however, for that very same place is continually 'challenged' by ever new processes, ever new subjectivities, in a continual struggle among them to try to resolve the paradoxes of modernity by granting meaning(s) to space (Minca 2001, Oakes 1997).

Accordingly, we cannot conceive of (the 'real') Bellagio as a place in the singular—as a fixed identity, as an original essence upon which the Las Vegas 'replica' draws. And its inhabitants are quite well aware of this differentiation: as one local community leader noted to me, in fact, Bellagio, the 'place' of the Bellagini—and Bellagio, the 'place' visited by the tourists—"are two entirely different places" (Leoni 2000) (again, if we conceive of place as the site of subject formation, as the point of intersection between actors and deposits of collective memory). The Anglo-Americans who make up the bulk of Bellagio's visitors come to the lakeside town already knowing it, in a sense—as the crystallization of a set of meanings that frame that 'exclusive Italian lifestyle' so highly prized by the global service classes. Gianfranco Bucher (2000), owner and manager of the exclusive Grand Hotel Villa Serbelloni, perhaps the prime purveyor of this prized tableau, was quite candid in his discussion with me: "they [the visitors] come here to just be, to sit and stare. To experience the *dolce vita italiana*. Not to see something, but to see themselves here." The British and American tourists that come to Bucher's five-star hotel strive to "re-create the Grand Tour" according to the owner. They are rich professionals and seek a quite different Italy from the 'pizza and gondola' crowd. The slogan of the hotel? "Il Tempo Ritrovato" (finding time again)—time to enjoy the Italian lifestyle that has become an integral part of a certain international class's *habitus*.

The Bellagio, Las Vegas, was undoubtedly constructed in part to evoke some of the traces of *memory* that exist within the Anglo-American cultural universe of the Bellagio experienced at the Grand Hotel Villa Serbelloni (and its numerous lakefront contenders). Steve Wynn and his design team were, in fact, guests of the hotel (non-paying guests, I should add) and spent several weeks with the establishment's staff, examining in detail the daily operations. Not only this: Wynn and company were also welcomed by the Bellagio municipal authorities themselves who generously provided the entrepreneur with detailed historical urban plans, as well as building specifications for the lakefront promenade. "What we expected," Bucher told me, "was of course something in return, perhaps some publicity for our town." They never heard from Wynn again, or anyone else from the resort, however. Not that business is suffering from this slight; the hotel has been booked at full capacity and has extended its opening season by one full month.

Others in town see the matter differently. Luca Leoni, owner and manager of the Hotel du Lac (Figure 7.3) (just down the street from the Serbelloni) and

Figure 7.3 Shoreline view including the Hotel du Lac, Bellagio, Italy (photograph by the author)

current president of the Bellagio Hotel Owners Association, has been to Las Vegas several times and thinks that "the *Bellagio* is gorgeous." "But it has nothing to do with our town—the landscape is completely different," he notes. "So there is no problem." He is critical, in fact, of the Municipality's pending lawsuit against the resort. "What is the point—we should rather all go there and learn how things should be done. Wynn has done a great job."

Certainly, the opening of the Las Vegas resort, in addition to the inflow of new tourists, has sparked a whole series of contradictory dynamics in Bellagio. These dynamics are very similar to those described in Oakes's Chinese village, albeit characterized by a rather different power dynamic between the observers and the observed, since in Bellagio the cultural and economic 'distance' between the two appears much more negligible. I would like to pause upon this question of power, however, for it is here that we should focus our attention, as it is power—the power to appropriate images, icons, places and to enclose them, eliminating all of the dialectical relations that 'make places'—that lies at the heart of the 'Bellagio question.'

Some of Bellagio's lakefront shops certainly cater exclusively to the tourist gaze, such as the rows of gentrified glass ornament shops ('I Vetri di Bellagio') and silk boutiques ('Made in Como'). And the new ferry terminal entrance is being built in an art nouveau style reminiscent of the Las Vegas Bellagio. Yes, Bellagio is changing and perhaps even becoming, in part, a caricature of itself. But, as Hans Bucher (2000) would sardonically tell me, "places change." What matters, however, is who are the 'owners' of such change?

If we think of place as both a material and metaphorical space—and thus also the product of its dialectical description, a product of always uncertain coordinates—then, just as in any other modern place, we can theorize the encounter between the tourists/spectators in the 'original' Bellagio and its more or less aware protagonists/objects-to-be-observed as essentially 'open.' Open in the sense that it always leaves some room for the subjectivity of the observed—an opening which, according to the particular context, may be more or less equitable (and here, political or perhaps even ethical considerations step in), but an opening which always remains; a subjectivity which we should never entirely disregard, unless we accept the responsibility of fully essentializing, and thus erasing, the Other/the observed (although the tourist experience purportedly always consists in trying, at least in theory, to preserve, to safeguard, its existence). The observed's existence as a modern subject depends, in fact, upon the recognition of her/his subjectivity, upon an acceptance of modernities other than the one we can envision from our privileged positions.

But what goes on within the Bellagio, Las Vegas? Where does this negotiation between diverse modern subjects occur? What role has been assigned to the memory of a place called Bellagio in the construction of the Bellagio resort? It is here that I would like to focus my attention: not on an analysis of the relationship between the Vegas resort and its Lombard namesake but, rather, on the dynamics underlying the construction of the idea of place, of a lifestyle *in* place.

Bellagio, Lombardy, is, obviously, a place. But, I would argue, the Bellagio, Las Vegas (as all of its promotional material stresses without fail) is also a place—in the meaning that we have given to it thus far, for within its space we also witness the encounter of diverse modern subjectivities: that of the tourist looking for luxury, Italian style, that of the Central American service worker who spends the day polishing the brass plaques of Via Bellagio, but also that of Italian visitors like myself who find themselves confronted with a celebration of 'Italianness' that they find difficult to recognize as their own—and who feel that, somehow, they've been had.

The principal reason for the confusion is not due to the fact that the two places share the same name (a relatively frequent occurrence) but, rather, due to the fact that the American Bellagio interprets the Italian Bellagio according to an entirely *internal* logic, the global service-classes' discourse on Italy. Who, then, are the observed subjects here, and who are the observers? Can we still speak of the typically modern contradictions of tourism (such as those described by Oakes (1998) in the case of the Chinese village) when a key element has gone missing, that is the inhabitants themselves, who should constitute at least one of the protagonists of that discourse on place?

Beyond Bellagio

The Bellagio, Las Vegas, is not a simple reproduction of place; it is much, much more. The Bellagio resort represents, in fact, an extraordinary materialization of a sign, of a globalized landscape, not so much because it loosely reproduces landscape traits that are considered 'typical' (by a certain public) of a northern Italian lakeside resort, but rather because the place called Bellagio has come to mean something entirely different by now. It has become a brand-name image associated with brand-name landscape which, in turn, evokes a certain timeless Italian *dolce vita*, most eloquently personified precisely by the Bellagio resort.

The tourists who come to be photographed at the Bellagio, or who pose in front of the resort's lake, do not do so because they necessarily recognize the traits of the place which the landscape purportedly evokes but, rather, because by now, that atmosphere, that music, has become a seductive and all-encompassing global icon. One goes to the Bellagio to see the idea of Bellagio; to see oneself *within* that idea of Bellagio, with the real Lake Bellagio just over one's shoulder (real, since there is no other Lake Bellagio—the one in Lombardy is Lake Como—but Como means nothing in this context). It is the idea of a *lifestylein place* that draws visitors into the Bellagio, visitors eager to see themselves within that idea and all that it symbolizes. But what is most curious is that *it is the very same idea of a lifestyle inplace that draws visitors to the 'real' Bellagio* (as the Hotel Villa

Serbelloni's manager perceptively noted, the desire to see themselves part of the "imagined picture")!

It is beyond the point, then, to dismiss the Bellagio resort as simply an extravagantly luxurious tourist *mise-en-scene*, lavishly garnished with lights, sounds and glistening brass. After all, there is no copyright on place names. Every place has the right to choose its own (although the erstwhile Mayor of Venezia, philosopher Massimo Cacciari, on the occasion of the opening of The Venetian resort, just a few minutes down the Vegas Strip, provocatively threatened to sue for violating copyright on the city and its landscape). What I would like to suggest is that what is most disturbing about the Las Vegas complex, rather, is precisely *the use of the memory of Bellagio* through the stylized reproduction of its idealized landscape— disturbing (from my Italian perspective) not so much because the Vegas Bellagio was reproduced without obtaining the opinion of the Bellagini and is thus a rather poor reproduction of a place of which we (as Italians) feel we possess a 'privileged' reading, but disturbing because the place called Bellagio, by de-localizing its dialectic, *cancels* precisely that subjectivity which, however repressed, oppressed and super-objectified, remains the central element of tourists' interest for a place. In other words, the Bellagio was (re)founded as a place through a fundamental act of 'forgetting'; by forgetting the very bases of the constitution of any place, that is the inhabitants; the putative 'producers' of that landscape and of that atmosphere that we usually seek out in our tourist experiences.

The Bellagio in Las Vegas is a place which has 'stolen' Bellagio's name and landscape and re-signified them, purporting to maintain a tenuous connection to their 'originals,' though in reality transforming them beyond recognition into something fundamentally different, into an exhibition of representations of Bellagio about which the Bellagini have no say. Within the Bellagio, we witness the veritable construction of a landscape *ex novo*, a veiled privatization of that landscape through its sublimation; the constitution of a place within which the negotiation of meaning has already been decided (and yes, it is a place—just ask the cleaning women scrubbing down the immaculate rooms, or the scuba divers who daily free Lake Bellagio of pesky algae). It is a place that can go bankrupt and close shop. It is a place which, to produce meaning, does not have to contend with the ties of memory.

But, above all, it is a closed space that pretends to be open. I do not intend 'closed' in the sense of many other so-called postmodern enclaves. The Bellagio is closed more metaphorically than physically; closed because it has killed off all negotiation of meaning, because it does violence to memory. And the power of this closure lies precisely within the ambiguous character of its spatial boundary—the famous lakefront sidewalk-*corniche*.

The public–private nature of the space of that strip of sidewalk is a perfect metaphor for the nature of the place that is the Bellagio. Everyone is welcome inside, though only according to certain rules—rules, however, which are not the rules that every community or institution tries to enforce within spaces considered of particular significance (for example, as symbolic of a certain historical memory, of a determinate social-political order, etc.). The rules here are the rules of a commodified global *lifestyle in place*—rules that are binding, even oppressive at times, but which rely upon, at least in part, the consent of the Bellagio's patrons and their acquiescence to the Bellagio project. Strolling along that glistening sidewalk, with the latest in Italian contemporary music filling the air, one becomes aware of having breached a magic threshold, an invisible confine which, precisely due to its invisibility, is so powerful. A no-man's-land between everyone's space and a space which belongs to someone; a space which, in order to function, must be both controlled down to its most minute detail and left physically open—or else the game is up.

And the game works precisely because there is no modern subject who can lay claim to this no-man's-land; there is no subject here who would conduct a legitimate fight against tourist commercialization (as inhabitant); no contender in the struggle for meaning. The observed objects are stripped of subjectivity because they are merely signs of signs, representations of representations, or else they are the observers themselves, joyfully paying tribute to this temple of the dollar just to be able to portray themselves with its golden icons in the backdrop. It is an inherently violent space, because it leaves no way out, if not desperate 'tactics' (see de Certeau 1984, Goss 1993a), which, nevertheless, lack any form of popular legitimation, especially the legitimation granted to the inhabitants of any place.

Within the Bellagio, there is no local population and thus none of the dialectic through which the place called Bellagio—or any other place—comes to be. Here, place has been constituted on private ground, the process of signification decided a priori, just as the programmed emotions evoked at regular half-hour intervals with the silky tones of "Con te partirò" for the benefit of the observers/observed on their balustrade/stage overlooking the lake.

The Bellagio may be a dream of place, to be experienced and admired by a middle class aspiring to the global service-class lifestyle. But it is a deadly dream, because upon reawakening one might realize that the would-be inhabitants are none other than the spectators themselves, destined to reflect only themselves, as though in a wall of mirrors, in the closed space of a stage(d) place which reverberates with the macabre images of a memory that can only be bought, never possessed.

8 Seducing global capital

Reimaging space and interaction in Melbourne and Sydney

C. Michael Hall

The seductions I have in mind do not happen by chance, they are not unplanned events. Instead, these seductions are usually well considered beforehand; they are prepared, presented in an appropriate atmosphere and introduced by the right words. So it is with cities as well as people. Cities are increasingly being promoted, developed, designed and built in order to attract and seduce parties ranging from investors to tourists. This chapter discusses the notion of cities seducing capital within the context of urban reimaging processes and the conscious attempts of cities to sell themselves in order to attract investment, employers and tourists. It then examines urban reimaging strategies in the cities of Melbourne and Sydney and the manner in which they have consciously tried to create themselves, and compete with each other, as 'world cities' of Australia.

Urban imaging

Culture, the economy and landscape are closely entwined, as are their practices and processes. The urban landscape—which includes the built environment and its material and social practices, as well as their symbolic representations (Zukin 1991: 16)—may be read as a text (McBride 1999). Based on the premise that "landscapes are communicative devices that encode and transmit information" (Duncan 1990: 4), researchers examine the 'tropes' which communicate this information, and how they are read by those who come into contact with them. Tropes are signs and symbols into which various meanings are condensed. They include items in and of the built and physical environment—such as buildings, monuments, public spaces, trees and parks—and also signs, slogans, relationships, brands and even language(s) associated with the landscape under study. The latter include the languages of consumption and production associated with specific places and spaces (Hall 1997, Jackson and Taylor 1996). As a 'text,' therefore, landscape is multidimensional, lending itself to many interpretations. As a conceptual tool, reading the landscape not only illustrates the ideology of the landscape (Cosgrove 1984a: 15), but also can illuminate the way it may "reproduce social and political practices" (Duncan 1990: 18). For example, Chris Philo and Robin Kearns (1993) identified the many ways in which the identity of the city is constructed and how the contemporary city reveals its different spatial conditions.

Issues of ideology, identity and representation have become central to analysis concerned with how the city is packaged as a product to be sold (e.g. Kearns and Philo 1993) and as a space which is specifically designed and constructed to encourage particular forms of leisure-oriented consumption. Contemporary urban imaging strategies are typically policy responses to the social and economic problems associated with deindustrialization and globalization and associated economic restructuring, urban renewal, multiculturalism, social integration and control (Roche 1992, 1994). Urban imaging strategies tend to have several aims, including attracting tourism expenditure; generating employment in the tourist industry; fostering positive images for potential investors in the region, often by 'reimaging' previous negative perceptions; and providing an urban environment which will attract and retain the interest of professionals and white-collar workers, particularly in 'clean' service industries such as tourism and communications (Hall 1992).

Urban imaging processes include the development of a critical mass of visitor attractions and facilities, including new buildings/prestige/flagship centres (e.g. shopping centres, stadia, sports complexes and indoor arenas, convention centres, casino development); the hosting of hallmark events (e.g. the Olympic Games, the Commonwealth Games, the America's Cup and the hosting of Grand Prix) and/or hosting major league sports teams; the development of urban tourism strategies and policies often associated with new or renewed organization and development of city marketing (e.g. 'Absolutely, Positively Wellington,' Sheffield: City of Steel, Cutlery and Sport); and development of leisure and cultural services and projects to support the marketing and tourism effort (e.g. the creation and renewal of museums and art galleries and the hosting of art festivals, often as part of a comprehensive cultural tourism strategy for a region or city).

Urban imaging strategies are therefore conscious attempts by places to seduce. In particular, not only do they seek to develop something which is attractive, but in so doing they aim to package specific representations of a particular way of life or lifestyle for consumption. The utility of urban imaging strategies for urban growth coalitions is that they consciously aim to create new urban places. Areas of waterfront and derelict land are replaced by the condominium, the cycle way and the retail complex, replete with cafés, restaurants and fashion houses, from which the poor are typically excluded.

Whether we are using culture in the sense of being indicative of a "particular way of life," or "as a reference to the works or practices of intellectual and especially artistic activity" (Williams 1983: 90), culture is becoming commodified and bought and sold in the global marketplace. Urban cultural policies are used to generate artistic and 'high' cultural activity in order to attract visitors and to make the city an attractive place to live for middle-class and white-collar workers and business, while wider notions of cultural identity are also being used to attract investment, visitors and employment. The use of cultural images to attract visitors is not new. It has been around for as long as tourism. However, the effects of urban place marketing may be pervasive within specific places targeted for consumption, because the notion of selling places implies trying to affect not only demand through the representation of cultural images, but also the construction, design, manipulation and management of the supply side, e.g. those things which make up a community's life, into a package which can be 'sold.' Such actions clearly have implications not only for how the external consumer sees places but also for how those people who constitute place are able to participate in the making of their collective and individual identities and the structures which sell place (Hall and Hodges 1997).

It is no coincidence that in this time of dramatic shifts in the character of contemporary capitalism that the 1980s were characterized as the decade in which consumers were taught "how to desire" (Bocock 1993, York and Jennings 1995: 44 in Pawson 1997: 17). The essential problematic of the new urban places is that causal spatial determinism has become juxtaposed with what Michael Keith and Steve Pile (1993: 25) describe as the "malleability of the symbolic role of landscape." Rather than seeing cities purely, even predominantly, as a site or place where practices and processes happen for all sorts of reasons, the urban landscape has become appropriated by urban growth coalitions and the leisured middle classes as a signification of certain social and cultural values—it is reified as both product and location (Ravenscroft 1999). For producers, an essential means of achieving this has often been by "romancing the [urban] product" through the use of brands (Pawson 1997). Branding is a way of seeking to add value to commodities including services and places. It is the communicative umbrella that seeks to simplify the complexity of a city in order to assist its consumption. A successful brand creates distinctiveness in the marketplace, including the highly competitive place marketplace. As Eric Pawson (1997: 17) notes:

> It is an investment in product quality at the same time as seeking to create more illusory associations to appeal to specific groups of consumers in the local spaces of globalised capitalism. Both branding and advertising are inherently spatial practices, used by producers in the expansion and differentiation of markets.
>
> In the case of urban reimaging, marketing practices, such as branding, rely upon the commodification of particular aspects of place, exploiting, reinventing, building, constructing or creating place images.

Consuming in the city

Given the shifts in the nature of production noted above, it should not be surprising that in the same way that the nature of urban production is regarded as having changed then so has the nature of urban consumption (Glennie and Thrift

1992). For example, in earlier work I reported "greater fragmentation and pluralism, the weakening of older collective solidarities and block identities and the emergence of new identities associated with greater work flexibility (and) the maximisation of individual choice through personal consumption" (Hall 1988: 24). In this context it therefore became apparent to those who were seeking to reimage the city that the desires of those who consume have to be accorded far greater prominence. In a world of consumers, differentiated by social segments and lifestyle niches, it is not only the inherent qualities of urban products that matter, but also their symbolic meaning (Pawson 1997). These attributes are encapsulated in successful brands "which, through careful management, skilful promotion and wide use, come in the minds of consumers, to embrace a particular and appealing set of values…both tangible and intangible" (Interbrand 1990: 6). Brands therefore demonstrate the falseness of trying to separate production from consumption (Bell and Valentine 1997). As E. Laurier (1993: 272) observed:

> To build binary opposites is to make one dependent on the other, and so there cannot be consumption without production…it is apparent that they merge in many places and that each process certainly does have effects on the other…even if they are causal or may never ever be explicable.

Figure 8.1 Harbourside retail complex in Darling Harbour, Sydney, 1989 (photograph by the author)

Figure 8.2 Southbank retail complex, Melbourne, 1995 (photograph by the author)

Walking through the retail spaces of waterside Sydney and Melbourne (Figures 8.1 and 8.2) you can see the reification of capital before your very eyes. Consuming is meant to be 'fun' in these festival market-places. Am I having fun? No. Yet there are many people walking through here, purchasing and returning to experience these spaces. Is it because they are 'new' or because they are something to 'do', some-where to go on a Sunday afternoon? Yet if they do come here they are not elsewhere. In cities that are rich with recreational spaces and places in which you do not have to consume—well, except perhaps for that ice cream—people are flocking here. Consumption or the possibility of consumption has become a form of leisure. Yet this consumption of an economic landscape and the practices and processes of capital which are revealed both in its surface and which are experienced within this space reveals the nature of cities and seduction. People are persuaded that this is where they should be on a Sunday afternoon. Trading hours have been revised to allow this to happen. Spaces are made fun by the provision of free entertainment (of the right sort), and it is the place to meet people (of the right kind—no poor people here). We consume therefore we are. However, in experiencing this space I wonder if my cynicism or analysis (or are they not one and the same?) is not misguided: people look genuinely happy—have they been seduced or are they just free of the postmodern melancholy of academics who have read too much Foucault (without reading Lukes or Marx) and who actually believe that people cannot make up their own minds?

We therefore come full circle: urban places are increasingly attempting to seduce but, if the argument regarding the relationship between consumption and production is correct, people equally want to be seduced when they consider their destination options.

Historical/contemporary settings

The historical legacies of colonial settlement in Australia have set the broad framework for the trajectory of city development and the evolving contexts of urban consumption and production. Sydney was the founding city of colonial Australia with the arrival of the first fleet from Britain in 1788. Although Sydney struggled initially, by the early nineteenth century it was a relatively prosperous outpost of empire that served as the capital of the colony and later state of New South Wales. Melbourne was established in 1835 with the settlement of the colony of Victoria commencing the previous year. Melbourne also struggled with its early development, but the discovery of gold in the 1870s soon provided for the creation of a magnificent Victorian city. The colonial context created a legacy that is still felt in the two cities to this day. As Peter Murphy and Sophie Watson (1997: 1) observe, "One way Sydney is constructed is in relation to Melbourne, its denigrated 'other': Melbourne is dull, Melbourne is serious, Melbourne is full of wowsers, in Melbourne it rains all the time." As capitals of separate colonies they vied for migrants, investment and the attention of the British motherland from an early age, often through the creation of urban space. Following the creation of the Commonwealth of Australia in 1901, colonial rivalry was transformed into interstate rivalry with both states seeking to reinforce the relative economic significance of each city.

For most of the twentieth century it could be argued that the rivalry between Melbourne and Sydney was played out as much on the sports field (cricket to be precise, as they played different codes of football) and in the machinations of federal politics than in formal place promotion and construction. However, from the early 1970s on there was a significant shift in the relationship between the two cities and the nature of place competition and construction. The oil crisis of 1973 witnessed the beginning of the end of a period of economic certainty in Australia and the first round of economic restructuring began in the wake of the removal of tariffs and duties. The growth of the aviation sector also served to reinforce the processes of globalization and led to increased place competition between Sydney and Melbourne. As Geoffrey Blainey (1983: 342) observed:

> The most important reason why Sydney…was able to challenge Melbourne as the financial hub was simply its slight advantage in proximity to East Asia and North America and its possession of the main international airport. Sydney's advantage was a few ticks of the clock but that was a big advantage.

The economic and political changes beginning in the early 1970s, which led to increasing circulation of global capital, had both cause and effect in the development of urban reimaging strategies and the creation of conscious cities of seduction. Melbourne and Sydney began overtly to seek to attract such capital and to retain capital that was already in place. In order to do so, not only were certain policy settings put in place to favour the location of capital, e.g. favourable tax advantages or government grants to particular companies, but cities began to shape themselves in such a way that symbolic capital served to attract and integrate financial and investment capital in a manner that structured the 'new' inner city landscape of convention centres, casinos, festival marketplaces, heritage housing, museums, entertainment complexes, casinos, sports stadia and events. Interestingly, the symbolic capital of the inner city was primarily funded by the public sector—often in public–private partnerships—rather than the private sector which often contained some of the most vocal advocates for the creation of these new inner cities. Such measures also highlight the manner in which the city, or to be more precise the urban core and the redeveloped inner city, aimed to seduce not only international capital, investors and visitors but also the local people themselves. For without the support of local growth coalitions and the seduction of the local populace, the creation of overt landscapes of seduction becomes difficult indeed. Many of these new places of the inner city were in fact old. However, the workingmen's terrace housing of the nineteenth century became the fashionable, heritage housing of the 1970s and 1980s. These places of the inner city—the Darlinghursts and Balmains of Sydney and the Carltons and St Kildas of Melbourne—became socially constructed as desirable not just through individual decisions but through the actions of government planning regulations, the media and real estate developers.

In the case of Melbourne and Sydney the desire to 'place themselves on the map' as a 'world city' has therefore been tied in to the broader processes of globalization and localization as well as the overall state of the economy. In walking through the cities of Sydney and Melbourne we are not just experiencing a surface landscape, rather we are also experiencing the result of numerous decisions which are influenced if not ultimately determined by capital.

Indeed, one can note that in the imaging of these two cities, economic conditions have had a major impact on reimaging strategies—with greater attention being given to the construction of urban space through events, casinos and waterfront development proposals at times of economic recession than during periods of economic growth (Hall and Hamon 1996). Such developments do not occur in isolation: they change the spaces around them and are often deliberately encouraged in order to do so. The old order of urban space occupied by the poor is to be replaced by new spectacular and ordered space in which there is little to surprise the visitor or local but which nevertheless is visited because the promotional material tells the visitor that this is where the real urban lifestyle is to be found. For example, Melbourne's unsuccessful bid for the 1996 Olympic Games was followed by Sydney's successful bid for the 2000 Olympics. In the case of Melbourne, the bid was tied to the redevelopment of the Melbourne Dockland area, while the key feature of the Sydney bid was the redevelopment of the former industrial site and waste dump at Homebush Bay on Sydney Harbour as the main games stadium complex. In both cities, the bidding for events by state governments reflected development of new cultural, leisure and tourism policies focused on attracting visitors to the city and broader urban redevelopment programmes seeking to develop cultural, housing, leisure and entertainment complexes in waterfront areas (Hall 1998).

However, in examining the two cities we should also note that the attempt to reimage the cities has not been a continuous, even process. In the same way that development is uneven in spatial terms so it is also in government policy settings. Indeed, place promotion and reimaging have moved in stages which reflect the broader condition of the economy. The 1970s witnessed the first stages of place competition in the form of promotional campaigns to attract investors and the redevelopment of the lower class housing for the new urban rich. The 1980s were the period in which the first stages of inner city redevelopment started to occur. By the late 1980s and early 1990s the development and 'rejuvenation' of inner city industrial and waterfront areas came to be connected explicitly with the hosting of events and the introduction of new forms of leisure and entertainment such as casinos and IMAX theatres. The city core and its immediate hinterland had gradually become transformed into a retail–leisure–tourism–business theme park for the attraction of further investment, the creation of an appropriate middle-class environment of conspicuous consumption and the provision of further tourism opportunities. Such a description of change is not just an account of urban change at the academic level; it also reflects the changing experience of walking through the city.

Figure 8.3 Darling Harbour, Sydney, 1996 (photograph by the author)

Figure 8.4 Southbank complex, Melbourne, 1998 (photograph by the author)

Later in time, still the same place? I keep coming back though. To Darling Harbour (Figure 8.3), well I justify that because I can claim I'm monitoring urban change and the Olympics. But Southbank (Figure 8.4)? Perhaps this is slightly different. I can see rowers on the river, real people entering and leaving the train station. It's also a relatively short walk along the river, near the park and not far from the MCG (Melbourne Cricket Ground)—the holy of holies. It also has some great restaurants and memories. Perhaps that's why I like the space. There's actually some sort of personal relationship to it. It's still a space designed to encourage people to consume but somehow I can suspend my disbelief here. I still prefer the streets of St Kilda or Fitzroy—they are messier and more interesting, this is still the planned space of consumption but somehow it's not that bad.

However, the casino has now come along and dominates the area. I have lots of pictures but why should I reify capital anymore? I love walking elsewhere in Melbourne, I love the footy, but you know I don't take pictures anymore. Somehow it seems to change the place when I take a photo. The seduction is all in the mind, take a picture and you have lost the mystery. Similarly give it a brand, and try and sell places and spaces and you have also lost the mystery. Ironic really isn't it? The more one tries to seduce the harder it is to draw those in who are cognizant of what you are doing. And if you do draw them in you have to wonder who is seducing who?

"Share the Spirit. And the Winner is...?" (Sydney Organising Committee for the Olympic Games brochure, nd)

In the 1980s and 1990s, Australian state and city governments spent increasingly large sums of money to attract events with the belief that it would ensure that their cities would be seen as 'world class' and attract a high media profile and, supposedly, international capital. However, such strategies did not always win full local support. For example, there has been substantial debate over the economic benefits of hosting the Qantas Australian Formula One Grand Prix in Melbourne. According to Tourism Victoria (1997: 24), the 1996 Grand Prix, which was a centrepiece of the Victorian government's events strategy, had an estimated overseas audience of more than 300 million people and total attendance was 289,000 over the four days; it also yielded the Victorian economy A$95.6 million and created 2,270 full year equivalent jobs. By contrast, a thorough evaluation of the 1996 Grand Prix by Economists at Large and Associates (1997: 8), for the community-based Save the Albert Park Group, concluded that the claim of A$95.6 million of extra expenditure was a misrepresentation of the size of the economic benefit, that the gross benefits were overstated or non-existent, and ironically, compared to what might have been achieved, Victorians were poorer where they could have been wealthier, if the government had chosen a more 'boring' investment than the Grand Prix.

A similar example of the costs of seduction is Sydney's hosting of the 2000 Olympic Games. The bid for the games had as much to do with Sydney and New South Wales politics as with a rational assessment of the economic and tourism benefits of hosting the games. The Sydney Games appeared designed more to assist with urban redevelopment and imaging than with concern over the spread of social, economic and environmental impacts (Hall and Hodges 1997). Nevertheless, until the recent confirmation of long-alleged (Simson and Jennings 1992) corruption and scandal within the International Olympic Committee (Evans 1999, Magnay 1999, Stevens and Lehmann 1999, Washington 1999), the success of the Sydney 2000 Olympic bid was highly regarded by much of the Australian media, government and industry, as having the potential to provide a major economic boost to the New South Wales and Sydney economies. The initial economic impact study, undertaken for the New South Wales (NSW) government by KPMG Peat Marwick, suggested that the net economic impacts as a result of hosting the Games would be between A\$4,093 million and \$4,790 million, and between \$3,221 million and \$3,747 million for Sydney (KPMG Peat Marwick 1993). However, in January 1999 the NSW Auditor-General, Tony Harris, calcu-lated the readily quantifiable cost to the state government of hosting the Olympics at \$2.3 billion, which was approximately \$700 million above the \$1.6 billion figure included in the 1998 state budget. Presumably, the greatest benefit of the Olympics is seen in terms of employment; an Arthur Andersen report noted that the Olympics will create 5,300 jobs in NSW and 7,500 jobs Australia-wide over a 12-year period (Power 1999). However, this still ranks as an expensive job creation exercise.

Despite the Olympic movement's claim about the social value of the 'Olympic spirit,' the Games have been more about the spirit of corporatism than the spirit of a community. The Olympics are not a symbol of a public life or culture which is accessible to all local citizens—even though the multicultural nature of the Sydney citizenry, including indigenous Australians, was an integral component of the Olympics bid. The Sydney Olympics, along with the increasingly event-and casino-driven economy of Victoria and the other Australian states, are representa-tive of the growth of corporatist politics in Australia and the subsequent treatment of a city as a product to be packaged, marketed and sold, and in which opinion polls and focus groups are a substitute for public participation in the deci-sion-making process. In the case of the Sydney Olympic bid, the former Premier and key member of the bid team, Nick Greiner (1994: 13), argued that "The secret of the success was undoubtedly the creation of a community of interest, not only in Sydney, but across the nation, unprecedented in our peacetime history." The description of a 'community of interest' is extremely apt, as such a phrase indicates the role of the interests of growth coalitions in mega-event proposals (Hall 1997). The Sydney media played a critical role in creating the climate for the bid. As Greiner stated:

> Early in 1991, I invited senior media representatives to the premier's office, told them frankly that a bid could not succeed if the media played their normal 'knocking role' and that I was not prepared to commit the taxpayers' money unless I had their support. Both News Ltd and Fairfax subsequently went out of their way to ensure the bid received fair, perhaps even favourable, treatment. The electronic media also joined in the sense of community purpose.
>
> (1994: 13)

Greiner's statement begs the question of 'which community?' and 'whose city'? Certainly, the lack of adequate social and housing impact assessment prior to the Games' bid and post 'winning' the Games indicates the failure of growth coalitions to recognize that there may well be negative impacts on some sections of the community. Furthermore, in terms of the real estate constituency of growth coalitions such considerations are not in their economic interest. Those which are most impacted are clearly the ones least able to affect the policy-making and planning processes surrounding the Games and the reimaging process.

The hosting of the Olympic Games in Sydney, and Melbourne's successful bid to host the 2006 Commonwealth Games, are both associated with the redevelopment and reimaging of the two cities in order to attract capital and investment. However, in the late 1990s and early 2000 a new stage in the reimaging process was reached. The reimaging of Sydney and Melbourne is now not just a matter of selling bricks and mortar, or even attractive urban views, it is now also a matter of selling lifestyles. Both Melbourne and Sydney, and particularly the former, now promote access to 'café culture' through wine and food opportunities, restaurants, colourful markets full of local and foreign produce, which can now be accessed through guided tours and lessons in how to cook the meat, fruit and vegetables that you bought that morning. Certain areas having a concentration of restaurants which reflect their local migrant populations are now markets with authentic 'ethnic foods'. Inner city suburbs, such as Fitzroy in Melbourne and Paddington and Balmain in Sydney, have been transformed from working-class neighbourhoods into gentrified middle-class suburbs representing 'authentic' city lifestyle, promoted in lifestyle magazines and television shows as goals of contemporary homemaking. Moreover, even some suburbs retaining their ethnic mix are now finding themselves drawn into the range of authentic 'colourful' tourist opportunities which have been commodified for visitors (Hage 1997). Marketers and tourism authorities construct such 'discovery' of place. This is not a natural-istic process of discovery; instead we are experiencing the production of surprise as a product—three surprises an hour and we have a tourism product.

Everything now seems to be a commodity available for consumption, but of course it has to be a commodity of a certain type with a given symbolic value which reinforces preferred lifestyles and representations of identity. These commodities are not evenly distributed in urban space. They are presently dispro-portionately located in the inner city areas, with many of the outer suburbs forgotten by those who seek to reimage the city even though that is where the majority of residents live. Nevertheless, these commodities still seduce. They are produced and/or packaged to seduce the visitor, to attract international capital, albeit for increasingly short periods of time, and to seduce the locals. Because at a time when restructuring and change seems to be the norm and when employ-ment becomes increasingly 'casualized' and insecure, so it is that political elites need to be seen to be doing something. New brands, new developments, the hosting of events and the creation of new leisure and retail spaces are all signs that something is being done. However, for those who are able to visit both cities on a regular basis, we can note that even though they are trying to reimage themselves as places which are different, they are also looking increasingly the same. Melbourne's Southbank is to Darling Harbour as the new Olympic Homebush Stadium is to the Colonial Stadium and the Melbourne Sports Centre. Both cities also have new legal casinos. David Harvey's (1989) concern over the 'serial monotony' of the redeveloped and reimaged city is reinforced. As Sharon Zukin (1991: 221) observed, the city is a site of spectacle, a "dreamscape of visual consumption." It is therefore perhaps to be expected that at a time when little competitive edge can be found for either city, in terms of the reconstruction of their physical space, emphasis is then placed on the lifestyle opportunities they offer to those who are able to afford them.

Murphy and Watson (1997) describe Sydney as a 'city of surfaces,' which also applies to Melbourne. However, while the spaces are the same the places are different. Place has a distinct location which it defines. Space, in contrast, is composed of intersections of mobile elements with shifting often indeterminate borders (de Certeau 1984, Larbalestier 1994). As Stephen Daniels and Roger Lee (1996: 5) suggest, "reading human geography...is a complex and critical act of interpretation" and as readers, we are engaged in interpreting writers' inter-pretations "of worlds which are already construed, or misconstrued, by meaning or imagery." Fortunately, city places and spaces are continually read "in various ways, by a variety of people pursuing a variety of endeavours, in walking, in working, in reading, in speaking, in all or any of our everyday practices" (Larbalestier 1994: 187). While space is being constructed by urban growth coalitions in the desire to reimage the city, places remain open to negotiation and interpretation.

Despite the conscious development of space in a manner that aims to seduce the visitor and the investor, places can still resist commodification. Indeed, it is the very complexity of places, with its dense social networks, which ultimately makes them interesting though not always easily accessible for the visitor. In experiencing the city I do not want to be seduced in a contrived space; rather I want to be surprised. Despite the efforts of successive state and municipal governments Sydney and Melbourne still have the capacity to surprise. However, such surprises tend not to occur in the planned spaces of the redeveloped water-front, rather they happen in the back streets and the suburbs. Yet these are the 'messy' spaces which the real estate speculators and developers have not yet reached and where successive politicians' grandiose monuments have not yet been realized. In promoting the city to the wider community, including those who live in the suburbs and those who cannot afford inner city living, and those who are seeking more than the casino and the riverside restaurant, it is to be hoped that those places will continue to thrive.

9 Urbanscape as attraction

The case of Guangzhou

Ning Wang

Cities are attractions for both tourists and instrumental travelers. However, for a long time, it has been taken for granted that cities, especially industrial cities, are primarily destinations for instrumental travelers, and of far less interest to tourists, especially holiday tourists. As a result, while cities are actually major destinations for many tourists, their touristic functions have been under-acknowledged in theory. The city has been dominated by a romanticist bias or "way of seeing" (Berger 1972) that has stressed the dark side of urban modernity and depreciated the touristic significance of cities. Although such a situation has begun to change in relation to the discourse on the arrival of the post-industrial age in which tourism and service play an unprecedented role (e.g. Bramwell and Rawdding 1996, Law 1993, Page 1995, Roche 1992), urban tourism still awaits paradigmatic or post-romanticist shifts. This chapter is an attempt to urge academic efforts in reaching this goal.

While much of the literature on urban tourism is mainly about urban tourism in *Western* cities, this chapter chooses to examine urban tourism in the cities in *developing* countries. Based on a case study of urban tourism in Guangzhou, a city located in the south of China, this chapter suggests that, after two decades of economic reforms, a certain distinctive seductive power has emerged in China's coastal cities. That is to say, these coastal cities have come to symbolize the achievements of the project of modernization and hence China's emerging modernity. Indeed, they have become the landscape of modernity in China. For domestic tourists in developing countries, urban tourism signifies greater meaning than for their counterparts in developed countries. For example, as a result of the unbalanced rate of development between the coastal and the hinterland regions and between cities and the countryside in China, the large coastal cities have become the first sites where modernity has emerged. Thus, urban tourism in this context indicates a tribute paid by domestic tourists from the countryside and the hinterland to national symbols of successful modernization. It is a rite of passage through which domestic tourists go on pilgrimage to national symbols of emerging modernity (cf. Graburn 1989, MacCannell 1973).

In this chapter I take an ethnographic approach, first discussing the social background of Guangzhou, and then the emergence of the new urbanscape of the city in relation to the well-known economic reforms in China. The analysis concludes with an assessment of the theoretical implications of new urban tourism in Guangzhou for tourism research and urban studies more broadly.

City and history

Guangzhou is the capital city of Guangdong Province and, with over 7 million people, is the economic and cultural heart and soul of the Pearl River Delta region—one of the most economically developed areas in China. Guangzhou is renowned historically as China's southern gateway and an important port city, with a history of over 2,800 years. As the most important historic port in southern China, Guangzhou is also reputed for its commercial tradition and a long-standing tradition of immigration and migration, having taken in many waves of northern Chinese migrants and being the source of many ethnic Chinese who now reside all over the world. Many families in Guangzhou continue to maintain close ties with their relatives in Hong Kong and abroad.

Close to Hong Kong and Macao, and still serving as a vital channel of communication with the rest of the world, Guangzhou has long been exposed to the influence of Western media, especially TV programs, popular music, and lifestyles emanating from Hong Kong. When China's reform and open policies were set in motion at the end of 1978, Guangzhou was one of the first places where the old taboos inherited from Chairman Mao's era were broken down. At that time, new reform policies were introduced experimentally in Guangzhou, along with thirteen other coastal cities. As a result of the past twenty years of social and economic reforms, and also partly due to a higher level of integration with the Hong Kong economy, Guangzhou and the Pearl River Delta have become one of the most economically developed and dynamic areas in all of China. It is no small wonder that Guangzhou is often presented by the central government as one of the best examples of successful economic reforms in the country and a 'window' on the achievements of the reform policies. Within China, Guangzhou signifies 'emerging modernity' writ small; although such modernity in Guangzhou is not comparable to that of North America, Western Europe, or even Hong Kong, it does offer a hope and a dream relating to the project of modernization in China.

After more than twenty years of reform, the economic achievements of Guangzhou are obvious. Guangzhou is now one of the three largest urban economies in China, along with Shanghai and Beijing. The per capita GDP and the average disposable income per person in Guangzhou are the highest in the country. Because of its rapid economic development, Guangzhou is full of opportunities for both local residents and outsiders. Millions of rural residents from the hinterland have migrated into this city to look for jobs and other opportunities. They are the so-called 'currents of rural labor forces' and 'floating population.' As a result of both economic achievements and the sharp contrast between Guangzhou and the hinterland, the economic power of Guangzhou has been incorporated into the place identity of Guangzhou as a major dimension of the city's character.

The economic achievements have also enhanced and upgraded the status of the local culture and lifestyle of Guangzhou. In the past, Guangzhou was characterized by its *marginal*, albeit distinctive, local culture (i.e. *Lingnan*, or Cantonese culture). The formation of this cultural marginality was, historically, due to its remote distance from Beijing, the capital city of five Chinese dynasties. As a result, the hegemonic cultural power of the central feudal empire on Guangzhou was relatively weak, which allowed Guangzhou to develop its own local culture, including a culture of dwellings, of cuisine, and of clothing. At the same time, however, from the perspective of the north, the local culture of Guangzhou was marginal to the extent that it was not positively valued by the residents in the north; its influence was narrowly limited (cf. Dun Woo 1998: 1). Even two decades ago, the local culture of Guangzhou still ranked on the margins, partly because of the city's weak economic and political position during the Maoist era. However, thanks to the reforms and the consequent economic 'miracle,' Guangzhou has witnessed an enhancement of its *cultural* identity. The local culture of Guangzhou has been revalued as one of the 'superior' subcultures nationwide. For example, a Cantonese person who spoke Mandarin with a Cantonese accent might have been laughed at in the north of China in the past, but now (s)he is respected and sometimes emulated, as if Mandarin with a Cantonese accent is the fashion. Lifestyles of Guangzhou have also become widely imitated and followed by many people in hinterland areas. For example, soon after they emerge in Guangzhou, new furniture designs, styles of domestic decorations, ways of celebrating holidays, and restaurant menus are imitated by people elsewhere. Thus, as a result of economic power, Guangzhou has gained a position as a lifestyle leader, and its local culture has attracted nationwide attention. (This condition might be considered somewhat similar to the influence of Los Angeles in popular culture, Guangzhou's sister city in the United States.)

With the development of its economy and the associated enhancement of its cultural identity, Guangzhou's tourism industry has also experienced rapid growth. Guangzhou is one of the top three most visited cities in China, along with Beijing and Shanghai. In 2000, Guangzhou received nearly 23 million visitors, which included 18.79 million domestic arrivals and over 4.2 million international arrivals (Guangdong Tourism Bureau 2001). The income that these visitors brought was over 41.5 billion yuan, and the income of domestic tourism was 29.12 yuan (US$1 is roughly equal to 8.28 yuan). Tourism is now an integral and important part of the urban economy of Guangzhou and the city, together with a few other cities, was awarded the honor of 'China's Excellent Tourist Cities' by the State Tourism Bureau in 1999.

Emerging power of the new urbanscape

Why do people visit Guangzhou? This is a difficult question to answer, for different people have varying reasons and motives for their journeys. With regard to the motivation of travel to Guangzhou, foreign tourists are different from domestic tourists, business travelers from sightseers, and seasonal immigrants from leisure travelers. However, there has been a corresponding and synchronous growth between both visitor arrivals and economic development in Guangzhou since 1978, and it is certain that one of the growing attractive powers of Guangzhou has much to do with the changing economic significance of the city. On the one hand, its stronger economic position brings a wider range of opportunities for the flow of people, goods, capital, and information. On the other hand, the stronger economy also leads to better development of the urban infrastructure and facilities and conditions that are necessary for the growth of tourism. Thus, although the drawing powers of the historical, cultural, and geographical heritage of Guangzhou cannot be denied, it can be suggested that a new seductive power of Guangzhou has emerged and has been growing most dynamically. This newly growing seductive power exists in the changing landscape of urban modernity, exemplified by economic prosperity, consumer spaces, and a new architectural landscape.

Such changes in the urban landscape of Guangzhou have been brought about by rapid economic development and societal reform. Indeed, one of the most visible achievements of the economic reforms there has been the rapid pace of change in the *urbanscape*. In 1987, I came to Guangzhou for training in English for half a year. At that time, there were not many new high-rises, not to mention skyscrapers. A few hotels, including the Baitiane, or White Swan, built in 1979, and the Huayuan, or Garden Hotel, represented the newest and most modern architecture in China, and were regarded as the important new attractions in the city. I myself was recommended by local friends to see the Baitiane and the Huayuan Hotels, and I did go and enjoyed the visit. I also took my wife to these architectural attractions when she joined me later on. There used to be controversy over whether domestic visitors (few of them at that time could afford such hotels) should even be allowed to enter the Baitiane Hotel for sightseeing purposes, for they were so numerous that they sometimes caused chaos and destroyed the carpets from overuse. However, the major investor of the hotel, Mr. Huo Yingdong, from Hong Kong, claimed that domestic visitors had the right to enter the hotel for sightseeing. Such an episode could never happen again in Guangzhou, for the landscape of architectural modernity has now spread to the entire city.

I left for the UK in 1988 and stayed there for nearly ten years. I returned to Guangzhou in 1998 to take up my academic position at Zhongshan University. On my return, I was amazed at the dramatic changes in the urbanscape. East of the old urban districts a totally new urban district, *Tianhe*, had emerged. This district was composed of many high-rise buildings, constructed with high technology and new materials. In the old districts, many newly built high-rises had replaced older houses. Along the banks of the Pearl River, which flows through the center of the city, contemporary high buildings and a number of newly built bridges constituted a spectacular riverbank landscape. In addition, the modern flyovers and heavy traffic also indicated the vitality and dynamism of modern Guangzhou. The emerging 'jungle' of a new high-rise skyline gave a different look and feel to this ancient city. I was interested in comparing tourist perspectives on the 'new' city with the older, more traditional attractions, as discussed below.

Guangzhou's architectural landscape continues its rapid pace toward modernization. The city's first subway line was opened in 1999, and the second was opened in 2003. The third one is under construction and a few more are being planned. A new airport was also under construction in 2001 to replace the existing Baiyun Airport, which opened in 2004. An elevated motorway that encompasses a good part of the inner city was completed in early 2000, while an outer circle motorway was finished in 2002. In addition to the new district of Tianhe, another new district, Zhujiang Xincheng, is emerging along the banks of the Pear River. Many more apartment buildings are being constructed or planned for construction. The many varied and constant construction projects make the entire city feel as though it is a construction site. The dynamic urbanscape of the city changes monthly if not daily. Even though Guangzhou was negatively affected by the Asian financial crisis (starting in 1997), urban construction did not cease then and shows no sign of slowing in the foreseeable future.

In addition to the spectacular new and creative architectural sights, another important dimension of the urbanscape of Guangzhou is the consumer landscape, which is composed of departmental stores, shopping malls, restaurants, commodities, and consumer lifestyles. This consumer landscape is the visible economic prosperity and commercial flourishing of Guangzhou. Well-decorated shopping centers and malls, the conspicuous display of commodities, the dazzling neon lights of shops and restaurants, and the conspicuous consumer lifestyles of the 'new rich,' all constitute an impressive landscape of consumer culture (Figure 9.1).

Moreover, the 'ethnoscape' (Appadurai 1990), a term used to describe the sight of the flows of people, has also appeared as an integral constituent of the urbanscape of the city. Business travelers, shopping visitors, event and conference attendees, friends-and-relatives visitors, rural immigrants, and tourists all constitute the landscape of new mobility in Guangzhou. Every year, millions of travelers pour into it for various reasons. Simultaneously, millions of local residents get out of the city for both holidays and instrumental travel. The flows of people moving in and out of Guangzhou make it one of the most bustling and busiest cities in China. What is worth particular mention in this respect is the 'immigrant armies' from the hinterland to Guangzhou and other cities of the Pearl River Delta. In the belief that "Guangdong Province is the best place to

Figure 9.1 Shopping Mall, Tianhe District, Guangzhou (photograph by the author)

make one's fortune," rural immigrants from all over China increasingly poured into Guangzhou and the rest of the province in the 1990s, as rapid industrialization created a huge labor demand. While supplying Guangzhou with cheap labor, the large number of rural migrants also created some social problems for the city. Rural immigrants were lucky if they got a job and were able to settle down in the dormitories which their employers provided. Many, however, were unable to find work and some despairingly returned home. Other jobless and homeless migrants continued to try their luck, roaming the streets in the daytime and congregating in spacious squares or the street corners at night. Some of them even committed robbery, burglary, and murder. These problems have been highlighted in the media, drawing widespread attention.

Simultaneously, changes in the visible urbanscape have corresponded to changes in the *imagescape* of Guangzhou. 'Imagescape' is coined here to refer to the symbolic landscape: the landscape that is represented in the media, tourism promotional materials, and on the Internet. In the past, Guangzhou was seen as a historical, cultural, and subtropical city. Relatedly, its symbolic landscape mainly consisted of scattered historical, cultural, and geographical sights. The major symbols that were seen as representing its place identity included the statue of the 'Five Rams' (related to Guangzhou's mythical origins), the Sun Yat-Sen Memorial Hall, the old Haizhu Bridge across the Pearl River, and other similar sites.

However, with the emergence of the new urbanscape, the image of skyscrapers (often that of the Tianhe District's Zhongxin Guangchang, the tallest building in Guangzhou) has come to be the new symbol of the place identity for the city. For example, in media, travel guidebooks, and souvenirs, the image of Zhongxin Guangchang, together with the surrounding buildings, is used to represent a new and modernized Guangzhou (Figure 9.2). Thus, in association with the changes in the urbanscape, a new imagescape of Guangzhou, symbolized through skyscrapers, has arisen. Yet the emerging new imagescape cannot be understood superficially. Rather, we can regard it as a set of signs and symbols that signify the total structural and cultural changes within the city. Thus, this urbanscape indicates not only the changes in the built environment, but also the city's levels of prosperity, liberty, opportunities, and moral tolerance, as well as its new challenges.

The new look of Guangzhou has also become attractive for many domestic tourists, though the appeals of the older attractions still exist. Official statistics indicate that 17.41 million domestic travelers visited Guangzhou in 1998 (Guangdong Tourism Bureau 1999a). This number, based on occupancy at a small number of large hotels, is certainly an underestimate because many domestic tourists stay in small hotels or with friends and relatives. In the same year, the

Figure 9.2 Zhongxin Guangchang or Citic Plaza in Tianhe District is the tallest building in Guangzhou (photograph by the author)

number of oversea tourist arrivals (including foreigners and travelers from Hong Kong, Macao, and Taiwan) was 3.03 million. Thus, the total tourist arrivals to Guangzhou in 1998 was at least 20.44 million.

Table 9.1 shows the number of visitors at selected major traditional heritage attractions, which are primarily visited by tourists rather than locals. The most frequently visited attraction is Yuexiu Park, which is a large public park built in traditional Chinese style with lush subtropical vegetation. It also houses the Five Rams statue and some other sites of historical significance. If Yuexiu Park's 10 percent visitation rate can be assumed to represent an ideal standard for assessing the popularity of an attraction in Guangzhou, then in comparison the other attractions listed in Table 9.1 all are poorly visited. As a matter of fact, according to my interviews with some of the managers of the attractions listed above, these heritage attractions (except Yuexiu Park) were usually visited by overseas tourists because they wanted to understand Chinese culture. In contrast, the number of domestic tourists visiting these sites is relatively small. The interest that most domestic tourists have in Guangzhou is not in its heritage, monuments, and relics.

Which aspects of Guangzhou interest domestic tourists? What do they do and where do they go once they arrive in Guangzhou? One possible answer may be that most of them are just instrumental travelers and have no intention of visiting the heritage attractions mentioned above to experience and understand the city as a destination. Another answer may be that most domestic tourists travel to Guangzhou not because of the lure of its heritage sites, but because of the lure of its urbanscape as a sign of China's emerging modernity. This latter answer is more plausible since instrumental travelers are also partial tourists who can undertake some leisure activities when engaging in more work-related activities (Cohen 1974).

As mentioned above, the urbanscape of Guangzhou consists of spectacular new architecture, shopping malls, department stores, and the flow of people,

Table 9.1   Visitors to selected traditional attractions in Guangzhou, 1998

Selected attractions	*Total visitors*	*Tourist arrivals (%)*
Yuexiu Park	2,199,800	10.8
Guangxiao Temple	1,156,428	5.6
Yuntai Garden	983,518	4.8
Liurong Temple	674,023	3.0
Ancestral Temple of the Chen Family	251,368	1.2
Sun Yat-sen Memorial Hall	196,933	1.0
Western Han Nanyue King Tomb Museum	179,003	0.9
Whampoa Military Academy	173,945	0.7

Sources: Guangdong Tourism Bureau (1999a, 1999b)

commodities, and images. This urbanscape signifies the economic achievements of modernization over the last two decades in Guangzhou. Jonathan Culler (1981) argues that tourists are 'semiotic armies' who travel in order to search for the signs and symbols of a destination's place identity. The same is true of domestic tourists to Guangzhou. They seek and collect the signs of its new identity, which are most typically exemplified by the architectural and consumer landscape. Thus, it is shopping tourism and sightseeing of the urbanscape rather than heritage tourism that most typically represent domestic tourism in Guangzhou. For example, going to see the Zhongxin Guangchang building and the Tianhe District's skyline, and to shop on the pedestrian streets of 'Shangxia Jiu Lu' (with its newly embellished pseudo-historical architectural motifs), 'Nonglin Xia Lu,' 'Beijing Lu,' or in the 'Tianhecheng Square' shopping center, are what domestic tourists most commonly do. When I first arrived in Guangzhou in 1998, I consulted friends about what was worth visiting. They recommended the Tianhecheng Square shopping center in Tianhe, where you could have a walk, look at things, eat lunch, and have fun in the center for a whole day. The center was a new experience and attraction for both locals and domestic tourists alike.

Change of representations

Two cultural factors helped to change the representations of Guangzhou's urbanscape. One was the popular media and the other was tourism, especially through promotional materials, guidebooks, and city travel maps. In the popular media, the imagescape of Guangzhou has undergone a shift. As mentioned earlier, Guangzhou was formerly represented in the media as a historical, cultural, and subtropical city. Nowadays, the newly emerged architectural urbanscape is added to the imagescape of Guangzhou and treated as its new morphological identity. The change of the physical urbanscape and the change of representations are interpenetrated into one another. Both constitute a dialectical process.

A special media episode helps to illustrate the changing representations of Guangzhou. In 1998, to celebrate the twentieth anniversary of the 'Reform and Opening Policy' in China, two TV series in the style of soap operas were produced for national distribution based on economic reform experiences in Guangzhou. Both were popular and highly successful. In the series *The Blowing of the Wind and the Ripples of the Waters* one program was titled "The Part of Advertising," a fictional story about several people who were involved in pioneering advertising enterprises in Guangzhou. Another title in the same series was "Piece of Stock," about the experiences of several of the first stockbrokers in China, mirroring the development of stock brokerage firms in Guangzhou. Both of these programs portrayed pioneering, exploratory, and experimental enterprises on the road to economic reform in Guangzhou. In the second TV series, the image of the Tianhe District, especially that of the Zhongxin Guangchang building, was used as a major backdrop. This was no accident. On the one hand, Guangzhou was regarded as a good example of successful economic reform in China. On the other hand, its spectacular new architectural urbanscape, especially that in the Tianhe District, helped symbolize China's emerging modernity and the future.

As a result of these media representations, a new place-myth of Guangzhou has been gradually created, one that is based on economic modernity. Spectacular skyscrapers, flyovers, shopping centers, and the dazzling display of commodities all act as a set of symbols that represents the new identity of Guangzhou. Through the presentation of the visible new urbanscape, Guangzhou is exemplified as evidence of urban modernity and economic prosperity, brought about by successful economic reforms. As a response, many domestic tourists journey to Guangzhou to witness and experience this emergence in person.

The new dimensions of place identity in Guangzhou have also been incorporated into the touristic imagescape of the city through tourist representations, especially those in promotional materials. Two widely used domestic Chinese guidebooks illustrate this. One is titled *Travel in China: Guangzhou* (hereafter TCG) (China Travel & Tourism Press, 1998) and the other is *A Guide to Tourism of Guangzhou* (hereafter GTG) (Guangzhou Tourism Bureau, 1999). These guidebooks presented pictures of the new urbanscape to communicate the image that Guangzhou is a modernizing city as well as a historical, cultural, and subtropical city. In the guidebooks, the new architectural urbanscape was presented as an attraction itself, along with monuments, relics, museums, entertainment parks, and festivals. Thus, TCG supplied 19 pictures of modern Guangzhou, among 99 pictures in total. Of 364 various sized pictures in GTG, 45 (12 percent) were of the new architectural urbanscape. Most of these pictures of new architecture were shown to convey the overall gestalt image of Guangzhou. Thus, they function to define the new context in which heritage sites are located. In addition, both guidebooks had detailed narrative descriptions and pictures of major shopping malls, department stores, restaurants, and cuisine, effectively the new consumer landscape in Guangzhou.

The making of this new imagescape of Guangzhou in the media and by the tourism industry is not an isolated process. They are both, to a certain extent, shaped and underpinned by wider societal and political processes and associated value systems. Since reform, the dominant ideology of Mao's 'class struggle' has been replaced with the project of modernization. This new ideology constitutes a new political value system. The dominance of production and productivity, for example, became one major criterion by which many activities have become assessed politically, economically, and morally, although there were also other criteria. This is exemplified in Deng Xiaoping's famous dictum (which used to be fiercely criticized by Mao), "No matter whether a cat is white or black, it is a good cat if it is able to catch mice." For Deng, the elements of market economics, i.e. 'cats,' can be assessed as 'good' if they can catch the 'mice' (i.e. can enhance the forces of production), no matter whether they are 'white' or 'black,' communist or capitalist. Thus, economic indicators became one of the underlying political values, which informed a wide range of social practices. Media, publications, and education had to embody this system of values. As a result, economic achievements brought about by the reforms were highlighted in the media and other forms of discourse, for the purpose of confirming the 'correctness' of the reform and opening policies. In this way, 'economic miracles' that can be found in the built environment are highlighted in both the media and tourism representations. Thanks to its outstanding reform achievements, Guangzhou received intense media (and political) attention. In the process, the new media image has itself become integral to the touristic identity of Guangzhou.

Theoretical implications

Undoubtedly, without the new architectural urbanscape, Guangzhou would still be worthy of visiting and touring. Indeed, it has long attracted tourists due to its local cultural appeals, including that of the world-renowned Cantonese cuisine. In addition, urbanity is itself an attraction for tourists, especially shopping tourists from surrounding areas and the hinterland. However, the rise of the new urbanscape in Guangzhou has gained an additional drawing power for the city. In reality, for many domestic tourists from the hinterland, to see the modern urbanscape is an eye-opener, which can itself be sufficient reason for tourism. Although, methodologically speaking, it is difficult to sever this additional drawing power from other aspects of the seductive power of Guangzhou, the former can be analytically assessed for its theoretical implications.

Except for some of the world's major tourist cities (e.g. London, Paris, San Francisco, Hong Kong), cities in general, and urban architectural landscapes in particular, have long been under-acknowledged as tourist attractions in the social science literature on tourism (Law 1993, Page 1995). As previously mentioned, cities, especially industrial cities, have often been regarded as the origins of tourists rather than as tourist destinations. By contrast, nature and countryside have been seen as the ideal spaces for tourism (cf. Dumazedier 1967). In such a romantic approach, tourism embodies 'escape attempts' (Cohen and Taylor 1992, Rojek 1993a), attempts to avoid the malaise of modernity through a search for authenticity from natural, rural, and primitive destinations that have not yet been 'polluted' by modernity (cf. MacCannell 1973). Most cities are the settings from which one escapes (Cohen and Taylor 1992, Dumazedier 1967, Dann 1997, Rojek 1993a). Even tourism promotion for most of the urban regions in the United States tends to avoid elements of urbanity while emphasizing in travel brochures the natural environment in the surrounding areas (Hummon 1988).

By contrast, urban tourism to Guangzhou reveals that the city can be a significant tourist attraction. While this chapter is not aimed at developing a general theory of urban tourism, it is discussed here in relation to the context of modernization in China. In the case of urban tourism in Guangzhou, cities are attractive to tourists not merely because of their better leisure and cultural facilities and resources in a general sense, but also because of their symbolic functions: that is, they are national symbols of modernity.

In China, modernization and economic development are spatially unbalanced. Coastal China had a better infrastructure and more technological capabilities than western regions when the reforms began in late 1978. In 1980 four coastal 'special economic zones' were granted development policies that favored the development of market economies. Some of these policies were later extended to fourteen coastal cities (including Guangzhou), allowing these zones and cities to become the first developed regions in China. For example, Shenzhen, one of the special economic zones in China, used to be a very small town of only 30,000 people adjacent to Hong Kong's New Territories in 1980. After twenty years of rapid development, it has become one of the most modern cities in China, with a large urban economy and a population of over 4 million. By contrast, the more distant hinterlands and the western regions, which had fallen behind the coastal areas in economic terms well before the reforms began, are now even further behind economically. Although these areas are presently being favored by central government policies, the gap between them and the coastal areas will be difficult to erase. However, for domestic tourists, especially those from the remoter parts of China, the new urbanscape of Guangzhou is horizon-broadening, because it symbolizes a new world and a wishful goal for themselves. Modernity is no longer a foreign or Western phenomenon, but now has a Chinese face. Modernity is now a domestic reality, a home phenomenon, a landscape that domestic tourists can travel to see and experience freely.

Touristic destinations are thus not a natural gift offered by the would-be God of Nature, nor fully a result of tourism planning. They are rather the spaces that are socially constructed in relation to a series of socially defined dichotomies: the modern and the pre-modern, the contemporary and the primitive, complexity and simplicity, the city and the countryside, the developed and the developing, and so on (cf. Lanfant 1995: 2, N. Wang 2000). Tourist geographies are thus, in part, the result of social and cultural 'classification' of spaces in terms of a set of modern touristic values. Two categories of such spaces are very basic: the space of home, work, and daily responsibilities and the space of difference, leisure, and freedom. These two categories of space can be exchangeable between 'us' and 'them.' 'Our' spaces of home may be 'their' tourist destinations and, conversely, 'their' spaces of home become 'our' tourist spaces. Human society thus creates double perspectives from which both 'we' and 'they' see each other's spaces differently. The various definitions of human spaces from each other's perspectives lead to two fundamental flows of people in opposite directions. If we have already emphasized the significance of the flow of urban dwellers to nature and the countryside for the purpose of seeking refuge, then it is high time that we acknowledge the significance of the flow of countryside (and suburban) tourists to cities for the purpose of 'eye-opening' or 'horizon-broadening' experiences of contemporary modernity.

Conclusion

If, generally speaking, cities represent for leisure visitors sources of fun, pleasure, excitement, and opportunity, then for domestic tourists from the hinterlands of China, cities, including industrial cities, have particular meanings in addition to these. This may be true for some other developing economies, but is probably due more to a combination of underdevelopment and strict communist era rule. Economically strong cities, such as Guangzhou, which have emerged in the age of reform in China, are frequently visited not only because there are more opportunities and resources for instrumental visitors, but also because they represent emerging modernity for domestic tourists. In this way, the modern architectural urbanscape in Guangzhou has become a new symbol of the city's place identity, a symbol that signifies the coming of modernity to Guangzhou and China as a result of economic reforms and openness to the outside world.

Such an architectural urbanscape has come to form a new urban context, which also embeds the earlier appeals of local heritage, culture, and the geographical features of Guangzhou. Such a re-contextualization of the existing touristic resources in terms of the newly emerged architectural urbanscape offers a new way of seeing Guangzhou as a touristic city. This re-contextualization reflects actual structural and cultural changes in Guangzhou, which are also symbolized in the imagescape of media and tourism promotional materials. In short, with the advent of the new urbanscape, Guangzhou has undergone an identity shift within the context of modernization during last two decades.

10   San Francisco and the left coast

Carolyn Cartier

> There is only one question remaining after Tuesday's election—why is the rest of the country out of step with the Bay Area?
>
> Carl Nolte (2000)

In a review of television specials set in San Francisco, cultural critic and college professor Stephen McCauley (1998) writes about the city as a place of myth and legend. His students claim having spent past lives there, a mode of orientation he finds otherwise reserved for places like "ancient Egypt, Machu Picchu, Petra; rose-red city half as old as time." He observes, "San Francisco has achieved mythic status among the disillusioned and disenfranchised from all over. Surrounded by water and frequently shrouded in fog, it has become, in our collective imaginations, a kind of real-world Oz"—a fantastical place where people actually live. Continuing the metaphors, he evaluates the public television documentary "The Castro" as a neighborhood history of a marginalized group whose community has become so successful that many of its businesses cater to tourists; the whole neighborhood "has taken on something like a theme-park atmosphere: Gayworld." Perhaps what disorients McCauley is the sight of people at leisure on these streets—just as likely they're not tourists.

This vantage is from the US East Coast, whose leading newspaper, *The New York Times*, dependably represents San Francisco with appreciation as a city to be taken seriously for its landscape iconography, cosmopolitan ethos, and cultural institutions. As the veteran *Times* writer R.W. Apple (1999) reported for the "Weekend" (not "Travel") section of a Friday edition, "more than any other, this is the city that Americans fantasize about." He continues:

> For me San Francisco has always been a paradox: a thoroughly worldly place, yet remote and somehow detached from world capitals like New York and London; a little indifferent to the mundane business of trading shares and casting votes; in other words, a bit smug.... Yet San Francisco looms large in the imagination of everyone who knows it, myself most decidedly included. Like Paris, Venice and Hong Kong, it is a city without peer, a city of myth and magic.

He complains about the weather, but is compelled to conclude, "nothing, in truth, not fog or rain or fire or quake, ever quite masks San Francisco's beauty or quenches its exuberance." He has been in San Francisco long enough to realize that the fog frames and intensifies landscape, lending movement, light, and spectacle to the hills, bridges, and built environment.

Apple finds the city's character as a 'town of tolerance' owes to the Beat Generation, the Haight-Ashbury, the popular music scene, and the large gay and lesbian community. He compares San Francisco's elite world of the arts—the opera, symphony, museums, and the new public architecture that houses it all— to those of New York City's institutions. He also effectively dismisses Los Angeles (and thereby the Getty Center, the Los Angeles County Museum of Art, and other heavyweight institutions) by reaffirming San Francisco's early twentieth-century role:

> San Francisco is on something of a cultural tear lately, cementing its position as the West Coast's cultural capital, although Los Angeles is by no means the pothole full of philistines many San Franciscans see when (if) they take the plane south. The San Francisco Opera is the nation's second largest, after the Met, and it has the largest budget of any West Coast arts organization.

Apple's cultural institutional view of San Francisco is akin to a worldview based in Manhattan's upper 'sides,' the establishment of the upper east side enlivened by the west side's cultural political sensibilities. His discussion collapses the continental proportions of geography into the 'flyover,' the noun and verb that places San Francisco within easy range of New York City. In this estimation, San Francisco meets the standards of the New Yorker's imagination as a weekend destination. In fact, among tourist 'feeder markets' to San Francisco, New Yorkers tend to make up the second largest category of domestic visitors by state (7.7 percent) after those from California itself, where visitors from Los Angeles (8.5 percent) comprise the largest proportion (SFCVB 1999: 26).

Ideas

Tourism is the largest revenue-generating industry for the City of San Francisco, and is the second largest employer, after financial services (contrary to the worldview from New York). The economic dimensions of San Francisco's touristed landscapes manifest not so much in total jobs or comparative industry growth, but in consumption and the revenue-generating power of the industry for city government. More than 16 million people on average visit San Francisco each year, which is more than 16 times the local population, including nearly 3 million international visitors, and they spend enough money to return to the city about half a billion dollars (*TWSF* 2001). With the exception of hotels and some transportation and local tour services, tourism and local leisure activity overlap considerably, in retail consumption, the arts, a highly articulated restaurant culture, and transport. So tourism in the city, with the exception of visiting particular sites like Fishermen's Wharf or Alcatraz Island, merges into local cultures of leisure consumption, producing what David Crouch (1999a, 1999b) calls 'leisure/tourism' as touristed landscapes of local lifestyle. Like other places whose touristed landscapes prevail prominently over time and trends in the world imagination, in spite of mass tourism, understanding San Francisco's enduring profile in the tourist imagination depends on sorting out place geography: enduring themes of seduction, the contexts of landscape formation, the comparative and contested place identities of locals, migrants and tourists—and ultimately the ways these themes converge in complex, constructed meanings.

The conceptual orientation of this analysis engages some of Dean MacCannell's ideas about the spatial structure of tourist places. A resident of San Francisco, MacCannell recognized that his 'landmark' study of the tourist might well have been situated there rather than Paris. But the case for Paris reflects its more substantially documented history and its role in urban modernity; it was also a major stop on the Grand Tour. Why these cities (and London and Rome) made the cut for MacCannell's (1976: 59) project owes to "a fully elaborated touristic complex which was not yet repetitive or encrusted with commercialized attractions." San Francisco possesses this quality: a city of distinct landscapes both lived and toured, a city whose touristed landscapes emerged from scenes of daily life. In MacCannell's language of structural places, San Francisco presents itself to the tourist as a set of districts—Fisherman's Wharf, Chinatown, North Beach, the Mission, the Castro, Soma, and more—differentiated and brought together by areas, arteries, and means of connectivity—Market Street, Mission Street, Grant Avenue, Union Square, cable cars, the waterfront. One doesn't 'know' the city, but apprehends some of its qualities via landscape encounters in these discrete places. But these districts or places and the areas that connect them are also residential and working areas, places where people live and enact lifepaths of work and leisure. It is this accessibility to the apparent 'back region' of Erving Goffman's (1959) conceptualization that MacCannell recognized as ultimately attractive to the tourist—the opportunity to get close to the 'real' otherwise behind the scenes. This element of tourist desire—encounter with the 'background' of the un-staged and the potentially intimate, and ideas and realities about widespread opportunities to access it in San Francisco—accounts in some part for the city's seduction of place.

Yet the idea of the touristed landscape seeks to intervene in the 'foreground' and 'background' of touristic space. Following the metaphors, the touristed landscape is the scene of an intermediate ground where the 'real' and the 'staged' may merge, and from one moment to the next. It is this space of mixing that is so appealing to the tourist because potential boundaries separating 'front' and 'back' regions are less apparent or absent, offering access to the unmanicured realities of the local scene. In the 1960s visitors sought this experience in the Haight-Ashbury; in the 1980s the Castro became a vital scene of street leisure for both gays and straights, what David Bell and Gill Valentine (1995: 7) have called one of the most "*visible* gay communities" in the world; in the 1990s the dot.com boom gentrified the Mission district, which was quickly discovered by visitors seeking the edge. Certainly we can locate in San Francisco aspects of the front and back region; but the overall experience offers visitors a full range of social and economic issues and events enacted on the street and places of consumption, whether gay culture in the Castro, daily marketing on Stockton Street in 'Chinatown,' or homelessness at the Civic Center. If anything demystifies San Francisco for visitors, it is the 'real'; the realization that the city of myth and imagination is full of places where the real plays out—un-themed, unconstructed, and not very pretty. San Francisco has its many elements of boutique-town-ness, but it also has areas where tourists socialized to constructed, themed environments find themselves utterly affronted by the realities of the place. But such unscripted elements are also the city's assets, making visitors feel as through they have really been 'somewhere,' why they then subsequently participate in the reiteration and repetition of San Francisco's profile of superlatives.

This chapter assesses San Francisco historically, first as a destination of desire and significance in the developing western United States, and a place whose transhistorical draw, based on periodically booming regional economies, contributes to constructing the idea of San Francisco as an iconographic center in the global imagination. (For example, it is typical in Asia, the region where I usually work, for people to think that 'Silicon Valley' *is* the San Francisco Bay Area, a notion which blurs any distinction between the city and the regional economy and whose size gets closer to approximating the significance of the region's high-technology industries in the world economy.) As a city historically settled by waves of migrants and capitalized by new elites, San Francisco early gained its profile as a center of diversity, yet structured by an ambitious political economic leadership boasting high style and based in speculative economic activity. Wealth spilled over into the built environment, the arts, and, above all, lifestyle priorities, in which performing one's values and tastes through activities, dress, and leisure consumption has been a leading characteristic of the city's place identity and visual landscapes since its early days. This lifestyle orientation is arguably what has opened up the space for social tolerance—as long as those social forms fit within a range of accepted cultural political positions and their implicitly, and sometimes explicitly, designated locations.

This analysis of touristed landscapes treats San Francisco as a place of distinctive appeal as well as some of its constituent landscapes. For example, survey data collected by the San Francisco Convention and Visitors Bureau show that tourist interest in San Francisco is based on visual perceptions of its environmental qualities, and on sensory dimensions of leisure experiences, including restaurants and weather. While San Francisco's visual landscape drives the tourist imagination, what has also driven San Francisco is a dynamic political culture, spurred by a history of positions contrary to conservative US federal governments, geared toward lifestyle quality, and characterized by diverse and cross-cutting alliances often times bridging class, sexuality, and race: this is the San Francisco of the 'left coast.' Where tourist narratives such as Apple's are widely rehearsed and repeated, they also work by masking the complexities of social processes that have formed the city's communities and political cultures, longer term processes that elude visitor focus. The city's culture of activism is also an arena for identity formation, worked out and perceived at all scales from local to global, and its axes of identity formation are situated around place or neighborhoods and their dimensions of class, race, and sexuality, as well as around lifestyle positions and leisure activities more generally. Lifestyle priorities and their arenas of articulation are arguably more significant than race and ethnicity as leading markers of identity formation in San Francisco, a position that I discuss below through local perceptions of political values, sport, and cultures of masculinity.

The culminating ideational point is how to characterize the San Francisco Bay Area's 'development landscape' (cf. Cartier 1997, 1998, 1999b, 2002) and its 'social formation' (Cosgrove 1984a)—what I call here the 'spectacle of environment.' As the tourist understands, distinctive ideas about the city's landscape aesthetics and environmental resources, as well as its cultural economy, characterize the evolution of San Francisco. The city's human–environment relations orient to land and water, topography and the Bay, the role of the waterfront in the city's economy—in sum, a certain natural environment of landscape development. Vistas in San Francisco—a city whose real estate development platted out land geometrically and gridded over a series of hills—offer vertical, stunning viewscapes of architecture and the Bay, natural and built environments. Whereas in the middle of Manhattan the pedestrian or motorist is hard pressed to be reminded of the larger environmental situation, traversing San Francisco streets and hills opens up lines of sight to the water, to views of the Bay and its region. This command of the viewscape, to be constantly reminded of the locale's stunning arrangement and proportions (its extraordinary *fengshui*), is the optic of understanding the San Francisco Bay Area as "an important case of regional capitalism grounded in the wealth of nature" (Walker 2001: 167). This is not the simpler notion of capitalist spectacle, in which society is simultaneously so seduced and dulled by the extraordinary possibilities of the visual, the possibility of the consumption of things, that alienation defines its condition, or the spectacle as 'festival' (cf. MacCannell 1992). It is closer to Guy Debord's (1994: 1) "immense accumulation of *spectacles*," the geography of a city and its region as the vast institutional and technical apparatus of contemporary capitalism, a spectacle of geographical display, of the wealth of nature laid out before you, the spectacle that could symbolize and bind together the body politic in rehearsing the political and moral order of the city-region. This contemporary spectacle works out through images, discourses, and gestures of local 'greatness,' mediating social relations; this is about how the San Francisco Bay Area's local CBS television affiliate, KPIX, gets away with its slogan, "the best place on earth."

Land and water: environment and development in landscape formation

The hills around San Francisco Bay shape perspectives of landscape. Like Hong Kong and Rio de Janeiro, San Francisco sits astride a ria coastline, a coastal plain of drowned river valleys, which creates conditions for the world's finest natural harbors and dramatic viewscapes. In the historic view from the Pacific, Angel Island in San Francisco Bay and the Berkeley Hills to the east made the entrance to the harbor effectively invisible by forming a seamless horizon. As a result, San Francisco Bay, in colonial history, was discovered relatively late in 1769 by land, as a Spanish expedition headed up the San Francisco Peninsula (Costansó 1911). The superlatives emerged in the early expeditionary parties:

> Indeed, although in my travels I saw very good sites and beautiful country, I saw none which pleased me so much as this...there would not be anything more beautiful in all the world, for it has the best advantages for founding in it a most beautiful city with all the conveniences desired, by land as well as by sea, with that harbor so remarkable and so spacious, in which may be established shipyards, docks, and anything that might be wished.
>
> (Font in Camp 1947: 4)

The Spanish had located the finest natural harbor on the eastern Pacific.

San Francisco has arguably been a transhistorical mecca since its urban settlement origins in the Gold Rush. Before the bridge, the narrow entrance to the Bay was known as the Golden Gate and it formed the first rite of passage for miners on their way to the foothills of the Sierra Nevada. The gold mines hit their production peak in 1852, followed by a silver mining boom in Nevada in the 1860s and 70s. The wealth was spectacular and conspicuous, but was it this lush?

> With the 1870's, San Francisco entered upon its Arabian Nights era of sultanic palaces, fantastic towers, gold dinner services, silver balustrades and doorknobs, Oriental carpets, Chinese brocades, marble, brass, glass, lutes, flutes, trumpets and kettledrums, marquetry, parquetry, Pompeian frescoes, South African diamonds, rainbow apparel, extravagance and extravaganzas. The

> Silver Era far surpassed the Golden Days of Forty-nine in spectacular incident and display. The river of silver streamed over the city and engulfed it.
>
> (Altrochhi 1949: 157)

As a consequence of wealth accumulated by fortunate newcomers and locally invested,

> a metropolitan and cosmopolitan city sprang up at San Francisco, a city that telescoped half a century's growth into a year…and was within five years a financial rival of New York and, in some respects, a cultural rival of Boston.
>
> (Bean 1968: 149)

San Francisco's per capita wealth had become the highest in the nation, unevenly distributed in the extreme, its culture distinctively male-gendered, and proudly anything goes. (The female population of San Francisco did not begin to approach parity with the male population until 1950.) This landscape scenario, of a rich nascent industry, the places it produced, and its consequential impacts on the cultural economy of San Francisco, would play out again in the late twentieth century in the rise of the global computer industry, centered south of San Francisco in Silicon Valley.

Since its origins, as a settlement built across hills and confined to a peninsula that maps out at just under 7 by 7 miles (11 by 11 km), the city's economy has been bound up with the waterfront and its transbay connections; almost as significantly, its visual environmental culture has embraced the Bay. In 1776, the Spanish located the presidio at the northern tip of the peninsula, and later sited the plaza to the southeast near the water's edge on the bay side. After 1846 it was called Portsmouth Square, named for the ship commanded by John B. Montgomery, whose name was given to the beachfront. Now a public plaza marking the eastern edge of Chinatown, Portsmouth Square is over a mile from the Bay's shore. Waves of reclamation and speculative real estate development proceeded incrementally east from Montgomery Street, which marks the financial district. Skirting that, the reclaimed shoreline has formed the city's port in a garland of piers and terminals. Unlike Oakland, where acres of rail and warehousing yards separate the central business district from the waterfront, San Francisco's high-rise business district commands the Bay and its vistas.

The topography of the Bay Area was also centrally at stake in the project to complete the first transcontinental railroad, which bound San Francisco to the larger region. (Its construction also drew another wave of migrants, Irish and Chinese, among whom the latter became the single largest minority population in San Francisco.) Even as there was no question that the railroad was headed for San Francisco, engineers knew that bringing the rail line up the San Francisco Peninsula was far less efficient than a terminus on the eastern side of the Bay.

Arguably the most consequential event in the history of Oakland, the Central Pacific selected the east bay city as the terminus in a deal that deeded waterfront land directly to the railroad, and which San Francisco was unwilling to match (Daggett 1922: 87–8). The ultimate destination of the transcontinental line pointed to San Francisco as the first city of the West Coast, while its penultimate destination, Oakland, underscored how San Francisco would both depend on and anchor its larger regional economy. Travelers were compelled to make the final leg of the journey to San Francisco by ferry. Construction of the Golden Gate Bridge and the San Francisco–Oakland Bay Bridge, completed in 1936 and 1937, respectively, established the region's iconographic infrastructure and fundamentally networked the Bay Area, making the North Bay and the East Bay integral regional communities.

The waterfront was also the basis of the economy, as the Port of San Francisco anchored industrial development and trade through the first half of the twentieth century. The political culture of the port, though, was arguably more consequential in the city's formation of place, and an arena for its evolving cosmopolitanism (Cartier 1999a). The port led union formation in the United States; San Francisco dockworkers organized as early as 1853, and its sailors formed their trade's first union in 1866 (Camp 1947: 279–81). In the process, San Francisco became the labor capital of the west and internationally distinctive for its union activism, to the degree that "at the turn of the century San Francisco was the first port city in the world to be known as a 'closed-shop'" (Camp 1947: 405). Labor activism also ultimately prevailed over other modes of community organization: the 1934 longshoremen's strike saw the breakdown of traditional barriers of craft, nationality, and race as it ballooned into a general strike (Nelson 1988: 133). Traditions of democratic activism, inter-racial community formation, and worker rights movements in the San Francisco Bay Area stem from this period, and thus arguably the maritime-based labor movement bequeathed to the city its transhistorical culture of progressive political activism. The maritime community was also the scene of gay culture, which was historically located around the waterfront and its bars (Stryker and Van Buskirk 1996). By the third quarter of the century, unionization was in decline nationwide, and, in the process of shifting to container shipping, San Francisco lost its cargo-handling role to Oakland. With its direct rail connections, Oakland became the leading container port between Seattle/Tacoma and Los Angeles/Long Beach; consequently, many of San Francisco's piers fell into disuse and the harbor front atrophied.

But the postindustrial shift soon worked through the waterfront landscape, led by the conversion of the great brick Ghirardelli Chocolate Factory in the 1960s, west of Fisherman's Wharf, into a retail center. It became a national example of adapting historic industrial buildings and listed on the National Register of Historic Places in 1982. Adaptive reuse of the California Fruit/Del Monte Cannery, between Ghirardelli and the Wharf, soon followed. The complex was once the largest peach cannery in the world, and its restoration as The Cannery, another boutique shopping scene, furthered the revival of the historic built environment. In the process, former 'back region' spaces of work and industry opened up to settings for the new leisure consumption. In 1978, the 'urban entertainment complex' (Hannigan 1995) arrived at the San Francisco waterfront at the eastern edge of the Wharf with the transformation of Pier 39 into a tourist shopping destination and open-air entertainment center. By 1989, *USA Today* named Pier 39 the third most visited attraction in the country; in 1992, the London *Observer* named Pier 39 the third most visited attraction in the world (Pier 39 2002). While the San Francisco Convention and Visitors Bureau ranks Fisherman's Wharf as the city's number one attraction, the Pier 39 corporation claims the spot for itself, counting an estimated 10.5 million visitors each year. In the process, Fisherman's Wharf has become a monument to more typical tourist consumption, distinguished by the proliferation of carnivalesque shops and a wax museum. The district's fishing and seafood industry also contracted, though the classic restaurants remain. So the 'back region' of fishing production and the touristed landscape it fostered has departed the Wharf, now increasingly "encrusted with commercialized attractions." This transformation in style has undermined the area's draw for locals, while its predictable entertainment complex landscape has substantially engaged mass tourism (Figure 10.1).

The touristed overdevelopment of Fisherman's Wharf stands in contrast to the city's western waterfront and coastline. In 1972, the world's largest urban park, the Golden Gate National Recreation Area, was established on both sides of the entrance to San Francisco Bay. At two and a half times the size of San Francisco, it encompasses 28 miles (45 km) of coastline and diverse built and natural sites, including the Presidio, Ocean Beach, Alcatraz Island, Stinson Beach, and the Muir Woods National Monument among many others. Park historians assess how the Golden Gate National Recreation Area (GGNRA) contravenes the usual assumptions about a national park. As a "compilation of urban greenspace and rural lands surrounding San Francisco's Bay Area, it reflects the growing tensions in the National Park Service about the purpose of a national park designation," in part by presenting "one of the most complicated management challenges in the entire national park system" (Rothman and Holder 2001). Instead of the federal agency dictating terms to surrounding communities, in San Francisco the Park Service had to respond to demands of the multidimensional Bay Area. What makes the GGNRA different may also make it an archetype for national parks in the twenty-first century. Instead of being remote from population centers, its urban location presents new challenges about connections: between different kinds of places and landscapes, more and diverse constituencies, and possibilities for integration with many communities. Above all, the GGNRA has conserved land around San Francisco Bay, maintained coastal access, and capped in perpetuity commercial development on both sides of the Golden Gate (Figure 10.2).

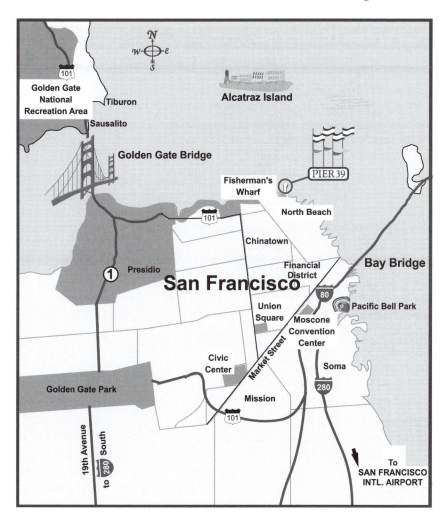

Figure 10.1 Tourist map of places of visitor interest in San Francisco

Source: © 2003 Pier 39 Corporation All Rights Reserved

The GGNRA responded to two important conceptual dimensions of San Francisco's touristed landscapes: the spectacle of environment, by opening up direct access to the land–water nexus of nature–society relations; and the 'intermediate ground' of the touristed landscape, by further opening up the 'back region' of Alcatraz Island and former military buildings at the shoreline and in the Presidio. The role of the park's accessibility in the regional social formation also arguably elides elite hegemony in other landscapes and places of the city. Over 13 million people visit the GGNRA annually, which makes it not only the most popular among 10 national park sites in California, receiving four times as many visitors as Yosemite National Park, but the most popular leisure destination in the

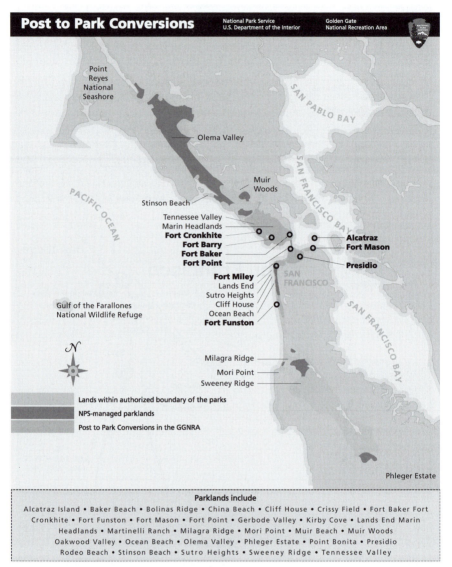

Figure 10.2   Post to Park Conversions, Golden Gate National Recreation Area

Source: Courtesy Golden Gate National Parks Conservancy

state overall, outranking Disneyland (State of California 2003). Of course local and regional residents make up the substantial part of total visitors.

In October 1989 the Loma Prieta earthquake hit, damaging the double-decker Embarcadero Freeway along the southeastern waterfront. The decision to tear it down effectively reopened the waterfront and Market Street's terminus at the Ferry Building. Light and space reappeared where previously the dark, hulking concrete mass discouraged pedestrian activity. In the process, the Ferry Building was renovated, high-speed ferry services renewed the site's function, and the waterfront south of the Bay Bridge erupted in a new high-rise, high-rent residential community. Construction of the new baseball stadium for the San Francisco Giants at China Basin, known as Pac Bell Park (a neoliberal toponym, after the local telephone company, Pacific Bell), provided the southern anchor for the revitalized area. The city spent a million dollars on mature Canary Island palm trees to frame the Ferry Building and line the street-car median along the Embarcadero, igniting debate over whether they looked "too LA." In the process, for the first time since the pre-industrial era, the city achieved unobstructed public access on all sides of the peninsula. These open-space priorities, sanctioned by success of the GGNRA, are also representative of ideas and realities about environmental amenities characterizing the political culture of San Francisco—and the 'left coast.'

The left coast

Its landscapes purportedly draw the tourists, but San Francisco's contemporary place seduction depends on its cultural political representations as the heart of the left coast. Ideational and material, the left coast has come to denote the regional geography of politically left-leaning trends in the United States. In other words, the phrase left coast is not regularly used in San Francisco, or in the west about the west; it is a toponym of comparison to the east, a term of 'place othering' or 'placism' from eastern vantages. For some it is a basis for the enduring draw of the west, as a region symbolic nationally of less traditional social norms and political conditions. For the other end of the political spectrum, it is a regional put-down, a handy discursive crucible of regional disparagement. For example, in 2002 when Congresswoman Nancy Pelosi was confirmed as minority leader of the US House of Representatives, critics attempted to dismiss her as "a 'San Francisco Democrat' who represented a latte-sipping, tree-hugging, Left Coast city that was light-years from the American mainstream" (Tempest 2002). At the regional scale, California as well as urban centers of Oregon and Washington all have some claim to being part of the political left coast, but California, and especially urban coastal California, stands out, with San Francisco as its center. It is a place that apparently breeds a distinctive politics whose Democrats must even be singled out from the rest of the Party. As William Safire (2001) has somewhat unwillingly accounted for it,

> In the past decade, California has become the most Democratic state in the nation. It has a Democratic governor, two liberal Democratic senators [both from San Francisco], and both houses of its legislature are firmly in Democratic hands. Not for nothing is it called the Left Coast.

In this essay, Safire complains about Californians vilifying energy deregulation (it is a pre-Enron collapse harangue). But in a more interested moment he addresses the term's etymology. The earliest citation he can turn up is the title of a 1977 record by the southern rock and blues band Wet Willie, "Left Coast Live" (Safire 2000), which was recorded at The Roxy on Sunset Boulevard in Los Angeles in 1976. Otherwise he cannot find sustained political usage of the phrase until the 1990s. But in the late 1980s around Cambridge, Massachusetts, the locally educated were glad to label Berkeley-types 'left coast'; that was the first time I heard it. Around the same time, in San Francisco, Rich Deleon was working on *Left Coast City* (1992), and encountered the phrase in political discourse (pers. comm.). A search of newspapers also turns up the phrase in *The Boston Globe* in the late 1980s, though not in a political sense but simply in a sports column to refer to where the Red Sox tended to lose regionally ("The Left Coast has been cruel in recent years") (Shaughnessy 1989). The timing and the place seem to point to urban northeasterners bringing 'left coast' into common usage by the 1980s as an irreverent toponym for their more politically charged regional counterpart on the left-hand side of the national map.

The idea of San Francisco as the center of the left coast, and Boston as one place where the phrase has been comparatively and irreverently deployed, is suggested in information on social capital formation in the United States. Deleon (2002) examined survey data from the first national comparative assessment of social capital, on over 3,000 people in 40 communities, for trends in comparative political participation and political ideology: in several categories, San Francisco ranked highest, followed by Seattle and Boston. In terms of political ideology, nearly 72 percent of white San Franciscans were glad to identify themselves as 'liberal,' which is nearly 30 points higher than populations in the next closest cities, Boston and Seattle. San Francisco also scored highest in political protest activity, followed by Seattle and then Boston. San Francisco is also the outlier in the category 'not religious,' followed, in this case, at some distance, by Seattle. (In the sociological literature, more than education or income, secularism correlates with liberalism.) On the question of support for immigrant rights, white San Franciscans surveyed showed far greater positive unanimity than the same group in any other major city (higher by at minimum 15 percent), including Seattle, Boston, and Los Angeles. Whether 'true' or not may not be the fundamental question: San Franciscans (in 2000, 49.6 percent of San Franciscans were white) choose to represent themselves as more politically left than residents of any other city in the country. Social science research is trying to pin down what is the 'left coast' because there is something to it.

California's voting patterns typically map out the differences between political positions in the Bay Area and elsewhere in the state. Results of both statewide and federal elections show, time and again, that at least four counties in the Bay Area consistently vote more liberal and socially oriented lifestyle positions than any other geographical concentration of counties in the state. San Francisco (city and county), Alameda (including Berkeley and Oakland), Marin (the coastal county north of San Francisco), and Santa Cruz (on the coast south of San Francisco, and location of another University of California campus) counties together consistently demonstrate the most left-of-center voting positions (Nolte 2002). In presidential elections, no Republican candidate has carried San Francisco since Dwight D. Eisenhower in 1956. San Francisco is also the West Coast center of national anti-war protests, which assumes its own inscribed landscape: assemble at the Ferry Building and march along Market Street to rally at City Hall.

The reality of the left coast has a substantial basis in social and labor movements and the migration of alternative lifestyle seekers, artists, writers, and progressive social thinkers to the west. In the early twentieth century, socialist and communist parties were both active in California, especially through the labor movement (Schwartz 1998). San Francisco's edgy literary lineage originated with Gold Rush satirists and continued through the Beat generation, centered in San Francisco's North Beach district around Jack Kerouac and City Lights Bookstore. The popular music scene—centered in the 1960s at the Fillmore Auditorium and dominated by the Grateful Dead and Jefferson Airplane (later Jefferson Starship), and in the 1970s by Santana, the Doobie Brothers, and more—was staggeringly rich. At Candlestick Park in 1966, the Beatles made their final public appearance. Berkeley, both the university and the town, was a world center of alternative politics and culture by the 1960s, lending intellectual heft to the cultural movement centered in the Haight-Ashbury district of San Francisco, which culminated in the 'summer of love' in 1967, the national climax of the Hippie Movement. The issues and activism that made '1968' a metonymy of the era's political culture played out distinctively in San Francisco and Berkeley. Between San Francisco and Los Angeles, California's cities and universities have been centers of innovative and alternative social thought to the degree that East Coast-based university faculty will refer to their West Coast counterparts, not always admiringly, as 'social theory left coasters.' Through it all, San Francisco has stood for the alternative of alternatives, drawing like a vortex all those who would seek its shelter from norms. What could better represent the liminality (Urry 1990: 11–12) of tourist experience? Likely moving to San Francisco, seeking liminal as lifestyle.

But the idea of San Francisco as the center of the left coast has been destabilized—at least in San Francisco—by the 'dot.comization' of the city. The dot.com boom of the late 1990s spilled over from Silicon Valley, impacting the real estate market and bringing a new wave of speculative capital. Apartment rents more than doubled in the second half of the 1990s (Huffman 2001), and what were relatively affordable artists' lofts disappeared from the south of Market as the municipal government sanctioned the district's development as Multimedia Gulch

(Redmond 2001). The Great Schism of 1999 was a high point of contrast between the city's new residents, propelled by silicon- and stock market-derived fortunes, and longer settled San Franciscans who found the *arrivistes* less motivated by the San Francisco-style culture of cool and left politics than a 'culture of speed' and its associated practices and *matériel* (MacCannell, this volume). Apparent flocks of SUVs descended on the city, charging around and challenging a long taken-for-granted spirit of urban environmentalism. By contrast to the upheavals of a generation ago, when Mayor George Moscone embraced the gay community, "it's no longer the sight of two men kissing on the street...that drives some people bonkers—it is 20-somethings in $70,000 sport utility vehicles talking to their brokers on cell phones" (Weiss 1999). Residents complained that the pedestrian environment—where crossing mid-street is a gesture historically respected by drivers—had deteriorated and that the down-to-earth quality of life declined in general. The impact was greatest in the neighborhoods on the city's southern side near access points to Interstate 280, which affords the most direct and fastest commute to the suburban cities and office parks of Silicon Valley. As one writer aptly put it, "becoming an economic suburb of Silicon Valley is hard for San Franciscans to take" (Weiss 1999). And long-time residents want to know, "to what extent can these strangers...be adapted to our culture?" (Deleon in Weiss 1999). That culture accepts alternative lifestyles, but not wrapped up in the new materialism's illegible symbolic politics—a challenge to the local social forma-tion. After the burst of the NASDAQ bubble in April 2000, the dot.com crash revealed the folly of controversial city development policies that pushed out the low-rent arts community south of Market and left empty office-lofts in its wake (Redmond 2001). Nevertheless, real estate prices in San Francisco remained near the highest in the country, fueling the city's seductive profile. The computer industry had grown to the point that Silicon Valley and San Francisco were no longer to be economically distinguished. By the 1990s, the position of the Bay Area's information economy in the world economy was so substantial that the value of goods shipped through San Francisco International Airport surpassed the combined value of goods shipped through all Bay Area seaports (Sims 2000).

Gender, masculinity, and the city

The combination of innovative mainstream achievement on the national scale and international economic achievement on the global scale only enhances San Francisco's seductive appeal in the globalizing imagination: the city of the unique and the different combined with the city of spectacular professional and economic distinction. In the 1980s this theme even played out on the sports scene. The rise of the practically glamorous San Francisco 49ers, as the most successful team of the era, and, as led by Joe Montana, by many measures the most successful quar-terback in pro-football history, contravened the profile of place the city had earned nationally in the 1970s as the capital of cults and murdered city officials. "Behind Montana's leadership…the 49ers in one magical season went from door-mats to world champions and transfixed a fragmented city" (Garcia 1995). The 49ers brought home their first national championship in 1982, four years after the Jonestown massacre in Guyana, which was followed by the slayings of Mayor George Moscone and the openly gay Supervisor Harvey Milk, by Supervisor Dan White, in their City Hall offices. Ideas about the culture of tolerance had shat-tered, a nostalgic notion.

Through the 1980s, the city's renewed sense of community integration and positive national recognition was actually attributed to Joe Montana. Upon his departure from the team, local newspapers recorded widespread, laudatory recognition.

> He did more than any one person to change the city of San Francisco…and made people around the country reevaluate the way they viewed San Francisco. And psychologically, he had a huge impact on how people here felt about themselves.
>
> (Garcia 1995)

Local sports writers reflected on it: "Montana became the Bay Area's signature athlete. Is there anyone really close? The team grew up with him, and so did a city's infatuation with football…. Montana's cool was taken as an almost mystical quality" (Keown 1997).

> The facts as they exist are pretty remarkable, but the Bay Area's feelings for him transcend quarterback ratings, winning percentages and even Super Bowl rings. Through dint of his…sense of style (aware all eyes were on him at all times without ever letting on that he knew), he grafted himself to San Francisco.
>
> (Ratto 1997)

"Montana's popularity was heightened because his personal style matched that of the city's—understated, graceful and self-effacing" (Garcia 1995). Joe Montana embodied cool; he performed it, and people recognized it: a low-key, elegant physicality, one version of San Francisco style. His fan base transcended class, gender, and sexuality (news articles also reported his following in the lesbian community) and gave the Bay Area a team tailored to local distinctions; contrary to local history, football popularity surged.

Widespread media coverage of the 49ers also re-scripted popular meanings of place and sexualized culture in San Francisco, and associated ideas about place gendering opened up to diverse masculinities. As R.W. Connell (1998) empha-sizes in writings about masculinity and globalization, plural masculinities actively construct in the context of particular social relations, locales, and histories, and often represent contradictory desires and conduct. These different local masculinities—gay culture and pro-football culture—together arguably maintained seductive tension in the landscape and its representative meanings. From one perspective, these masculinities offered up enticing extremes of male-gendered possibilities; from another, in the city itself, they also merged in local imaginations. As San Francisco's leading popular columnist, Herb Caen (1994), noted one day: "Onward: It has now been established that the football star who frequents the Castro is not a 49er." Apparently people wanted to wonder. By 1998, ideas about such possibilities had become explicit. The sports television channel ESPN aired a special, "Outside the lines: the world of the gay athlete," which featured discussion about rumors surrounding the sexuality of various athletes, including post-Montana 49ers' quarterback Steve Young. The gossip made Young consider whether to do commercials for VISA: one would feature him giving 49ers receiver Jerry Rice a bouquet of flowers; in another, Young and a woman dine in a restaurant with Rice, and when the woman says the restaurant is chilly, Young places his coat over Rice's shoulders. As the three depart in a horse-drawn carriage, Young says, "you look great tonight," and the woman and Rice reply simultaneously, "thanks" (Slusser 1998). These are not questions about whether Young is gay or bi; they are representations of negotiating local culture and identity. Ideas about sexuality are heightened in San Francisco, where near-mainstreamed gay culture—the number of gay and lesbian couples in San Francisco is the highest in the United States, more than four times higher than in Los Angeles, Washington, DC, or New York (Black *et al.* 2002: 64)—mixes into and sometimes mixes up heterosexual norms. VISA's commercial spots clearly appealed to the sexual tensions at play in the city's leisure landscapes: gay culture in San Francisco would negotiate the otherwise hyper-masculinized culture of pro-football on its own terms.

The so-called Satangate episode was the moment of collision for San Francisco's leading popular cultures of masculinity. In 1997 the 49ers franchise was angling for political support to build a new stadium that would replace the often foggy, wind-rattled Candlestick Park on the city's southern bay shore. Team owners hired an openly gay political consultant with a splashy local persona to promote the cause with voters. One of his organized events was his own 50th birthday party, which featured male strippers and "sex acts performed before many of the city's most powerful figures" (*The Washington Post* 1997). Local talk shows and newspapers debated the event and its fallout for weeks, as "San Franciscans...fretted over the damage done to the city's image nationally" (*ibid.*). Even before the birthday bash dissembled into 'Satangate,' the crux of the problem was that the proposed stadium plan included a casino, which the local anti-stadium campaign successfully argued would not be welcomed in San Francisco. While the 49ers' then owners (widely discussed as from the Midwest) saw a gold mine in the casino (utterly "encrusted and commercialized"), located between the airport and the city proper, locals knew, like tourism industry professionals, that San Francisco's tourism industry fundamentally had no need for such stylistically inappropriate development.

San Francisco's tourism industry

The reality about the tourism industry in San Francisco is that it is neither systematically promoted nor studied much by its professionals: its highs and lows reflect the vagaries of the economy and otherwise it largely takes care of itself. The San Francisco Convention and Visitors Bureau supplies tourist information about the city and assists in convention planning, but it has never had to work hard to attract tourists or maintain the industry. Its survey data are slim and only in 1998 did it undertake a focus group-based study to gain information beyond the usual quantitative data on visitor place of origin, length of stay, local destinations, expenditures, and the like (Bratton 2002, pers. comm.). Neither has the City of San Francisco historically maintained any kind of office of economic development, and analyses of the service sector in general are not well developed. In effect, San Francisco's bounty of historical assets and cultural and environmental amenities has continued to yield the region an enviable and largely unplanned regional economy (Walker 2001).

As a diverse group of service sector industries, economies of tourism vary in definition and scope. The California Economic Development Department defines tourism as part of the 'visitor industry,' which is one of San Francisco's three leading service clusters, in addition to 'finance, investment and dealmaking,' and 'communications and media content.' It includes transport, travel services, hotels, restaurants, retail trade, arts and entertainment and related business services, and is the city's second largest employer, accounting for 12 percent of all jobs in 2000, after the financial sector, which employed 21 percent of all workers (Sims 2000: 1–6). While tourism is not the city's leading source of jobs, it leads in contributions to city revenue. In 2000 it provided $474 million in local taxes and fees, among which the hotel tax alone yielded over 50 percent of the city's general fund and paid $24 million for arts and cultural programs. In that year over 17 million people visited San Francisco, spending nearly $20 million a day (*TWSF* 2001).

By a variety of industry measures, San Francisco is, about nine years out of ten, ranked as the top urban tourist destination in the United States: "In travel surveys it tops all other North American destinations for scenic beauty and restaurants, and ranks exceptionally high in terms of its uniqueness, variety of things to see and do" (Sims 2000: 3–49). But when tourists try to account for it, their terms of reference extend beyond descriptions of sites and scenes. What they love about San Francisco is all the ways the city's environment makes them feel; their human–environment relations transcend the experiential for the psychological and the visual for the emotional. Many are utterly transfixed: in San Francisco, tourists feel renewed "happiness," the sense of "romance," even "ecstasy" (SFCVB 2000: 16). They translate feelings of freedom into reinvigorated ideas about life possibilities, as if the city bequeaths to them good fortune. Descriptions of their feelings suggest even spiritual terms of well-being and centeredness. On these terms, visitors' emotional orientations arguably represent more than place imaginations; they are seductions of place, in the ways that seduction draws and compels, whose power to effect subjectivity draws from sensory experience and embodied liminality—intensely legible for the subject but nevertheless a subtle complexity irreproducible in themed environments. Tourism researchers conducting focus group surveys assessed these conditions as their most outstanding findings:

> Possibly the most surprising find of this project was the degree of conviction and emotion with which focus group panelists spoke when discussing the feelings they have while in San Francisco. Panelists naturally had different ways of articulating these feelings, but they all seemed to be trying to find a way to express something similar. They described feelings akin to intense happiness, deeply felt well-being, or joy and freedom. Several panelists described the feeling of being in San Francisco in almost spiritual terms, saying it made them feel "complete" or "grounded."
>
> (SFCVB 2000: 16)

Researchers attributed these tourist subjectivities to two categories of embodied experience: "extraordinary physicality" and "delightful complexity." More than scenic beauty, visitors wanted to explain their experiences in San Francisco through multiple sensory modes of place, including the intensity of natural light, crisp marine air, smells of foods, vertical sense of space, and encounters with diverse people. Researchers also found visitors commonly expressed a quality of vibrancy, though this "most frequently mentioned aspect of San Francisco's physicality is also the most difficult to describe…it's tempting to say that the city emits some magical energy that enlivens the soul" (SFCVB 2000: 14). Panelists attributed the city's complexity to its diverse neighborhoods, restaurants, and activities. "When the topic [food] was introduced in the focus groups, the panelists invariably became, for a moment, almost giddy" in their comparisons of diverse cuisines; for many, attempting to explain the finer points of dim sum was a bonding moment (SFCVB 2000: 15). As if straddling cultures hemispherically, they found San Francisco to be "America's most European city" and the "gateway to Asia." These tourist perceptions of San Francisco echo the representations of the mythic city and more, as visitors, time and again, account for their encounters with San Francisco's constituent places through ideas about embodied, sensory experiences. Their imaginary geographies invoke global proportions in which San Francisco is neither east nor west, but a pivot of cultural geographic encounter, a place of cosmopolitan possibilities on the edge of the continent—where Asia is the 'far west' and the 'right coast', as US westerners say, is 'back east.'

Conclusions

Complaints about touristed landscapes are common among San Francisco's residents, fans, and critics who have occasion to 'other' the tourists. Their obvious presence—140,000 strong on any given day, in a city that has never sustained a population over 800,000—in efficiently consuming the place, can undermine ideas about the city's more elusive magnetism. This place dialectic, between visitors and locals, in urging upon each other discourses and performances of the seduction of place, suggests the power of San Francisco landscapes in local, national, and global imaginations: it is simultaneously the coolest city in the country, an overrated boutique town, the gay capital of the United States, breathtakingly beautiful, a veritable campsite for the homeless, an address that represents having made it in the millennial economy, and a symbol, always, of alternative possibilities. Its native population tends toward smug about the merits of the place, a position of performativity easily adopted by members of its newest class of the speculatively rich. Residents constantly parse the city's meanings, its neighborhoods, and the quality of daily life. Ideas about the left political landscape are practically taken for granted: to be a Republican in San Francisco represents, for the majority, little more than a position of curiosity because that affiliation is so token, more of a throwback to an earlier era than a threat to the prevailing political agenda. Following the logic, it is a matter of time before the rest of the country catches up with the progressive politics of the Bay Area.

But the city's historic reputation for tolerance is part of its mythic image, belying a political character of local elite preferences for upscale landscapes— more 'spectacle of environment'—underscored by mid-twentieth-century racist housing policy, which made San Francisco as common as other US cities. Residential segregation was a serious problem, exemplified by the case of Willie Mays, the San Francisco Giants' baseball star, who in 1957 was initially denied a home loan for a property in an exclusive neighborhood (Broussard 1993: 240). In the 1970s the San Francisco Redevelopment Agency subjected the Fillmore district to urban renewal, which slowly but ultimately gentrified Fillmore Street and environs, in the process largely compelling the departure of that area's African–American community from the city. Knowing San Francisco's history of housing and redevelopment policy means knowing that the city's leadership and its elite-property development alliance exported its own inner city problems out of town (e.g. Wagner 1998). For most residents, this history is conveniently forgotten or unknown, and contemporary rights to the city are based almost

completely on financial means to afford the place, and as the basis of lifestyle politics. Such events of intolerance do not figure in the constructed meanings of San
Francisco that sustain the city's image nationally and globally, which is a narrative
the city's visitors and residents, in large part, both desire.

San Francisco's future as a highly sought-after touristed landscape is reasonably
assured. With the exception of the over-commercialization of Fisherman's Wharf,
its leisure districts have significantly retained their intermediate, mixed ground of
characteristics, affording accessibility to the local scene. Its economy has shifted
almost entirely to service sector and high-technology industries, including
communications, media, and biotechnology. Its airport was substantially
upgraded in the 1990s and houses, in some terminals, changing art exhibits
worthy of visitors with no need to fly. An extension of BART, the Bay Area Rapid
Transit system, has served the airport since 2004. Once sidelined by major
conventions as too small for their meetings, The Moscone Convention Center,
enviably located downtown adjacent to the Modern Art Museum, designed by
Mario Botta (the only Botta building in the United States, funded by local philan-
thropists), has added another building. The city's enduring place seduction will
depend on the continuing emergence of cultural forms, in popular expressions of
daily life and thought, lifestyle alternatives, as well as maintenance and enhancement of landscape quality, especially visual and physical access to the water, and
places for performativity of lifestyle choices and aesthetic experiences. Whether
the city's latest round of speculative capital will open up new sites for these possibilities remains to be experienced.

Part III

The beach

The prodigious enrichment of the emotions provided by the sea-shore...transformed the means of expressing desire. A new intimacy became established between the walker and the elements. The ocean was no longer merely a sublime spectacle to be gazed at from the top of a cliff or a picturesque scene to be framed from the height of a viewpoint offering a clearly separate, dominating vision. The dialogue with the waves or with grottoes suggests that the spectator's position...should be abandoned. It ushered in desire for close contact as a prelude to an imaginary merging with the scene.... All the ways of using the shore were modified as a result.

Alain Corbin, *The Lure of the Sea* (1994: 172)

11 Tourist weddings in Hawai'i

Consuming the destination

Mary G. McDonald

Hawai'i, long the top choice of honeymoon travelers from the United States and Japan, is now a leading destination for weddings themselves. Bridal couples from east and west marry in Hawai'i to frame their ceremony within a romantic exotic. Wedding firms have created two different products for the two markets: a wedding on an exotic tropical beach for North American couples and a wedding in an exotic Christian church for Japanese couples. When tourist weddings were few, ceremonies took place on beaches and in churches where local couples might marry. The success of Hawai'i's wedding firms, however, has caused a shortage of the sublime: a construction boom is underway across Hawai'i creating privatized 'natural' landscapes and simulated sacred spaces for tourist weddings.

The tourist wedding industry transforms subjects while consuming and creating place. The industry produces a double displacement (Crang 1996b) by moving social rituals to distant stages and by overwriting prior social landscapes at the destination with new tourist space (Figure 11.1). The growth of tourist weddings can be understood in part by focusing on the tourist as consumer, but must also include economic interests creating the bridal tourism market and carving out new tourist wedding spaces. This study examines bridal tourists' choice of Hawai'i, the industries that 'process' them, and the ways wedding firms gain ground in local landscapes.

Consumers who choose offshore weddings employ the distance in two ways. First, their journey is a variant of the liminal space–time so important in ritual transformations. They venture beyond known bounds, they leave old rules behind to celebrate only themselves, they surrender themselves to a new unity, and they return home transformed. But second, they use the margin not as the dark grove of many initiations, but as a place for their self-conscious and self-monitored wedding performance. The place itself will appear importantly in photos and videos for friends and family, marking the personal style of the couple. The place will confer status, marital and otherwise.

Subject-centered views of wedding tourism are ultimately indexed to the tourists' home culture. Tourist behavior and experience somehow 'works' in a functionalist way. But this leaves a great deal of ground unexamined. That ground

Figure 11.1 One kind of sacred space becomes another: a Japanese wedding party gathers for photos at the crypt of Hawai'i's King Lunalilo in the center of Honolulu (photograph by the author)

includes the workings of a tourist wedding industry itself, with its ability to commercialize rituals and sacred spaces in conjunction with vacation tours. This is an industry that beat Appadurai (1986: 3) to the discovery of "the commodity potential of all things." Firms sell ceremonies for ground rent. Wedding venues probably count among the overlooked spaces of daily life in which "moral hetero-sexuality" is routinely performed (Hubbard 2000). Controversial new wedding chapels in Hawai'i embody ambitious profit motives indulging the moral and simulating the sacred.

The industry acts not only on its customers, but also on the destination itself. Metaphorical summaries of the destination as margin or stage insufficiently capture the active production of space to accommodate tourist weddings. A place-centered view is needed, trained on firms, bulldozers and neighborhoods at the 'margin.' This study will review new spatial practices of bridal tourism, its agents in Hawai'i, and their use of the islands.

The growth of bridal tourism

Wedding ceremonies are now available on a worldwide menu of identity-through-consumption. The wedding and tourism industries call their shared growth niche 'bridal tourism.' Competition among places for bridal tourists abounds. The newsstand issue of *Destination Weddings and Honeymoons* offers "weddingmoons" in Bellagio, Krabi, Punatapu, Ixtapan, and Quepos. Theme parks and shopping malls host weddings as the ultimate in visitor attractions. The Internet invites couples to wed in Kenya or Sri Lanka; the bride can be rowed to her wedding by Fijian warriors. Wedding rituals now fit into the "pattern of everyday consumption that renders us more and more like tourists as we purchase not products but representations and experiences" (Oakes 1999: 309). One's wedding has become an opportunity to shop for an exotic experience.

In industrial societies, romantic love and heteronormative marriage play out on topographies of accepted places for courtship, weddings, and honeymoons. Precisely coded landscapes and the crossing of spatial boundaries are essential to the experience of romance (Illouz 1997: 115, 137). A couple's decision to wed sets in train movements through space that formalize the rite of passage. The couple chooses a wedding site within constraints in their legal and religious landscapes. The bride may determine setting and style, in that the teleological imagination of love and marriage remains specifically gendered (Giddens 1992). Family and friends travel from afar. The wedding is enacted within ritual microrules of spacing and timing. The wedding party proceeds to a reception site. The couple departs on a honeymoon then returns to cross the threshold. No matter how abstracted, contemporary weddings retain a structure of symbolic movements comprised of ceremony, tour, and return home to new civil status. This is a highly normative social script, not legally necessary in any industrial nation today, yet followed by many in some order of events, even by those who otherwise partake little of religion or ritual.

These movements evolved with the rise of "the romantic love complex" that changed relations between parents and children, set partners on a long-term life trajectory, and gave primacy to the marital relationship (Giddens 1992: 41–4). Travel associated with love marriages was never random, however, but socially patterned. Weddings long remained situated in the social milieux of the bride or groom's church and home. The act of elopement breached socio-spatial convention yet established its own common paths. Destinations for quick and easy weddings such as Gretna Green and Las Vegas attracted streams of couples in what Urry (1990) calls "travel for." The honeymoon that was a private circuit of family visits in centuries past became a twentieth-century public appearance of newlyweds at popular honeymoon destinations, most famously Niagara Falls (Dubinsky 1999, Shields 1991).

Post-World War II baby boomers were the first to move their wedding ceremonies to sites that expressed transcendental inspiration. Rising age of marriage, independence from parents, sexual revolution, dual incomes, second marriages, and prior travel allowed couples to think creatively about wedding venues. Couples chose parks or inns where family and friends could congregate and relax. Such alternative choices may have shared the original spirit of romanticism:

rejecting social conventions, valuing nature, and pursuing self-expression. Yet as with the nineteenth-century romantics, the latter-day counter culture of "modern autonomous imaginative hedonism" inevitably underwrote its own forms of consumerism (Campbell 1987: 77). Travel brokers and hoteliers began packaging resort weddings as products. While the destination wedding appeals to a "sense of distinction" (Bourdieu 1984), it has become another commodity form.

Romance has been a vital part of the place-image through which Hawai'i markets itself as a destination (Goss 1993b). Hawai'i's state population of 1 million residents is accustomed to seeing and serving lovers, honeymooners, and second honey-mooners among the 7 million visitors Hawai'i receives each year. In the year 2000, over 46,000 tourist couples staged their wedding ceremony while vacationing in Hawai'i, as shown in Table 11.1.

Bridal tourists are clearly involved in the larger social processes MacCannell (1989) identified: attraction, ritual, sacralization, framing, enshrinement, mechanical reproduction, and finally social reproduction. In one sense, social reproduction in tourism was never so obvious: Hawai'i's tourism board prays these unions will be fruitful so couples "come back with their children" (Tang 1999).

The margin consumed as wedding site

For the principals, the wedding in Hawai'i is a product they imagine and purchase from afar. The liminality of the destination wedding is part of its efficacy. The importance to all rites of passage of a time and space of separation was empha-sized by Van Gennep (1960) as early as 1909. As a transit to the limen or margin, the tourist wedding allows the couple to expect and claim experiences of communitas, freedom, harmony, unity, and authenticity (Turner 1967, 1969). The liminality of the exotic setting is always one of the attractions of the tourist desti-nation (MacCannell 1989, Goss 1993b) and is a vital signifier allowing consumers to see and feel romance (Illouz 1997: 142). The wedding ceremony at the margin asserts the couple's independence, their distance from conventions. The couple may be betwixt and between in terms of marital status, but they can have both

Table 11.1 Weddings in the state of Hawai'i, 2000 (number of couples)

Legal marriage, one or both Hawai'i residents	9,209[a]
Legal marriage, both non -Hawai'i residents[b]	16,146[a]
Ceremonies for Japanese tourists, unlicensed	30,000[c]

[a] Office of Health Status Monitoring (2001).

[b] Almost all residents of USA and Canada, small numbers of Europeans and Latin Anericans.

[c] Endo (2000), Matsuo (2001).

anonymity and exposure, both rejection and observance of ritual, in their passage to the limen and back.

The visual correspondence of the destination to a spatial ideal type is also part of its power to effect the transformation. The place must satisfy the tourists' romantic gaze (Urry 1995) in a double way, both as a landscape they find inspiring and as a frame in which they idealize themselves in wedding photos. The picturesque is all important; weddings like tourism are designed to "end in photograph" (Sontag 1977). The tourist destination already provides many varieties of "bodies on display" for tourists' visual and affective consumption, mirroring the couples' own imaginations (Desmond 1999). In the tourist wedding ceremony, the tourists themselves are the stars, but they grant the image of the place itself a strong degree of performativity, or power to act upon them. Austin (1962) argued his speech act theory of performativity through the famous example of wedding vows. The concept of performativity was extended by Derrida and Butler into theories of identity produced through continual iteration and citation (Derrida 1982, Butler 1990, 1993). In this spirit, Parker and Sedgwick (1995: 11) argue that marriage is effected less by a couple's vows and more by the "surrounding relations of visibility and spectatorship" during the ceremony and ever after. Marriage, they find, is one long processional on an "invisible proscenium arch that moves through the world" commanding constant witness, continually bringing itself into being. In this light, the tourist wedding is not a self-transformation but an intersubjective one: the bridal picture from Hawai'i will circulate among future witnesses to sustain the facticity of their marriage.

The margin is only available to tourists through the construction of space that "can be called a stage set, a tourist setting, or simply, a set" (MacCannell 1989: 100), and through repeated additions to allow more tourists to use similar space and gaze upon the same object (Urry 1995: 136). The tourist's liminal path at the picturesque destination wends its way through a rearranged lifeworld of the host community. But to the host community the destination is not margin, it is home. Widespread construction of tourist stages leaves less backstage and finally no off-stage—invisible proscenium arch, indeed. A full understanding of tourist weddings must begin in the spirit of Urry's double argument in *Consuming Places* (1995): that consumption takes place at specific sites, and that economic interests continually restructure and commodify locations. Both processes are at work in Hawai'i. Practices that may be read as stylization of life for consumers may have erosive effects on cultural values taken more broadly (Sayer 1997: 25), so studies of tourist performance must ask who built the stage, and on top of what or whom?

Hawai'i's wedding industry at the crossroads of exoticas

Wedding tourists rely on wedding brokers to instruct them how and where to get married at the destination. Potential wedding tourists find many firms ready to help them marry in Hawaiʻi, via advertisements in English or Japanese. Wedding firms serve primarily North American couples or Japanese ones, and promote different styles of weddings to the two sides of the Pacific. Couples arrive expecting to consume a wedding in a utopia counterposed to, yet specific to, the home culture. The wedding 'set' has already been constructed in their imagination. In broadest aesthetic terms, for North American couples this is an oriental fantasy and for Japanese couples an occidental one.

North American couples seek weddings in 'natural' settings of beaches, gardens, and waterfalls outside the urban core of Honolulu. Many naturalistic wedding sites are modified to signify 'tropical paradise' to the Euro-American romantic imagination. The South Seas visual landscape of paradise has been a mainstay of Hawaiʻi's romantic place-myth. The 1937 Paramount film *Waikiki Wedding* with Bing Crosby sold the idea of romance in Hawaiʻi to a prewar generation. In 1958, the tune of a love song in Hawaiian got an English lyric and a new title, *The Hawaiian Wedding Song*. On the lagoon of Kauai's Coco Palms Inn, American hotelier Grace Buscher invented the Hawaiian wedding as a product for North American tourists (TenBruggencate 2000). Her hotel became the setting for Elvis's wedding in the 1961 film *Blue Hawaiʻi*, and tourist weddings have taken place on the site ever since. New resorts copied visual icons of a tropical romantic Eden and simulated village life, engineering beaches, lagoons, and waterfalls to correspond to North Americans' image of a romantic wedding and honeymoon setting (Bulcroft, Smeins and Bulcroft 1999). Many years of advertising by the Hawaiʻi Visitors Bureau constructed a place imagery of paradise and engaged consumers' desires (Goss 1993b). Hawaiʻi's tourist industry constructed Hawaiʻi as an imagined "place-for" (Shields 1991: 9).

North Americans were the main customers for Hawaiʻi's tourist wedding business until the 1990s. Their numbers continue to grow. In 1993, 7,771 North American tourist couples married in Hawaiʻi; by 2000, over 16,000 couples did so. North Americans married across all the islands: 7,722 couples on Maui, over 4,000 couples on Oʻahu, about 3,400 couples on Kauai, and over 2,000 couples on the Big Island of Hawaiʻi (Office of Health Status Monitoring 2001). North American couples seek small private ceremonies in outdoor settings, and they find these through many small firms in the wedding business. North Americans come from a do-it-yourself wedding culture in which couples purchase wedding components, choose the wedding venue, assign roles to relatives and friends, write the script of their ceremony, choose the music, and book their own wedding and honeymoon travel. If a wedding planner helps produce part or all, the role of the firm is to stay as invisible as possible.

This relaxed and self-produced wedding style combined with Americans' imagined Hawaiʻi has become the wedding North Americans project onto Hawaiʻi. Americans who have never been to Hawaiʻi trust in a set of icons of the tropical Pacific to provide a spiritually sympathetic setting for their wedding. North Americans shop via the Internet for a wedding provider with a package that suits them, often making their own arrangements. Businesses with names such as A Very Simple Wedding, Simply Mauied, Affordable Weddings of Hawai'i, or #1 Hawai'i Weddings $95, hope for a call from the North American minimalists. The simplest wedding is often on the beach. Beaches are public property, but couples must obtain an event permit in person from the state for a wedding on the sand. Brides and grooms provide their own tropical costumes and may be barefoot. Flowers may be leis or blossoms in the hair. Music may be a tape recording, usually Hawaiian music. The photographer writes "Just Married" in front of the couple in the sand. For the post-Elvis generation, the newest firms are named Eco Weddings of Hawai'i and Extreme Weddings Hawai'i. Firms serving North American couples are geographically spread throughout the islands; few have addresses in Honolulu's hotel zone of Waikīkī.

Some neighbor island weddings take the form of 'destination weddings,' formal productions at resorts with attendants and guests from home. In the 1990s, hotels added weddingscapes in the form of gazebos and bowers in tropical gardens with view of pool, waterfall, sea, or all three. Wedding consultants joined the fold of resort services to provide ceremonies coordinated with lodging and dining at the resort. Hotels can exclude the public for extra privacy and security. A Lanai luxury hotel got help from the police to exclude and even arrest journalists on public roads and beaches during the wedding of Bill Gates in January 1994 (Taylor 1995). Hotels' limitations as wedding venues are their high costs and booking limits of one or two weddings per day. North American couples can wrap family, friends, wedding, and honeymoon into one destination wedding in Hawai'i if they have sufficient resources, but fewer than 10 percent of weddings for North American tourists are resort-provided.

Hawai'i state and local agencies have had an active hand in promoting Hawai'i as a wedding and honeymoon destination in North American markets. The island of Maui has marketed itself in North American bridal media for 20 years. The state of Hawai'i has recently been pushed by its hoteliers, small businesses, and advertising media to promote bridal tourism, the only niche it has ever targeted. The Hawai'i Visitors and Convention Bureau developed a small target budget within its $6.9 million dollar North America advertising fund to promote Hawai'i wedding tourism beginning in mid-1999 (Tang 1999).

To the surprise of Hawai'i residents in the 1990s, numbers of honeymooners and wedding couples from Japan surpassed those from North America. In 1982, just over half of all Japanese couples traveled overseas on their honeymoon, but by 2000, 97 percent did so (Lynch 2000). The number of Japanese tourist weddings in Hawai'i topped 30,000 in 2000, six times the 1989 level of 4,700 couples (Endo 2000: 112). Friends and family have accompanied these couples, as well. In 1999, 20 percent of all 1.8 million Japanese visiting Hawai'i came for a wedding or honeymoon (DBEDT 2001). Almost all the weddings of Japanese tourists take place in and around urban Honolulu on the island of O'ahu while couples and guests stay in Waikīkī hotels. Their weddings are very different from those of North American tourists, and these differences start in Japan.

The exotic wedding Japanese couples seek in Hawai'i is a 'chapel wedding,' a Western-style product of Japan's own wedding industry. Japan's legal requisite for marriage is a couple's registration as a new household in the vital records of their town hall, so a wedding ceremony has no legal import wherever it occurs. The Japanese state issues no marriage licenses authorizing couples to marry, nor does it license marriage officiators. Nevertheless, almost all couples choose a ceremonial wedding and reception that serves to unite their families. In the 1970s and '80s, weddings in Japan grew into extravaganzas of conspicuous consumption. Families wished to impress each other with their means, weddings secured important social patrons for the young couple, rules of hospitality required hosts to offer the very best, and wedding venues added more luxury and spectacle to ceremonies year by year (Kamata 1985, Kogawa 1985). By 1986 the average Japanese wedding cost 7.5 million yen, or 1.6 times the average annual household disposable income (Horioka 1987).

The 1990s brought a sudden end to the extravagant times of the bubble economy. Many Japanese owners in Hawai'i sold their hotels at a loss, and Japanese tourism to Hawai'i fell overall. In Japan, many families had less money than before and hesitated to impose expensive social obligations on others. Weddings in Japan became somewhat simpler (Watanabe 1998, Japan Information Network 1997). Many couples decided to forgo the Japanese kimono, and instead chose a Christian-style 'chapel wedding.' Many non-Christian couples began requesting Christian churches to marry them. Ministers recognized that the Christian-style wedding was a fashion, but for the brief chance to evangelize and to collect handsome fees, many Japanese Christian churches obliged (McKillop 1999, Catholic Bishops' Conference of Japan 1992). Hotels and wedding palaces in Japan hastily responded to the trend by remodeling interiors as 'chapels' with architectural citations such as stained glass and altars. Wedding halls also increased their stocks of white wedding gowns, veils, gloves, and tailcoats to rent to brides and grooms. These venues also enhanced the staged authenticity of their Christian-style weddings by hiring Westerners as officiators. Real missionaries being in limited supply and given to preaching, wedding halls simply trained English teachers and other Westerners as 'talent pastors' to dress and act the part of minister. The hired minister reads lines in simplified Japanese and Japanized English, such as the word *ra-bu* for 'love.' The wedding hall prepares the script and instructs the minister to abide by the stopwatch timing of each ceremony (Zuercher 2000, Harris 1998). Guests sing a hymn. This is the syncretic European-style 'chapel wedding' that has already been "re-made in Japan" (Tobin 1992) and is now carried to, not simply found in, Hawai'i.

In the context of simpler elegance in the 1990s, the alternative of a wedding in Hawai'i was attractive for Japanese couples. An overseas ceremony could be socially less complicated, stylistically more authentic, and more economical. A Hawai'i wedding for a couple or for an entire entourage would cost no more than a wedding in Japan. A bad economy in Japan has been good for the Hawai'i wedding business. Many Japanese brides in the 1990s were older, more independent from parents, more concerned about personal and family enjoyment of their weddings, and attracted to an offshore wedding as a statement of a cosmopolitan personal style. Couples in love marriages were freer to marry abroad than those bound by traditional negotiations surrounding arranged marriages. Japan's "tourism production system" (Yamamoto and Gill 2002) combined overseas tours and wedding services into packages for a mass market. Following the pattern pioneered in Hawai'i, Japanese wedding operators began romancing places around the world for permission and sites to marry Japanese couples. By the end of the 1990s, the Japanese bride and groom could choose destination weddings from Disney World to Banff and from Bali to Paris. About 780,000 marriages are registered in Japan each year, 10 percent of them enacted overseas.

Overseas wedding sales begin in Japan. Prospective brides see images of destination weddings in Japanese magazines and television programs. Couples choose their overseas wedding package in Japan from among the offerings of tour companies teamed with wedding firms. JTB (Japan Travel Bureau), Jalpak (a Japan Airlines subsidiary), Kintetsu, Hankyu, and NTA are the top five travel providers of wedding tours to Hawai'i (Matsuo 2001). JTB has Wedding Plaza information desks at city centers throughout Japan. The 1998 Look JTB brochure for Hawai'i offered 35 wedding venues with photos and details about each location (JTB 2000). Alternately, a couple contracts with a Japan-based wedding firm offering overseas and domestic weddings. Bridal firms advertise, recruit customers, and sell every detail of each wedding before the couple leaves Japan.

When the couple reaches the overseas destination, they are in the hands of a branch of the Japanese wedding firm. In Hawai'i, these firms include Watabe Wedding, the largest provider of hotel weddings in Japan, established in Hawai'i in 1973, Matzki Inc. (or Matsuki in Japan), whose Wedding Emporium/Tuxedo Junction date from 1977, Bic Bridal Hawai'i, established 1992 as a branch of the largest operator of commercial wedding palaces in Japan, and Best Bridal, an operator of six wedding halls in Tokyo, incorporated in Hawai'i in 1999. The wedding day begins with makeup, hairdressing, costuming, and photos at the salon location or at the hotel room of the couple. The wedding firm contracts a white limousine to drive the couple to the ceremony location. The wedding fulfills the Japanese requisites for a chapel wedding. The bride wears a rented 'Shinderera' (Cinderella) white gown with a fitted waist, long full skirt, modest bodice, cap sleeves, tiara, veil, pearls, and gloves. The bride and groom expect a sanctuary with Christian symbols, a Bible, organ music, a hymn, a minister, vows, rings, and a chance to sign the wedding registry book. Few Japanese weddings take place outdoors or in the casual style of North American tourist weddings.

The Japanese tourist wedding in Hawai'i is precisely the one that North American couples are trying to escape (Figure 11.2).

The Hawai'i branch offices of the Japanese wedding firms must deliver the wedding experience that has been sold in Japan, must work with travel wholesalers to assure a steady flow of customers to Hawai'i, and must make business run smoothly in the host locale. These firms now conduct more than 30,000 couples through their wedding experience in Hawai'i each year, or over 100 weddings per day on O'ahu. The wedding firms require a great deal of cooperation from local wedding sites, churches, clergy, small businesses, neighborhoods, and governments in order to do business. The firms must do the ontological front work for the Christian-like wedding ceremonies that have no legal or religious significance. They must answer to puzzled clergy, congregations, wedding chapel neighbors, and tax officials in Hawai'i. They face indirect problems such as parking for the fleet of 500 owner-operated white stretch limousines that has grown on O'ahu to serve weddings (City and County of Honolulu 2000). Above all, since business has now outstripped the capacity of local churches, wedding firms need space that can host the one-at-a-time tourist activity of 'chapel weddings.' Firms have been busy creating new wedding venues in churches, homes, restaurants, and in the form of new commercial 'chapels' with views of the sea.

Figure 11.2 North American couples want a Waikīkī wedding in the sand beyond, but this Japanese bride and groom prefer white tulle, harp, and make-believe minister in the lobby of the Royal Hawaiian Hotel (photograph by the author)

Even while new wedding chapels have elicited protests in Honolulu's neighborhoods, the state of Hawai'i has been explicitly promoting wedding tourism in
bridal markets in Japan. The Hawai'i Visitors and Convention Bureau collaborated
with ad agency McCann Erikson in Japan to include weddings within the state's
"Aloha Magic" campaign (Hawai'i Visitors and Convention Bureau 2001), which
places large photos in two-page magazine spreads and on train station posters,
showing over 20 varieties of the 'magic' of Hawai'i aimed at a range of life stages
and lifestyles. Captions express an inner thought of the Japanese visitor implied in
each photo. Several of these ads target young women, known to be the decision-
makers as to where to marry and where to honeymoon (Matsuo 2001). One of
the Aloha Magic images is a panoramic photograph of a white-gowned Japanese
bride and groom taken through a wide-angle lens, such that the camera is implicit
in the image. The photo is captioned with the voice of the bride as she recalls,
"Everyone else was happy too." The advertisement sells the warm moment in the
sun, the past-tense documentary value of the photograph, and the idea of happy
family in the Hawai'i wedding trip.

Watabe Wedding: displacing rites

Japanese destination weddings in Hawai'i are largely a product of one company,
Watabe Wedding. Watabe was a kimono shop in Kyoto that began renting wedding
kimono to brides in 1953, incorporated as a wedding firm in 1971, and debuted
on the Tokyo Stock Exchange in 1997. With over 40 locations in Japan, Watabe is
the largest wedding dress rental firm, handling its own Avica label gowns manufactured at its Shanghai factory. Watabe coordinates weddings in upscale hotels
throughout Japan. Hawai'i was Watabe's first overseas location in 1973. Watabe
now has 20 offices producing Japanese weddings in the United States, Australia,
Taiwan, Guam, France, the United Kingdom, Italy, Holland, and Switzerland. It
has married over 100,000 Japanese couples outside Japan. Over half the company's
earnings are now generated by offshore weddings (Watabe Wedding 2001b).

Watabe Wedding advertises at bridal fairs, through Japanese magazines such as
Oz Bridal and Watabe's own website, which opens with the slogan "The journey of
eternal love can begin from anyplace on earth." Hundreds of pages of print and
electronic text describe destinations and wedding sites in detail. Customers are
assured that by selecting Watabe, they and their families will have no communication problems or other surprises on their wedding trip abroad. Watabe promises
on-site support identical to service in Japan. Watabe Wedding can take care of
you, your ceremony, your makeup, your guests, and can let you choose your
dress, tuxedo, and your menu beforehand from the identical selection in Japan
(Watabe Wedding 2001b). Web pages detail all the steps the couple will follow
from the moment they step off the plane at their overseas destination. "Step one,
call Watabe Wedding, in Japanese of course." For the return home, Watabe can

prepare souvenirs to take back to associates and arrange wedding receptions among friends back in Japan.

In Honolulu, Watabe Wedding occupies two floors of the Waikīkī Business Plaza. Marble showrooms hold 1,500 wedding gowns and 3,000 tuxedos for the nearly 20,000 weddings they produce each year. Hair salon, makeup stations, and photo studio are attached. Watabe's clientele has been almost exclusively couples from Japan, though Watabe recently began promoting its weddings to non-Japanese (Yim 2001). The cover of Watabe Wedding's Japanese brochure for Hawai'i contains a cartoon bride and groom gazing at each other, faced by two smiling onlookers, surrounded by church, limousine, shoppers, surfers, a hula dancer, cupids, Kamehameha I, Diamond Head, and Waikīkī Beach, all tied together with Watabe's trademark red ribbon. The two-page photo inside the front cover is a close-up of a blonde Western bride and groom kissing in an ocean-front setting. The words Hawaiian Wedding appear in English above the Japanese table of contents to the 53-page brochure (Watabe Wedding 2001a). Next comes a page of photos of Japanese bridal couples with Western children dressed as attendants, and a list of rules for overseas weddings.

Watabe Wedding's brochure promotes a chapel wedding in an authentically occidental setting. Photos and text are a tour guide to the nine steps the couple will follow on the day before the wedding and 15 stages of the wedding day itself. These include hotel check-in, consultation with Watabe, costuming, limousine ride, Hawaiian minister, chapel with views of trees and sea, rings, kiss, signing the wedding register of the church, a shower of flower petals on the church steps, photographs in the sanctuary and by the sea, and reception with the option of Hawaiian entertainers. The brochure offers options for rings, flowers, cake, reception food and drink, photographer and videographer. The text explains the role of bridesmaids and best men [sic] in the American wedding, and offers to rent additional costumes for these attendants. Watabe suggests a sample seven-day plan for a Hawai'i vacation that includes wedding and recreation for all family members, and lists 17 Japanese tour companies through which Watabe's travel packages are available.

Watabe offers 27 Honolulu wedding venues in churches, chapels, hotels, and restaurants. All of these require a site fee and not one is a beach or parkland. Some sites are available exclusively through Watabe. Watabe and other wedding companies get their wedding sites in two ways. The first is by negotiating the terms of a wedding package offered by existing churches or hotels that routinely host weddings. The second is by building or contracting a custom-built commer-cial chapel. Reception sites are also increasingly custom-built banquet rooms. Watabe has been actively building its own sites for both weddings and receptions. This is partly a response to the growing numbers of weddings and the need for sites, but Watabe is no doubt itself romanced by the greater profits possible in the vertically integrated wedding.

Churches for enacted weddings

Making place on O'ahu for 100 Japanese tourist weddings per day has not been easy. Neighborhoods and local institutions are filled with conflict over the use of churches and the siting of new wedding chapels. When Dubinsky studied the growth of Niagara Falls as a honeymoon destination, she found that land use change created only a few "neighborly difficulties" (1999: 185). Perhaps the tone of conflicts mellowed in Dubinsky's retrospective view. But for residents of Honolulu in the 1990s, the expansion of wedding tourism has brought painful controversy. The provision of the stage itself for happy tourist performances has alienated residents from each other and from their own place.

Churches already established in the local community might be the 'genuine' occidental sacred space Japanese couples seek, but commercial demand does not fit easily into local churches. In the 1970s a few local churches, through members' ties to Japan, began to make money by obliging Japanese wedding firms' requests for ceremonies. The model was tempting but difficult for other churches to follow. How should churches treat non-members, non-Christians, and couples without marriage licenses? Most churches insist that weddings be performed by authorized clergy and attended by their own wedding coordinator. Churches define what days and times weddings can take place. Sundays, though the most popular day for weddings in Japan, must be reserved for regular church services. The church typically provides sanctuary, minister, organist, and on-site coordinator. The wedding firm adds flowers, limousine, photographer, and wedding firm escort. The church package can cost the couple between $1,000 and $4,000, and the wedding firm adds thousands more for costume rentals, salon services, photography, video production, and reception.

Genuine churches have come to grief in the wedding business. One such church was one of the 1970s pioneers of Japanese weddings, Calvary by the Sea Lutheran Church, on Kalanianaole Highway on Honolulu's eastern oceanfront. The pastor of the church, Doug Olson, performed the weddings partly in Japanese for 20 years until the congregation objected in 1992. The church turned over wedding functions to church member Paul Tomita of UI Productions, who had been working on Japanese weddings at Calvary since 1978. Under the contract, UI produced 'church' weddings for customers booked by Japanese wedding firms and paid the church for each wedding plus $40,000 per year to Pastor Olson in lieu of lost income. By 1996, Calvary was performing six to eight Japanese weddings per day, 170 per month, or 2,000 weddings per year. Revenues from tourist weddings were $500,000 annually. This windfall attracted the attention of the Internal Revenue Service and state tax authorities. The state of Hawai'i questioned whether commercial weddings should be taxed in the same way as any other tourist entertainment. Largely through the case of Calvary by the Sea, Hawai'i laid claim to taxes on commercial weddings in the following way:

> The test is whether the primary purpose of the tourist "wedding" activity is religious or fundraising in nature.... For example if the ceremony performed is in fact a wedding (as opposed to, say, a reenactment of one)...then the activity will be considered religious and not income-producing. On the other hand, if the weddings are arranged, packaged and conducted through a commercial entity without church involvement other than making available the use of church premises, the "wedding" activity will be considered fundraising in nature.... In general, the honoraria or fee that the minister, priest, or officiator receives for performing the wedding ceremony is subject to the 4 percent general excise tax.
>
> (Department of Taxation 1997)

Accordingly, the state claimed that Calvary by the Sea Lutheran Church owed general excise taxes on Japanese weddings dating back almost 20 years. The church countered that weddings were a religious ministry, but because the minister could not communicate with the Japanese couple and spent less than eight minutes with them, the religious argument failed. In 1998 the church council voted to pay the tens of thousands of dollars of back taxes claimed by the state (Gillingham 1998).

The commerce of enacted weddings also strained the congregation of the Boston missionaries' first Honolulu church, Kawaiahaʻo Church. This historic church, built of coral blocks, stands close to the state capitol in the government district of Honolulu. It is a touristed site in its own right, as the scene of coronations and marriages of monarchs of the Kingdom of Hawaiʻi. The church's blue blood has made it one of the most popular choices of Japanese couples since 1984, when the church entered an agreement with the wedding company of Honolulu attorney Mack Hamada, Fantasia International Trio, Inc. Tourists and residents of Honolulu now see Japanese wedding parties on the steps of the church at the rate of one wedding per hour most days of the year (Figure 11.3). Kawaiahaʻo's commercial weddings grew to a $1 million per year business in 10 years. The organist quadrupled his salary with wedding fees to over $112,000 in 1995. But in 1996 money troubles erupted between Hamada and Kawaiahaʻo Church. The church terminated its contract with Hamada, who retaliated with a lawsuit; the legal battle lasted four years (Anwar 2001). Church staff and congregation then fractured over whether weddings should be managed directly by the church office or by an independent in-house Royal Wedding Center. The latter mode prevailed. One church member attributed the legal battles to "greed" under the pretty surface of the wedding business (Anwar 2001).

A home wedding? Not in my back yard

If churches are constrained as wedding venues, other authentic places for tourist weddings are stately homes with ocean views or picturesque gardens. North

Figure 11.3 Tour packages bring couples from Japan to enact weddings at Honolulu's Kawaiaha'o Church at the rate of one ceremony per hour. The state has encouraged wedding tourism, but the public is bemused by thousands of non-legal ceremonies taking place each month (photograph by the author)

American tourists will marry outdoors on private estates and farms, but formally dressed Japanese couples want to be inside a home with a spacious interior. Individual owners of large homes on O'ahu have offered their estates for weddings. In some cases the collapse of bubble-era property values impelled owners to try to offset declining property values through wedding business revenues. Attempts to place tourist weddings in elegant private homes have, however, inflamed quiet neighborhoods.

One example in the 1990s involved Honolulu attorney Rick Fried, who hoped to build a wedding chapel on his vacant lot. Fried entered a tentative agreement with Calvary by the Sea Lutheran Church, which was being investigated for tax liability. Under Fried's plan, Calvary would lease the land, build a chapel, and conduct Sunday religious services. Fried would use the chapel for commercial weddings the other six days, paying Calvary $100 per wedding as a site fee. When Calvary declined the deal, Fried approached the Prince of Peace Lutheran Church in Waikīkī. The Lutherans had sold their 10-story Laniolu Building on Lewers Street to Duty Free Shoppers and had been holding services in the Sheraton Princess Kaiulani Hotel for three years. Again, however, Diamond Head neighbors objected to the prospect of limousines bringing eight weddings per day in and out of Fried's driveway (Campos 1997). Fried shifted his sights for a wedding chapel to another property, the Walker Estate in Nuuanu Valley. He held an option to buy the 5 acre (2 hectare) estate from its Japan-based owner, Minami Group, and developed a plan for a 2,500 square foot (230 square meter) chapel. Neighbors protested the plan before their Neighborhood Board (Morse 1997) and Fried abandoned his second wedding site.

Home chapels in residential neighborhoods have opened, however. Two waterfront estates along Honolulu's upscale east shore operate as wedding venues. One is the Bayer Estate, built in the 1930s, a home listed on the state's Register of Historic Places. The owners are Richard and Susan Mirikitani; he a vice-president of Castle and Cooke and brother of a city council member, she the operator of the wedding business. Because their house is on the historic register, weddings are among the commercial activities allowed to fund the preservation of the building. The wedding business was approved by the city in 1998 as conforming to uses allowed in historic homes, but lawsuits by neighbors forced a review. Neighbors were especially concerned about limousines entering and leaving. After hearings in 2000, the city director of planning and permitting approved wedding use and so the Bayer Estate hosts weddings today (Lind 2000).

Another wedding site in the 'hotbed' district is an estate owned by sitting city council member John Henry Felix. Felix made his fortune in Honolulu mortuaries. To try his hand at weddings, he purchased an oceanfront home. In September 1999 he entered an exclusive agreement with Tokyo-based Best Bridal and allowed 20 Japanese couples to use it that month for a site fee of $300 per ceremony (Dingeman 1999). The home is in a residential zone, however, so the neighbors and the city have viewed the councilman's wedding business as illegal. Felix argued that weddings are a home occupation like swimming lessons. His own city council had ruled in May 1999 that weddings are a personal service, not allowed in residential properties. The City Department of Planning and Permitting cited Felix for conducting a commercial wedding operation in a residential zone and fined him $100 per day. Councilman Felix vowed to fight city hall and defiantly allowed weddings to continue in his home. Fellow council member Steve Holmes called Felix's wedding business "embarrassing" to other council members, reinforcing the "common view of politicians that we're a self-serving bunch.... John Henry is a very wealthy man. It makes no sense to me that he would do this" (Dingeman 1999). Felix's own constituents in East Honolulu appeared before their neighborhood board in January 2000 to complain about the wedding chapel business (Kuli'ou'ou/Kalani Iki Neighborhood Board 2000). Felix has allowed weddings to continue on his estate, and reported that he earned over $100,000 from weddings in 2000, while fines of over $40,000 for 1999 and 2000 remained unpaid (Dingeman 2001). In 2002, a church's purchase of a tennis club on the same oceanfront block brought the number of wedding operators to four in a row. One local news report called East Honolulu "a virtual strip mall of wedding businesses" (Roig 2002).

A wedding in a grand home appeals to Japanese tourists and is a low-overhead way for Japanese wedding firms to conduct weddings. Commercial wedding sites in residential neighborhoods have, however, met stiff community resistance. One neighbor of a proposed wedding site disdained the idea of a "Las Vegas-type wedding chapel" (Abedor 1998). Another feared if zoning rules loosen, "residential zones won't be safe from…bars (or) porno stores…. Unsavory elements will move in. Crime will rise and we will no longer be safe" (Ing 2000). Another resident, dismayed at the governor's plans to expand tourist space in Honolulu, suggested that the Governor's Mansion be turned into a tourist wedding chapel (Levinson 2000). The editor of *The Honolulu Star-Bulletin* concluded from these disputes that the city must "include 'wedding mills' with…drug treatment facilities and other activities that set the not-in-my-back-yard alarms ringing" (*Honolulu Star-Bulletin*, 25 July 2000).

Weddings at Saint Simulacra's

Each year the tourist wedding industry has needed more wedding venues, what the Hawai'i Visitors and Convention Bureau calls "hardware" (Matsuo 2001). Commercial chapels in commercial zones have been the answer. Wedding chapels have sprung up in defunct restaurants such as the former International House of Pancakes and the round revolving floor atop the Ala Moana Building. Watabe Wedding contracted for its first exclusive commercial chapel at one end of a Honolulu restaurant with a panoramic view of Diamond Head. The owner of John Dominis Restaurant and Chapel is Andy Anderson, developer, former state senator, former head of the state Republican Party, and, in 2002, Democratic primary candidate for state governor. His restaurant is in Kewalo fishing basin on a street of fish dealers, trucks, water, ice, and the smell of fish. The setting is atmospheric for a seafood restaurant, but hardly so for a chapel, so limousines bring wedding couples indirectly through one tree-lined back street redeveloped by the state.

New chapels have brought considerable investment to O'ahu, but alienation as well. In an effort to provide a chapel near Waikīkī, Tokyo Produce Company built Diamond Head Gloria Chapel for $5 million in May 1996 (Figure 11.4). Tokyo Produce is a wedding producer based in Shinjuku Ward, Tokyo, doing business as Gloria Bridal in Honolulu. Tokyo Produce purchased two adjoining corner lots totaling 27,000 square feet (2,510 square meters) on the first urban block behind Waikīkī's Kapiolani Park at the foot of Diamond Head. The property has a narrow frontage on Monsarrat Avenue zoned B-1 for Neighborhood Business, across from small stores and Unity Church, another busy tourist wedding venue. Streets on two sides of the Gloria Chapel property are narrow and residential, lined with homes dating from the 1920s on lots of under 5,000 square feet (465 square meters). Tokyo Produce got city approval

Figure 11.4 A Tokyo firm built Gloria Bridal Chapel near Diamond Head for weddings seven days a week. Couples arrive from their hotels in limousines that dwarf the bungalows in the neighborhood (photograph by the author)

for two buildings, Gloria Gardens Chapel and Club House. The white chapel with vaulted, steepled roof has a white cross above the entry. Viewed from the front door, the chapel seems to sit on the slopes of Diamond Head, and the same slopes can be seen through the narrow windows high behind the altar, though busy Monsarrat Avenue runs directly behind the nave. The two-story Club House holds offices, wedding gown salon, photo studio, and a 50-person reception hall. The chapel has done 1,000 weddings per year for Japanese couples, with small receptions for many. Gloria Bridal offers its own packages as a direct one-stop wedding firm, and wholesales ceremonies in its chapel to other wedding firms. Gloria Chapel does not sit easily in its neighborhood, however. Constant traffic of limousines, busses, and photographers' cars turned the narrow access streets into a 'parking lot' for the business, in the words of unhappy neighbors (Adamski 1999a). One venue for protest against the chapel by neighbors has been the Liquor Commission. Gloria has tried to obtain a liquor license to sell champagne or other liquor for receptions. It brought annual petitions to the Liquor Commission from 1997 to 2000, but each year was denied a license due to protests of a majority of residents living within 500 feet (150 meters) (Adamski 1999a, 1999b, Wright 1999). Even commercial zoning did not guarantee a happy fit in the Monsarrat neighborhood.

To expand, Gloria Bridal wanted a second chapel where it hoped to avoid the problems of close neighbors. It became a party to construction of a chapel on a rocky coastal promontory that angered another part of the island, however, alienating a second community. Gloria Bridal chose a site at Sea Life Park, a marine theme park that sits in a conservation district on one of the last undeveloped coasts of O'ahu. The theme park owner, Attractions Hawai'i, was adding new profit centers to several of its visitor attractions in the late 1990s, including all-terrain vehicle rides and North American-style waterfall weddings in fragile Waimea Valley. At Sea Life Park, Attractions Hawai'i reactivated a long-dormant plan to build a luau site on adjacent public land, but replaced the luau area with a wedding chapel and reception hall. Gloria Bridal agreed to become a concessionaire of Sea Life Park to operate weddings in the planned Saint Catalina Chapel. Attractions Hawai'i graded a level site into the steep rocky face of the hill in early 2000 without a permit and was fined for doing so. The company sought a building permit and faced questions from community members worried about the appearance of the prominent cliff above the highway. In May, the general manager of Sea Life Park, Wayne Nielsen, declared the bridal chapel would be an "accessory" to the existing theme park. "We're very sensitive to keeping the beauty of the mountains. We don't want to build anything that will block the view" (Roig 2000). The firm won a permit for two structures in July 2000 (Watanabe 2000). Construction alarmed the nearby rural community of Waimānalo. Seeing the results of the construction, neighbors complained to the press. One writer asked:

> How were permits ever obtained to destroy one of O'ahu's most beloved views? Whether this horrible hillside structure is for burgers or blessings for bucks, a wedding chapel is completely inappropriate for this location, and no amount of landscaping or paint can ever disguise it.
>
> (Henderson 2000)

The neighbors' surprise at the prominence of the chapel buildings, 25 feet (8 m) high and of 2,000 and 3,000 square feet (185 and 280 square meters), brought them to their Neighborhood Board to complain, "The chapel is huge and stands out as you come around the mountain" (Waimānalo Neighborhood Board 2000a). Gloria Bridal and Sea Life Park sent spokespersons to the November Neighborhood Board meeting to explain the chapel and to ask community approval of a liquor license. Residents blasted the appearance of the chapel and the idea of serving liquor on the site. Gloria Bridal's attorney for the liquor license application, Wayne Luke, himself a member of the license-granting Liquor Commission, received a unanimous "no" from the Neighborhood Board. One resident explained, "Many in the community see the chapel as a classic example of Sea Life Park deliberately misleading and deceiving the Waimānalo community, therefore the trust has been broken" (Aguiar 2000a). Facing staunch opposition from Waimānalo neighbors, Gloria Bridal withdrew its application for its liquor license for Saint Catalina Chapel in December 2000 (Adamski 2000), but revived its application and won its license in May 2002. Community members raised new protests against Gloria Bridal and against a Liquor Commission that "isn't listening" (Vorsino 2002, Lum 2002, Aguiar 2002).

Another commercial wedding chapel opened in 1998 on the shore of Kaneohe Bay, beside stone-walled Kahaluʻu Fishpond. The Aloha ke Akua chapel was built by World of Aloha as an exclusive venue for Watabe Wedding. This onetime property of King Kamehameha IV sits on 35 acres (14 hectares) of shoreline and is a registered historic landmark, but no community master plan governs Kahaluʻu Pond. The chapel's construction required a special shoreline permit to clear, grade, landscape, and pave the large parking lot. The chapel and its fishpond are visible from the road but sit behind locked gates and private property signs (Figure 11.5). Though the plan failed to get majority endorsement by the Neighborhood Board, it was approved by the City Council, and the developers reactivated the fishpond as a working ogo fish farm to enhance local legitimacy of the project (Kaleikini and Wong 1997). The chapel conducts six to eight one-hour

Figure 11.5 Displacing public access to the shoreline behind its new tourist wedding venue, The World of Aloha, Inc. showed no aloha; the sign reads "Private Property, No Trespassing." (Photograph by the author)

weddings per day, the last at 5 p.m., and received a liquor license based on the argument that champagne is a necessary "amenity" in the competitive world of wedding chapels (Aguiar 2000b).

Conclusions

Marriages are performed through spatial movements to and upon ceremonial stages. In consumer societies, wedding rituals and venues are now commodified in the particular spatial practice of 'bridal tourism.' Wedding brokers bring couples not to every destination, but to particular destinations where tourist infrastructure, romantic imagery, religious flexibility, and legal convenience coincide. Bridal tourism takes a particular spatial fix in Hawai'i.

Ceremonies in Hawai'i are neither legally necessary nor legally relevant for tourist couples, yet are a booming form of celebration and identity consumption. By choosing to marry offshore, a couple asserts a self-determination and self-stylization. Couples who know little of Hawai'i arrive for a wedding that is simultaneously private and in the viewfinder: through wedding photos that include icons of the romantic exotic, the couple creates an image of themselves that will in turn be consumed by others at home.

Place is performative, effecting the transformation of subjects, but not in its natural or unprocessed state. Even a place like Hawai'i is performed upon, rearranged by industries forging a visual correspondence between advertised product and wedding-in-Hawai'i venues. Hawai'i's two sets of images, places and styles of offshore weddings, for North American and Japanese tourists respectively, afford a clear example of the cultural specificity of the exotic in the eye of the beholder.

The marriage that makes all tourist weddings possible is that between the state and private capital, now cooperating in the $1.1 billion niche of bridal tourism. The state provides advertising, state and county land use approvals, and enabling business and civil regulations. Many public officials, their families, and their political donors have business interests in the wedding industry. Hawai'i's landowners, officials, and politicians have allied with Japanese investors needing to expand a particular built environment—storybook chapels—as the shop floor of their industry. Citizens have not always been happy to see wedding chapels appearing in their neighborhoods, and have often found their local state allied with the investors and cooperating in the displacements.

Cultural analysis of tourist weddings in the fullest sense must involve the question of the material production of venues at the destination. Concepts of the destination as margin and stage for performance too easily assume tourist space to be preexisting, available, and inert. From the vantage of the 'margin,' however, tourist space is none of these; it is invented, extracted, and reactive. How neighborhoods and nature become sets for the romance industries must be explained in terms of firms, the products they advertise and sell, and the places they restructure into consumable wedding venues. The last produces strains and conflicts within the host culture. Wedding firms facing local backlash produce ever more quasi-sacred spaces sealed from everyday life on Oʻahu. Local residents must ultimately forfeit cultural space of their own for another form of tourist consumption. The destination wedding appears to be rich, happy, transcultural displacement, but a place-conscious analysis will find alienation at the destination.

12 Maintaining the myth

Tahiti and its islands

Anne-Marie d'Hauteserre

Tahiti has a worldwide reputation as a tropical paradise yet actually receives few visitors. This chapter explores how images of 'Tahiti' unfolded from European explorers' descriptions—and how they have been appropriated by other destinations. Polynesians have participated in the creation of the images but, thanks in part to the French colonial legacy, they have resisted the transformation of their islands (their *fenua*) into a touristed destination. Its continued remote exoticism adds value to other 'destinations,' which exploit the myth for mass tourism.

Tahiti's identity has endured since its discovery as the ultimate seductive destination. Why then, despite its highly recognizable world image, does Tahiti receive but a minuscule number of tourists—just 6 percent of all arrivals in the South Pacific (including Australasia, Polynesia, Melanesia and parts of Micronesia)? Answering this question is approached by understanding Tahiti as an 'imagined world,' constituted by both Europeans' and Polynesians' historic ideas of the islands. While the myth of Tahiti placed it on the global scene, its actual landscapes have been, and continue to be, produced by the everyday practices of its residents. Its *image* has been maintained by its visitors and its residents, past and present, in the ways that Oceania has always been a region of mobility and cross-cultural mixing.

Histories of exploration and imperialism have shaped the culture and representations of Tahiti and its islands. However, we need to go beyond assumptions about the unidirectional influence of colonialism on colonized societies; basic and important continuities in Tahitian culture, values and dispositions have characterized local society, even as they have transformed. Interaction with European cultures has not been fatal to Tahiti and its islands; by contrast, such interactions demonstrate ways that residents have maintained and recreated Polynesian ways of life in spite of colonial history. France formally colonized the Society Islands, the Tuamotu Archipelago, and the Marquesas to establish French Polynesia in 1842, but the continued political and economic presence of France, up to the present day, has, unlike many independent tropical island states, diminished the islands' need to rely on tourism development. The islands' strategic value to

France—the Tuamotu Archipelago was France's nuclear testing site—has resulted in significant financial subsidies and public services, which have reduced the need for local economic self-sufficiency and economic diversification and, in turn, prevented the need to commodify Tahitian culture for tourism.

Tahiti's allure as trope of the tourism industry now serves to sell other 'paradise' destinations. Tourism developers in other parts of the world have appropriated discursive representations of Tahiti to enhance their own competitiveness. In the course of being 'discovered,' Tahiti's landscape was transformed into a myth, which has ultimately led to the re-visioning of many other tropical locations anxious to develop "economies of pleasure" (Manderson and Jolly 1997). Tahiti's own continued existence as a set of lived places and as an exclusive destination also enhances the seductive mystique of the 'tropical paradise' narrative.

Locating Tahiti

Tahiti and its islands consist of about 1,400 square miles (3,660 square kilometres) of land extending over about 1.5 million square miles (4 million square kilometres) of the Pacific Ocean in the humid subtropics south of the equator. Only the high basaltic islands, such as Tahiti and the other Society Islands and some of the Marquesas, were farmed and yielded surpluses sufficient to support a differentiated population in the premodern period. On these high islands, steep, jagged volcanic peaks are surrounded by aquamarine lagoons inside fringing coral reefs. As one British naturalist rhapsodized, "One of the finest islands in the whole of the Pacific must be Moorea. The steepness of its jagged volcanic peaks, sometimes swathed in layers of mist, is a masterpiece of natural sculpture" (Mitchell 1989: 198) (Figure 12.1). The island has tall coconut-palm-lined coastal strips, offset in the interior by sharp ridges covered in contrasting patches of pastel fern and dark tropical vegetation. The atolls, where the high island has subsided and the coral reef surrounds the lagoon, comprise 84 of around 130 islands, but only 48 of those are inhabited. The Tuamutu Archipelago contains most of the area's atolls including Rangiroa, in which "the sense of remoteness and space that is part of the mystique of Polynesia is exhilarating" (Linden 1996: 160) (Figure 12.2).

Tahiti, as used in this chapter, stands for all of the islands that make up French Polynesia. The local tourism development office uses the phrase 'Tahiti and its Islands' to market the destination. The name 'Tahiti' is more resonant with most would-be travellers than French Polynesia, even though the colonial toponym correctly indicates the political and economic context of the islands. Many travellers also know of two other Society Islands: Moorea, a high island with spectacular cliffs, and Bora Bora, a high volcanic island surrounded by an outer coral reef, which was used by the US Army during the Second World War. For armchair travellers, photographs of Moorea and Bora Bora are commonly used to

Figure 12.1 Peaks of the island of Moorea (photograph by C. Cartier)

represent 'Tahiti.' These three islands have 87 percent of the hotels, comprising 4,305 rooms (Ministère du Tourisme 1999).

The name 'Polynesia' was coined by Charles de Brosses in 1756 to describe the multitudinous islands that had been discovered in the Pacific Ocean, by combining 'poly' with the Greek root for island, *nesos*. J. Dumont d'Urville (1832) first restricted its use to the area within the triangle from Hawai'i to Rapa Nui (Easter Island) and Aotearoa (New Zealand) during a meeting of the French Geographic Society in 1830 (Figure 12.2). The wider Pacific has since been regionalized on the basis of ethnographic differences of color, social organization and gender organization: Polynesians are categorized by lighter skin and with more hierarchical social organization and greater gender equality compared to Melanesians, in the southwestern Pacific, with darker skin and relatively egalitarian social organization but marginal treatment of women. But like many early 'trait geographies,' Polynesia as a cultural–regional category and these divisions of regions of the Pacific have been culturally constructed and tell us little about the regions themselves and especially processes of regional formation. Polynesians are the descendants of Austronesian-speaking Lapitans who established pioneer settlements 4,000 years ago in the region that includes Fiji, Tonga and Samoa. Polynesian skills in building and operating boats and elaboration of indigenous systems of environmental knowledge enabled navigation to places hundreds of miles away on the open sea, and count among humanity's major achievements. The sea has always been incorporated in the Polynesian definition of place (Teiawa

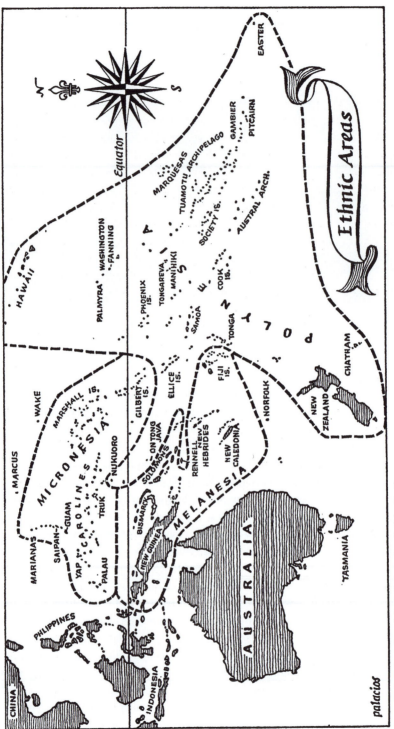

Figure 12.2 Polynesia and the 'ethnic regions' of the Pacific.

Source: Oliver (1989)

2001), to the degree that Hau'ofa (1994) has proposed a paradigmatic shift to perceive the Pacific Ocean as a highway rather than a barrier to Polynesians.

Origins of the myth: constructing the Tahitian paradise

The first European voyages of discovery renewed Western ideals, transmitted through generations by religious and humanist traditions, of founding a new, just and prosperous society and of rediscovering paradise on earth. Discovery of new lands revived numerous myths, such as the bucolic myth, the myth of Eden and of paradise lost, of the Golden Islands (of Hercules' fame), the myth of cornucopia and of the 'noble savage,' first developed by Vespucci in his famous 1503 letter, *Mundus Novus*. Such mythical visions evolved long before Tahiti was discovered, and through them Europeans attempted to make sense of new lands. "As the realities of America proved increasingly at variance with the vision of an earthly paradise…this vision was transferred to other regions less well known," in particular to the 'Southern Continent' (Hammond 1970). The idea of a still unknown continent in the southern hemisphere became a quest of most navigators, and a European alter ego to the temperate zone of the northern latitudes.

The representational construction of Tahiti as paradise in Western discourses is closely tied to the history of global exploration and discovery, in particular to the famous navigators of the eighteenth century, Louis Antoine de Bougainville and James Cook. Explorers and their witnesses "depended on each other's testimony in forging their narrative" (Kabbani 1986: 114), and in this process navigators celebrated the Pacific as a new mirror of European power, often in the name of science and reason (Scemla 1994). This world region, the Pacific, discovered last, became the first to be scientifically studied. However, descriptions of Polynesia have always revealed more about European symbolic systems and values than about Polynesian society and beliefs. As Augé (1997: 87) observes, "memory tends to cling more readily to myths than to historical facts." The Tahitian paradise does not reside in the Pacific, but in words, "words that bloom, words that melt pleasurably under the tongue like sweets, words that make one dream" (Michelet 1980: 62). Myths certainly also accomplish such real tasks as the justification of power, authority or political acts. Myth, adds Barthes (1986: 15), is a "word chosen by history. It cannot be deduced from the nature of things." Myth ignores history as well as geography.

Europeans interpreted Polynesian society through prevailing trends in Western intellectual thought. The Romantic era fantasized the natural world and led to the ascription of differences between the Polynesians and their visitors, who then proceeded to capture and maintain them. It was assumed that Tahitians instinctively lived in perfect harmony with their (to these early observers) fertile environment, and they became stereotyped as 'noble savages.' The Enlightenment mindset led Europeans to assume subject status while Polynesians were represented as objects, described as specimens of scientific enquiry. In the late eighteenth century, Julien Crozet (1783) explained the presence of Polynesians and their islands as living on the remnants of a drowned continent. Only the summits had escaped some volcanic cataclysm to become islands of refuge for the human survivors; they could not have navigated between these islands, considering their primitive technologies. Ignorance of Polynesian human–environment relations and social organization provided a kind of *tabula rasa* for explorers' plausible if extravagant idealizations of Pacific society (Moerenhout 1837, Spate 1988).

Legacies of explorers and travellers

Each century has had its particular obsession: during the eighteenth century men were fascinated with islands, as exemplified by the success of Defoe's novel *Robinson Crusoe*. Its narrative, like that concerning most island settings, confirms the castaway, "established in both his power bases, masculinity and whiteness, as the master of all that he surveys" (Woods 1995: 146). Even in the twentieth century South Pacific islands have maintained their mystical lure. Margueron (1987: 6) observes that "Europe seems to have never ceased to need happy islands." Towards that end, Louis Antoine de Bougainville found the happiest island of all.

Bougainville's circumnavigation of the globe from 1766 to 1769 brought France considerable prestige. His undertaking was not just a nationalistic or commercial venture; it was also animated by a genuine scientific curiosity, by the finer and admirable impulse of the search for pure knowledge (Wright and Rapport 1957). When Bougainville and his crew, in the first days of April 1768, after several months at sea and with empty stores, sighted a large island and saw that it offered anchorage, Tahiti did appear as the promised land. Its residents were quickly integrated into this mythical template: "as long as I live I will celebrate the happy island of New Cythera: it is the true Utopia" (Bougainville in Martin-Allanic 1964: 685). The name was used to pay tribute to the pleasures discovered on this island, since Aphrodite, the Greek goddess of love and beauty, is supposed to have arisen out of the sea at the island of Cythera. Bougainville (1771: 190–1) emphasized idyllic aspects of the setting:

> Nature had placed the island in the most perfect climate in the world, had embellished it with every pleasing prospect, had endowed it with all its riches, and filled it with large, strong, and beautiful people...[who] constitute perhaps the happiest society which the world knows.

A permanent element of the myth since its origins is the *vahine* (Tahitian woman), which has been woven through the works of explorers and travellers, authors and painters, from one era to the next. The myth finds strong erotic appeal in the encounter: the impossibility of the sailors on Bougainville's ship to respond when first faced by the nudity of Tahitian women (Taillemite 1978: 80). As Flaubert (1992: 85) confirmed, "This is madness, he said to himself, how could I ever touch her?...But by this renunciation, he set her upon an extraordinary pinnacle." In the first encounter, Bougainville (1771: 189–90) explains:

> the dugouts were filled with women whose faces were as beautiful as most European women's, and whose bodies were more beautiful. Most of these nymphs were naked...I ask you, how was one to keep four hundred young French sailors, who had not seen women in six months, at their work in the midst of such a spectacle?

Tahiti seduces because it has become a fetishized object of the male gaze, an erotic commodity (Baudrillard 1988d); and it seduces more because it represents tensions between Polynesian sexuality and Western norms. In Europe, nudity had become "a sign of the inferiority, even bestiality of wild men or savages" (Goody 1997: 220); concealment of the body was indicative of civilization, marking Europeans as superior. One of the attractions of the uninhibited nudity was (and remains) that it is not how life is usually lived: in Tahiti warmth and nakedness combined to create the conditions for sexual freedom, providing the perfect breeding place for white men's dreams (Woods 1995, Kaplan 1996). The fantasy of an erotically exotic female Pacific Other became a vehicle for Western libidinal fantasy.

Discourses of mythical representation were embedded in relational frameworks of power established between the centre and the periphery and reinforced through repetition and reiteration. Such assigned power differences appear to be natural, to have the "appearance of substance" (J. Butler 1993: 33). As Bougainville (1771: 189) continued, the women "particularly are distinguished by the regularity of their features, their gentleness, and their natural affability.... They are very hospitable, very gracious, and quick to caress." The accounts fixed on women's sexual availability and desirability while they reinforced prevailing gender and colonial discourses, "with the added cachet of scientific observation and intercultural persistence" (Dawson 1991). Labelling Polynesian women as sexually promiscuous became an identity-constituting performance, reinforced by participation. Authors like Pierre Loti (Julien Viaud), Somerset Maugham and others structured the myth through distinctively male-gendered experiences.

Philippe Commerson, a widely respected member of France's scientific elite, and whose interpretation of Tahitian life most influenced Diderot, elaborated on Tahiti as utopia. In regard to sexual practices, he proclaimed that:

> There neither shame nor modesty exercises its tyranny: the act of procreation is a religious one.... Some censor with clerical bands may perhaps see in this only the breakdown of manners, horrible prostitution, and the most bald effrontery; but he will be profoundly mistaken in his conception of

natural man, who is born essentially good, free of every prejudice, and who follows…the gentle impulses of instinct not yet corrupted by reason.

(Commerson 1769: 198–9)

Lieutenant John Elliot, who accompanied Cook, confirmed the vision:

Tahiti appeared to us to be the Paradice of those Seas, the Men being all fine, tall, well-made, with humane open countenances; the Women beautiful, compaired with all those that we had seen…with lively dispositions…. When we were on shore here we felt ourselves in perfect ease and safety.

(Elliot and Pickersgill 1984: 19)

Cook and his naturalists, Johann and Georg Forster, were careful to distinguish the social origins of the women who came on board ship. They claimed, contrary to Bougainville, that these women were of the lower class and that none were married, asserting that "Conjugal fidelity is exacted from her with the greatest rigour" (Forster and Forster 1968: 133). Dawson's (1991) interpretation is that both men were reinforcing and easing European gender conceptions for a European audience. Foster himself recognized that much of the trading of sexual favours was instigated by contact with Europeans and was practised to placate their violent intrusion (Pearson 1969). Many of the details reported by these explorers and their narrators illustrated European ideologies of territorial and global possessiveness; women could be sexually appropriated since they belonged to inferior men.

After the Bougainville voyage departed, the memory of Tahiti only exacerbated the crew's miseries, plagued by disease and threat of starvation, as all the islands they subsequently came upon seemed to be either barren or inhabited by people who were antagonistic to them. These difficulties did not prevent Bougainville from making useful geographical discoveries, such as recognition of the atomization of the Austral continent into innumerable islands and one small continent (Australia), and carefully drawn maps of the areas explored. Bougainville also brought back to France a Tahitian of princely heritage, Aoutourou, an apt representative of the noble savage: very tall, pale (aristocratic Tahitians avoided the sun), and handsome. He was formally presented to King Louis XV in 1769 (Hammond 1970). The popular excitement stimulated by his presence and by the descriptions of Tahiti added cachet to the scientific and political aims of the voyage.

When Gauguin (1949) sailed for the Pacific in 1891, he was adamant that Tahiti had remained an uninvaded, remote garden of Eden—the myth of Tahiti and its islands denied them any history. Gauguin's paintings, and the later visits to these islands by many literary figures, contributed to the maintenance of the Arcady myth introduced by Bougainville. Robert Louis Stevenson and his party, including his mother, his wife and her son, all quite accomplished writers, produced an endless stream of letters to the *New York Sun*, which sponsored the trip to the South

Seas in 1884. These letters gave detailed accounts of the life of the people of Tautira (a district at the opposite end of the island from Papeéte, Tahiti's largest settlement and colonial capital) which they described as an "earthly heaven" or "the garden of Eden." High culture's fascination with singular, elite figures of travel such as Melville, Zane Grey, Victor Segalen or James Norman Hall, who are noteworthy because of their contribution to the romanticized image of the islands, facilitated the continuing diffusion of what had become a "static utopia" (Sturma 1999). Such notions of untouched authentic 'primitive practices' tend to reify and value a specific moment in the ongoing history of a people (Clifford 1997, Crang 1999), placing them outside of time, irrevocably tied to a primeval past.

In the twentieth century too Tahiti appeared primitive to Western writers. Biddle (1968: 18), who visited Tahiti in the 1920s, wrote "of a still unspoilt although rapidly vanishing primitive society." Rowe (1949) visited in the 1930s, having "lived for long among Polynesians still in a fairly primitive state." Each of these men imagined he was savouring the last pleasures of Polynesian life before the 'happy islands' vanished. In this way, Tahiti is always just on the edge of extinction; its fugacity itself is seductive. Melville (the most famous of Pacific beachcombers), Loti and later Gauguin tragically celebrated its "gently dying out under the touch of civilized nations" (Loti 1878: 95). The population of Tahiti had actually dropped from 35,000–40,000 at contact to 8,000 by 1800 due to diseases introduced by European visitors and would not start to increase again until the first decades of the twentieth century (Campbell 1989: 153).

Using the image

There were of course imperfections in these paragons but such news had difficulty reaching European shores. Amazingly, the myth of Tahitian paradise survived even though Bougainville's *Newsletter* and some of the nineteenth-century accounts presented an accurate if unflattering vision of the people: they practised warfare, oppressive taboos, infanticide, human sacrifices on *marae* altars and slavery. Social order was hierarchical and its political organization tyrannical. The myth also survived in spite of the fact that the first encounter by Wallis of Tahitians required the use of deadly firepower over several days. European hostility and violence typically disappear from texts on Tahiti.

The image of New Cythera, launched by Bougainville, is now exploited by the modern tourist industry to attract visitors to (m)any tropical areas in the 'pleasure peripheries'. Although "the myth of Paradise is by now a thoroughly shop-worn cliché…virtually every travel brochure on the region contains similar images, no longer the exclusive preserve of Tahiti, which inspired them" (Douglas and Douglas 1996: 32). Destinations reconstruct themselves on the Polynesian palimpsest of mythic serenity and graciousness. The eighteenth-century explorers set the imaginative pathways for contemporary tourism, and fantasy thus fundamentally predates most journeys, shaping tourists' preconceptions and expectations.

The blend of literary and artistic images that Tahiti engendered through the centuries, "together with recent and more blatantly commercial imagery, has reinforced the view of tropical islands as paradise" (King 1997: 3). Bali advertises itself as "Paradise for a song" (Cumming 2000) and tour organizers' brochures about New Caledonia feature Tahitian tropes. In Hawai'i, "The more exotic and erotic part of the [Kodak Hula Show]," which has promised 'Hawaiian culture' for more than 60 years four times a week, "comes in the brief presentation of Tahitian dance and dancers...focusing the gaze on the rapidly moving hips, bare mid-riffs, and bra covered breasts [*sic!*], all moving to the intense beat of Tahitian drumming" (Buck 1993: 2). Augé (1997: 52) argues that the image is now applied to artificial paradises, like the Paradis Aquatique Tropical in central France, a "tropical island fringed with white sand, lapped by blue and warm waters, covered with coconut trees" where no coconuts fall from the artificial trees.

In consideration of such ersatz Tahitis, we should connect with Augé (1995), who has argued that meaning-saturated places are increasingly disappearing altogether, claiming they have been replaced by 'non-places' as the real measure of our time. With the commodification and predictability of the destination, actual destinations are somewhat arbitrary (Kirschenblatt-Gimblett 1998, Clifford 1997, Zurick 1995). Images have as much to do with an area's tourism development success as the local recreation and tourism resources. The real magic lies in that the audience is predisposed to believe in the illusions (Bourdieu 1986: 137). Realities of the Pacific are not well understood outside the region, and so Tahiti remains a mystery. The myth anchors resistance to disenchantment, resistance to knowing the political economic power dimensions that underlie these socially constructed tourist images, as they are integrated into and essential to global mass tourism. Tahiti's continued existence as an 'exclusivist and defensive enclave,' a destination too remote and exotic for most visitors, has in turn enhanced its mythical image. Tahiti is not utopia, but there is no other. The mystery that shrouds its socio-cultural landscape lends more credence to other fantasy paradises conjured by marketing. The power of the touristed tropical landscape is maintained by the endurance of 'Tahiti,' a transhistorical metaphor that continues to evoke aspects of both mystery and reality.

Maintaining a Polynesian landscape

If our human landscape is our collective autobiography, there are potentially as many landscapes as cultural ways of seeing. Historical processes of globalization, modernity and mythology have transformed the idea of Tahiti into a myth, which could, in time, stand as a shorthand statement of its character. Has the myth reduced Tahiti as a place to a stereotype? Singular privileged accounts—the ones the tourism industry has inherited from Enlightenment era explorers and many of their literary counterparts and descendants—need to be dislodged in favour of more multiple and diverse ones. What has been left out of discourses of colonialism and imperialism must be included, especially Polynesian participation in the continued imaginative reconstruction of paradise in the South Pacific.

The question remains, though, of how particular representations are asserted, how they survive across different local–global scales, and whether they are the result of a specific agency or a dialectical process by both residents and travellers (most Polynesians themselves are both). The transformation of spaces into places does require a conscious moment (de Certeau 1984). Polynesian agency, born of history and geography, has produced positions from which they have enunciated their own narratives and cultural practices. Polynesians knew, for example, how to placate Wallis when confrontation did not work: they offered shore parties with several very handsome young girls whose 'perfections' dissolved much of the hostility (Pearson 1969). They also understood the military superiority of these new visitors and resolved to welcome future ones with great festivity rather than resist, and to use this European strength to resolve internal conflicts (Toullelan 1991). Thus the landscape of perceived welcome and cornucopian abundance encountered by the first explorers was also a social landscape engineered by its occupants—not a generous unadulterated one provided by nature.

It is difficult to know what changes in Tahitian concepts of their place followed the contacts with foreign navigators. What is certain is that many of the inhabitants declared the island, on the basis of frequent European visits, to be "the finest part of the whole inhabitable globe" (Newbury 1980: 78). Tahitians were not just at the mercy of colonial forays. Visitors to Matavai, where much of the early history of European contact in Tahiti unfolded along its eponymous bay, were valued more for what could be assimilated into Tahitian culture than for what might change it (Figure 12.3). The Tahitians arguably spatially contained contact to Matavai Bay, as there were a variety of anchorages available around the island. Exchanges were not a European innovation in Tahiti and its islands, as confirmed by much recent archaeological work that demonstrates exchanges with other Pacific island peoples (Newbury 1980, Rolett 1996, Kirch 2000). It is also probably safe to say that most of the sexual encounters experienced by Europeans since the first encounters had to be purchased (Oliver 1989: 354). These correctives to the myth suggest that European impact was in part locally managed.

Polynesians have not transformed Tahiti according to the mythical palimpsest or a 'placeless' tourism destination. Their cultural landscape is an ordered assemblage of places and things, which act as a signifying system through which local social organization is communicated, reproduced, experienced and explored (Lippard 1997). Polynesian culture continues to represent a coherent picture of the inhabited world of Tahitians even as they continue to be critically aware of the way in which they are portrayed. Accordingly, they have developed the islands as habitat,

Figure 12.3 Matavai Bay, Tahiti; island of Moorea on the horizon (photograph by the author)

not as attraction. This is not to say that the Polynesian landscape is more 'authentic' or primitive. Tahitians have embraced modernity in their urban and economic spheres and participate in global networks of trade and communications. Tahitians remain both tribal and modern, local and worldly. They were not displaced by European discovery in the eighteenth century, and their contemporary existence and authenticity are not threatened by twentieth-century global rootlessness.

Change has been relatively gradual in Tahitian society, and it has been characterized by some continuity in practice and ideology (Baré 1985, Edmond 1997). For example, Polynesians conserved their language despite 200 years of foreign occupation. V. Smith (1998: 88) confirms that "imported theology subsisted in juxtaposition with traditional elements of faith, failing to dispossess the latter of their authority." Cultural practices have especially endured. Tahitian dance, for example, has maintained its deep social significance and is practised by all generations as well as for tourism (Figure 12.4). The main cultural holidays continue, especially the *heivas* (Polynesian cultural celebratory festivals), which occur in July, now to accommodate Bastille Day celebrations.

Heivas have a long history: Captain Cook described them as attended by young people "to dance and to make merry" and by all ranks of people (Marra 1967: 206–7). During the *heivas* a complex gamut of songs and dances is presented to an appreciative audience, which tourists are welcome to join but cannot monopolize. The celebrations include competitions, exhibitions and cultural re-enactments, scheduled in numerous venues. They culminate in recreated ceremonies at

Figure 12.4 Tahitian dancer
(photograph by the author)

reconstructed *Maraes*, reinforcing the functioning of the sites as symbolic centres of Polynesian cultural survival and revival. These periodic displays of culture permit Polynesians greater agency in their cultural portrayal than typical tourist venues, such as museum displays or cultural parks. Shows in hotels sometimes are cancelled because participants have more pressing matters to attend to than entertaining tourists. There are no plastic leis in Tahiti. Women weave fresh flower leis every day for resident use and for the performers at the central market. Departures by residents of the islands are marked with real shell necklaces because of international restrictions against fresh produce imports. So Tahitians have shielded themselves from "trinketization" (Khan 1997) of their culture, allowing the mythic image in whose creation they participated to endure

Tourism in Tahiti and its islands

The great irony of the enduring myth of Tahiti as desirable destination is that, on the world scale, and in comparison to other island destinations, hardly anybody goes there. The *Bulletin de la Société d'Etudes Océaniennes* published a special edition on tourism in 1922; the relative absence of tourists led the lead author to conclude that Tahiti had all the attractions for tourism but lacked organization (Sigogne 1922). In the 1930s, Jourdain (1934) pleaded for better organization of sailing tourism while recognizing that the distance from France (then an oceanic voyage of 30 days) might be a major impediment to its development. Sheer distance has been something of a transhistorical condition in Tahiti's tourism development. In the 1950s only 700 visitors a year made it to the islands, and visitor numbers only substantially increased with jet technology and the opening of the Faa'a International Airport in 1961. Visitor arrivals grew to 10,000 in 1962 and to 85,000 by 1975. But even by jet, Tahiti is a long haul. All of its major outbound markets lie at considerable distance (6,400 km to Los Angeles, 9,500 to Tokyo, 6,100 to Sydney, 7,800 to South America), which generates great expense. Landing in Tahiti does not bring one that much closer to most of its islands either. In 1997 Tahiti received 180,440 visitors, compared to 6.8 million for Hawai'i (GIE Tahiti Tourisme 1999, Ministère du Tourisme 1999).

Distance aside, Tahiti has chosen to escape from the tourist gaze: it has limited the number of visitors to its shores by refusing to become a mass tourist destination, even though tourism has been an economic activity for many decades. The Tourism Development Office advocates an 'exclusivist' kind of tourism, on the basis of the high costs of tourism development in the area and its privileged representation (GIE Tahiti Tourisme 1999). It targets rich visitors in search of uncrowded and remote exotic paradises where nature and culture (not just Polynesian, but also Chinese and French) are deftly mixed. Catering to a limited niche market reduces the need for mass tourism and commoditization of local culture. Investments have been recommended to facilitate more visits to outer archipelagos, including home stays, which would redistribute economic benefits, and some locals offer resident lodging; but the cost of living in Tahiti is high so even this alternative is not cheap. Polynesians believe that their *fenua* (home) is deservedly expensive to enjoy.

Imperialism and tourism

Tahiti has not embraced tourism as its leading development strategy but (or because) it still boasts a relatively high per capita income. Competitiveness as a tourist destination has not been a necessary strategic consideration. Tahiti's development has included tourism as one of several economic activities which are presently pursued: high-seas commercial fishing, tropical agricultural products exports (flowers and the *nono*, a native fruit with medicinal promise), black pearl farming and technological developments like solar power. Thus tourism is one link between the Pacific island nation and the global capitalist system, not the only one or the dominant one (cf. Linnekin 1997: 231). Its main link to the global scale is through its colonial ties to France, which has provided financial support for several decades. This 'benevolent' support has reduced the urgency of local economic self-sufficiency (d'Hauteserre 1999). Some have described it as a helpless addiction to the cash handouts accompanied by complete subservience to the French state (Mitchell 1989). But there is a dual nature in possession and exchange, and colonialism in French Polynesia has had a specific history and geography.

The French Pacific colonies were largely left to themselves in the nineteenth century: French Oceania was not the empire (Aldrich 1990: 331). The French presence in the Pacific in the nineteenth century was predominantly naval and missionary and its power was based on the mobility of its navy, whose commanders doubted the economic profitability of colonization in Oceania and favoured the interests of the indigenes over those of colonization (Aldrich 1990). France was little involved in trade. The formation of a French empire in the Pacific gave France a stake in the region for possible future exploitation of its resources but French colonization was episodic and slow. Missionaries and traders from various countries were the main colonizers. The colonial lobby had constantly to renew its strategies to obtain investments by the French government. A French steamship company finally started serving Tahiti in 1923. Not only did French colonialism differ from that of its rivals, but it was even observed that "Tahiti remains a colony to be created" (Aldrich 1990: 334).

The colonial relationship entered a new level of significance when nuclear tests were projected for the region. Nuclear tests took place in the Tuamotus beginning in 1966 and were finally terminated in 1996. During this time, non-military public spending averaged 24.6 percent of GDP in the 1960s, and 36.9 per cent in the 1980s (Henningham 1992). Subsidies will continue because of Tahiti's strategic and emotional interest to the French government. Chesnaux (1991) commented that "unlike other colonial relationships rooted in economic exploitation, this one, instead, is motivated by economic investment and national pride." In 1995, France spent 125 billion Pacific francs in the local economy, 150 billion in 1998 and 152.5 billion (US$1.5 billion) in 1999. A contract signed in 1996 guarantees continued payments at this level until 2006 (Mission 2000: 35).

Nuclear testing fundamentally contradicts images of paradise. The same remoteness that allows Tahiti's myth to endure explains in part its role in France's nuclear regime, which regards Oceania as a frontier space rather than a place. The myth of Tahiti has been useful to the French government as it veiled the military use of some of the region, just as the ideological work of any dominant myth is to naturalize relationships of power and domination. The small number of visitors reduced the number of prying eyes in all areas of French colonial activities.

France has accommodated some Polynesian aspirations for autonomy to continue its hold. Decentralization of the French state in the early 1980s and 30 years of passionate demands resulted in reclassifying French Polynesia from a colonial dependency to a *Territoire d'Outre-mer* (overseas territory) (TOM) in 1984, which provided the territory with autonomy (Faberon 1996). In 1998–9, the issue that most preoccupied the local government was the campaign to reform governance of French Polynesia to catch up with the progress made by New Caledonia, another French colony in the Pacific. French Polynesia would become a *Pays d'Outre-mer* (overseas country) benefiting from even more independence from the French government but without totally severing ties. The French Polynesian version of this new statute of autonomy, unfortunately, is high on symbolism but low in substance. France would still retain control in vital domains such as foreign policy, justice, nationality and law and order. Independence, though, contrary to the situation in New Caledonia, is excluded as a future option (Von Strokirch 2000). Polynesians are aware that independence would only bring them neo-colonial ties to the capitalist world and its fickle markets. The returns would be much lower and in turn engender a deteriorated standard of living. Although the majority of the population does not desire separation from France, deliberate exclusion of a referendum on self-determination undermines credibility of the decentralization process.

Conclusion

Tourism, a global economic enterprise, is a spatially differentiating activity. To draw tourists, destinations need to differentiate themselves and through application of marketing techniques, too often commodifying themselves into staged environments in the process. Tahiti, even though it has possessed a seductive image, an important global advantage in these times of competitive destination marketing, has resisted 'theming' and 'revisioning' itself as a destination landscape. It has not become a place of tropical homogenization even if its image of exotic tropical paradise has been widely borrowed.

In terms of becoming a commodified tourist site, Tahiti, the local, has been a "primary site of resistance to globalization" (Kaplan 1996: 160). Polynesians have never ceased to be agents of their own cultural practices and representations (Saura 1998), which has enabled them to maintain a Polynesian landscape in the face of global intrusion (colonial or touristic). Tahiti illustrates the conceptualization of a local place as "a negotiated reality, a social construction by a purposeful set of actors" (Ley 1981: 219), the actors here being the Polynesians and other local residents together with representatives of the French government. Foreign visits do reinforce Polynesians' pride in their homes, but the goals of the French government and Polynesian interests have sidestepped the need for tourism to become the chosen path of economic development. Tourism has thus only minimally reconfigured the region's cultural and physical endowments. The myth, thus perpetuated, can continue to be exploited elsewhere.

13 Uncertain images

Tourism development and seascapes of the Caribbean

Janet Henshall Momsen

The word 'Caribbean' conjures up Kodachrome images of azure seas with matching skies framing green palm trees along unblemished white sand beaches, awaiting Robinson Crusoe's footprint. It is a picture unchanged since the first white tourist, Christopher Columbus, wrote home from the Bahamas of vegetation lush like that of Andalusia in April, of large flocks of parrots, of sweetly singing birds and plentiful, exotic, heavily laden and aromatic fruit trees. From Cuba, he wrote to his patron King Ferdinand in 1492, "Sire, these countries far surpass all the rest of the world in beauty" (in Watts 1987: 1). Thus the region's first publicist sold the image of an Edenic, unspoiled paradise to attract investment and visitors half a millennium ago. Little since has diminished tourists' fascination with islands (King 1993). As the Acting Prime Minister of Trinidad and Tobago, the Hon. John Humphrey, said, "It may be argued that as it is now perceived, the Caribbean is the best brand name among tourism destinations." Yet the geography of the brand name is not entirely clear.

Understanding the complexities of Caribbean geography depends in part on understanding its colonial history. The four major colonial powers, the Spanish, British, Dutch, and French, and even the Danish, all had spheres of influence in the Caribbean (Figure 13.1). They transformed the islands into sugar plantations geared to the evolving world economy, and left a legacy of dual economies in which plantation crops remain the leading regional product while the internationalized tourism sector has become the leading foreign exchange earner. Historic agricultural production depended on intensive migrant and slave labor, and as a consequence, the contemporary Caribbean is the most densely populated part of the Americas. The region's landscapes reflect these conditions: often impoverished, densely populated island interiors contrast with coastal zones of more diversified economic development and sojourning tourists, overwhelmingly focused on the beach. The touristed landscape of the coastal zone draws visitors away from the social and economic realities of the islands at large and so the tourist gaze focuses on the 'seascape.'

This chapter explores several points of intersection between landscapes and seascapes, tourist imagery and tourism economy. For industry professionals, this

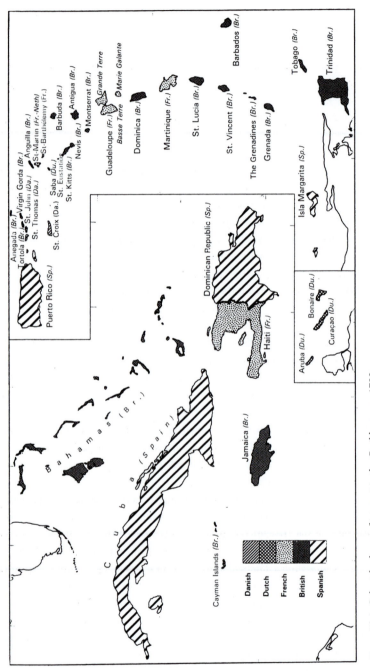

Figure 13.1 Colonial spheres of interest in the Caribbean, ca. 1780

Source: Stinchcombe (1995)

discourse focuses on the meaning of the region's 'brand name.' Jean Holder, Secretary General of the Caribbean Tourism Organization (CTO) (which has 34 members including mainland countries of Suriname, Guyana, Venezuela, and Mexico), said that

> the Caribbean tourism brand is one of the world's most sought after, and a most enduring product. It came as no surprise to me when some of our colleagues in Central America informed me that to be considered a Caribbean tourism product was something that they very much desired.
>
> (CTO 1998a: 26)

In Mexico, for example, in 1967 the Federal Program for Tourist Development adopted the Caribbean brand name as a deliberate strategy in order to encourage growth in the Yucatan peninsula (Mariñez 1996). The Program tapped the lucrative market for Caribbean holidays in this peripheral-region-with-tourism- potential: the Yucatan coasts have climate and littoral resources similar to Caribbean island resorts, especially white sand beaches and coral reefs, archeological sites, and prox- imity to the eastern and southern United States (Lee 1978). Thus the coastal village of Cancún, which had 600 inhabitants in the 1960s, was established specifically as a 'sun, sand, and sea' tourism growth pole for the poor and thinly populated territory of Quintana Roo, which became a state of Mexico in 1974. Cancún subsequently experienced dramatic in-migration and has become one of the hemisphere's leading destinations, attracting over a quarter of Mexico's tourists.

Clearly the Caribbean brand has a strong positive identity for tourism and tourism development. It is a leading world destination for 'sun, sand, and sea' or '3S' tourism. But is this brand any more than a set of signifiers? For some trav- elers, realities of place seduction in the region can belie the official marketing discourse. A few years ago, in Cahuita on the Caribbean coast of Costa Rica, as I savored the marijuana smoke and reggae beat emanating from every little bar, a local African–Caribbean resident told me that "tourists come here for the reggae." We both knew that they also came for the sun, white sand and warm sea, the hair- braiding, the West Indian food and, above all, for the drugs. However, Costa Rica's national tourist board discourages visitors to the Caribbean coast, spreading the word, erroneously, that it is an area of high crime. In this case, locals and critics interpret the government's warning as a reflection of institution- alized racism as Costa Rica's Caribbean coast was historically populated by black migrants from Jamaica and other islands and has a distinctively Afro-Caribbean culture. Has the 'branding' of the Caribbean contributed to constructing the region's places as versions of staged authenticity for the tourist imagination (MacCannell 1976), while attempting to marginalize others, like Cahuita, off the map? But first, another set of images the tourism industry might not want you to know about.

The fourth 'S' of Caribbean tourism

Increasingly this '3S' Caribbean brand image has incorporated a fourth 'S,' for sex, in a distinctive form. Sex tourism, in which poor and usually female young people provide sexual services for a fee to male foreign tourists, is one extreme form of relationship between host and guest. Prostitution of young women and even children in tourist areas is a major source of income for some of the local population, especially in the Spanish-speaking countries. This is an important factor for tourism in the Dominican Republic and in Cancún and is once again growing in Cuba as dollar-spending tourists become more common (Kempadoo 1999). In many of the Dutch islands of the region women from the Dominican Republic and Venezuela engage in these services (Martis 1999).

However, in the English-speaking territories, the specifically Caribbean phenomenon of 'beach boys' has emerged and is spreading even beyond the region. Relationships between local young men and visiting women have been termed 'romance tourism' by Deborah Pruitt and Suzanne LaFont (1995) rather than sex tourism, in that they may occasionally lead to longer term, post-vacation links. Rather than being motivated by immediate economic benefits, the men involved may be looking for a visa to enable them to emigrate or the status of sexual experience with an exotic 'other.' Joan Phillips (1999) also notes that men in sex work in Curaçao and St. Maarten see themselves as having a romantic liaison with a tourist, not as being prostitutes. In the Dominican Republic such men are known as 'Sanky-Pankies' (Cabezas 1999). They capitalize on the desire for 'exotic racialized encounters' and claim that the characteristics desired by tourist women are blackness, dreadlocks, youth, fitness, and dancing skills. Heidi Dahles (2002), in work on Indonesia, identifies how this cultural form is globalizing. Young Indonesian men in tourist areas have modeled their approach on this Caribbean style, especially on forms of dress and mannerisms including the so-called Rastafarian hairstyle. But young men's interest in foreign women causes resentment among local women and is seen as damaging to community sexual mores (Momsen 1994). It has also become common enough that tourists report it as harassment and a negative factor in visitor surveys (Barbados Ministry of Tourism 1997).

Despite this growing interest in a specific type of sexual encounter, tour operators' brochures clearly target heterosexual male tourists, or at least the patriarchal family, in that the icons, signifiers, and symbols used to mark the touristed landscape are overwhelmingly of women with perfect figures wearing skimpy swimsuits—no portrayals of muscular men are to be found (Kempadoo 1999). "That pleasure is associated culturally with women, rather than men, enables women's bodies—specifically, the image of the 'woman in a swimsuit'—to signify the particular pleasures of the beach holiday"

(Marshment 1998: 31). Sexual objectification of women is found in many aspects of the industry with female hotel employees in particular being told how to dress and comport themselves in much greater detail than is true for male employees. Women in the hospitality industry may be expected to flirt with guests in order to encourage additional consumption at the bar, for example (Griffiths 1999). The 'managed heart' identified by Arlie Hochschild (1983) is also a stereotype that symbolizes how women appear to be particularly suitable for providing a range of people-connected hospitality services. But the worker is also an agent in her own right: even menial hotel work may offer a pleasant air-conditioned environment, a smart uniform, the possibility of tips and small gifts, and the opportunity of meeting people from many different places; so it may be more attractive and pleasurable (and possibly safer) than the conditions of work in field or factory. Thus the landscape created as desirable and enticing for tourists may also provide aesthetic and social satisfaction for those whose work constitutes its production, even as sexuality constitutes an element of gendered economic relations used to promote enjoyment of the touristed landscape.

The development of Caribbean tourism

The globalization and democratization of tourism began during the 1960s, when a newly affluent middle class endowed with ever increasing amounts of disposable income and leisure time started to take advantage of relatively cheaper air fares, propelling tourism to its current status as the world's largest industry. Between 1960 and 2000 world international tourist arrivals grew from 69 million to 697 million (CTO 2000, WTO 2001). Stayover tourist arrivals in the Caribbean have generally grown even faster than the global rate, increasing from 4.2 million in 1970 to 20.3 million in 2000 (excluding Cancún and Cozumel), while cruise passenger arrivals rose from 1.2 million in 1970 to 14.5 million in 2000 (CTO 2002). In general, tourism growth in the Caribbean has fared better than the world average. On the whole, visitors are willing to pay for an up-market quality product permitting an annual growth in visitor expenditure of some 9.5 percent in the 1990s with average spending 31 percent above the world average, second only to average visitor spending in North America (CTO 1998a: 26). For many Caribbean countries tourism is the strongest and fastest-growing sector of the economy. The region is a 'must see' destination for tourists with high disposable incomes.

In addition to the foreign exchange earnings from tourism, the industry employs a major portion of the Caribbean labor force. It is estimated that in the region as a whole over 300,000 are employed in tourist accommodation establishments and if those indirectly employed are included, such as taxi drivers, sports operators, souvenir makers, and salespeople, then the total number of jobs generated by the tourism sector would be almost 900,000 (CTO 2000). These figures lead to the conclusion that 'by almost any economic indicator, the Caribbean is four times more dependent on the tourism industry than any other region of the world' (CTO 1998a: 5). This dependence is complicated by the fact that, because of the high level of imports needed to meet visitor demands, there is considerable 'leakage' of receipts from tourism. Furthermore, "like it or not, this dependence grows daily as other economic sectors are marginalized by fast-moving global political and economic events" (CTO 1997a: 8).

Tourism carrying capacities

The physical environment of the Caribbean adds its own layers of complexity. The region's islands form two general groups, the Greater Antilles, which encompasses the larger islands to the west, from Cuba to Puerto Rico, and the Lesser Antilles, which are further divided into two different subgroups. The most common grouping is the Leeward and Windward Islands: the Leeward Islands range from the Virgin Islands to Guadaloupe or Dominica, while the Windwards are to the southeast—exposed to the northeastern trade winds—from Dominica or Martinique to Barbados and Grenada. The other grouping treats the Lesser Antilles as two arcs across the eastern Caribbean. The more eastern of the two arcs, running from Anguilla and St. Barts through Barbuda, Antigua, and Martinique to St. Lucia, with an outlier at Barbados, is the drier group of islands and thus the prime area for sunnier beaches. Volcanic activity, higher elevations, and orographic precipitation characterize the islands in the western arc, which includes Montserrat, Guadaloupe, Dominica, St. Vincent, and Grenada (Figure 13.2). All of the islands have substantial tourism economies, though the smallest and driest have generally been the most sought-after destinations and consequently the most severely impacted.

The classical, paradisiacal image of an island, replete with natural resources, is belied by the realities of human impacts. Tourism is an invasive industry with widespread deleterious effects on the environment, especially on the smallest, most accessible island destinations (Briguglio *et al.* 1996). Caribbean island ecosystems are particularly vulnerable because of their small size combined with natural hazards such as drought, hurricanes, and seismic activities (Watts 1995). Montserrat, for example, has seen very little tourism since the 1995 volcanic eruption and continuing pyroclastic flows. There is now an attempt to restart tourism based on exploiting the volcanic disaster as a curiosity factor (Sobers 1999). The demands of rich visitors who expect to be able to take frequent showers and play golf on lush green well-watered grass, put pressure on water supplies in dry islands such as Antigua and Barbados. Limited sewage systems on these islands have led to near-shore pollution and other problems (Lorah 1995). On tourism-reliant Barbados, for example, untreated sewage, which damaged the

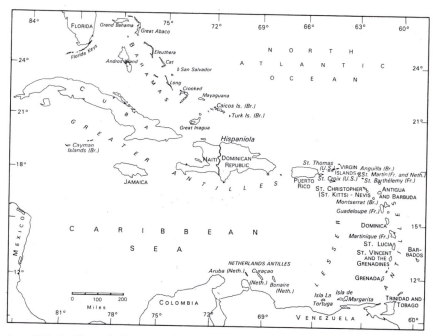

Figure 13.2 Caribbean island groups in the twentieth century.
Source: Stinchcombe (1995)

protective coral reef, has been blamed for beach erosion on its developed west coast. Beaches there were washing away at the rate of a foot a year, and in response Barbados scrambled to build more sewage treatment capacity, borrowing $7.3 million from the Inter-American Development Bank to combat the erosion problem (French 1990).

Growing awareness of the way in which tourism can destroy the very environment on which it is based has led to attempts to measure carrying capacities for the region. Caribbean islands have generally high tourism density ratios (the ratio of the number of tourists per square kilometer on any one day) and tourism penetration ratios (the number of tourists per thousand local inhabitants at any one time). Tourism penetration ratios are highest in the smallest islands of Aruba, the British Virgin Islands, and the Cayman Islands and tourism density ratios show similar patterns, being highest in Bermuda and Aruba (CTO 1996, 2000). Given the widespread focus on growth in numbers of tourists it is not surprising that most islands exhibit increases in both ratios in the era of mass tourism. However, some of the countries in the northern tier of the region with long-established substantial mass tourism industries, such as Bermuda, Puerto Rico, and the US Virgin Islands, plus tiny Anguilla, have had declining penetration and density ratios in the 1990s (CTO 1996, 2000), reflecting static or reduced numbers of visitors, suggesting that optimum visitor levels in relation to the size of both the island and the local population have been exceeded. Overall, the most mature destinations often with greatest dependence on US-based tourists are the ones where carrying capacity is reaching crisis levels, which is characteristic of the later stages of Butler's tourism life cycle model.

The tourism life cycle model formulated by R.W. Butler (1980) predicts an S-shaped curve with an initial stage of discovery of an emerging destination, a period of rapid development, promotion, and consolidation, and finally a stage of maturity and potential decline, characterized by decreasing growth and profits, high densities that alter the visitor experience and disturb the host population, and the increasing substitution of human-made attractions for degraded natural ones. "As the place sinks under the weight of social friction and solid waste, all tourists exit, leaving behind derelict tourism facilities, littered beaches and countryside, and a resident population that cannot return to old ways of life" (Pattullo 1996: 106). Jerome McElroy and Klaus de Albuquerque (1991) suggest that examples of all three stages are currently found in the Caribbean. The silver lining may be that the fear of the economic impact of the destruction of the touristed landscape may be finally forcing the Caribbean to take ameliorative action (West Indian Commission 1992).

Rescuing the landscape for and from the tourist

Tourism has been seen as a new form of plantation economy in the Caribbean in terms of its seasonality and its reliance on a narrow foreign market, and on foreign capital and managerial skills (Weaver 1988). But its effect on the landscape is very different in that tourism is concentrated in the land/seascape interaction zone and often has little noticeable impact on the interior of islands. Thus the tourist land-scape may be seen as a spatial mirror image of the traditional plantation landscape. In a touristed landscape the island land use pattern of concentric circles is typified by high levels of degradation in the outer rings of beach and coastal areas, while the core interior zones of the islands, furthest from the coast, are least impacted. There is also a further internal difference between core and peripheral islands of the same political unit with the peripheral islands, such as Barbuda and Tobago, developing a different type of tourism because of limited accessibility (Weaver 1998).

Several islands are experiencing significant environmental degradation as a direct result of tourism development. The damage to Antigua's tourist-impacted coastal zones has drawn widespread attention (Baldwin, this volume, Lorah 1995, Weaver 1988). "Antigua's beaches are being destroyed....Raw sewage and domestic wastes are unchecked. Agrochemicals, hazardous wastes and toxic wastes are not controlled. Not only is the foundation of tourism threatened, public health is endangered too" (Coram 1993: 166). Barbados is also having problems of beach erosion as hotels build groins to protect their own beaches and so interrupt the natural patterns of sand deposition. Coral reefs are destroyed by soil erosion (Momsen and Lewis 1990) as well as by overuse by tourists. Solid waste disposal and sewage treatment are demanding new methods of control. Fresh water in these coral islands comes from subsurface sources. As they are drawn down below sustainable levels the watertable falls, salt water invades and the groundwater becomes unusable, which has already happened in Antigua. By 1994 in Barbados groundwater resources were almost 90 percent used up or committed (Watts 1995). Perhaps even more worrying is the decline in water quality in the many rivers of the least penetrated, ecotourist-focused island of Dominica.

However, new directions in planning Caribbean tourism may ensure a future for the touristed landscapes of the region. After many years of intra-island competition for visitors, the Caribbean's regional tourism research organization, the Caribbean Tourism Organization, based in Barbados, is working to assist with differentiating between the attractions of individual islands. These emphases or types of niche tourism include ecotourism (Dominica), health tourism (St. Lucia), inclusive resorts (Jamaica and St. Lucia), and heritage tourism (Nevis and Barbados), while Cuba offers some of the cheapest package holidays. This new regional and post-Fordist approach to Caribbean tourism marketing, emphasizing differences and the recognition of the need for cooperation, is a reaction to the region's situation of unequal power relations of wealth, race, and class at different levels. This can be seen at the world economy level between core nation hotel and travel companies and peripheral host nation's tourist boards; at the state level between small, resource-poor, tourism-dependent islands and larger, wealthier Caribbean countries with more diversified economies; and at the local level between predominantly poor, black rural hosts and rich, white urban guests. The CTO now stresses sustainable tourism which it defines as "the optimal use of natural and cultural resources for national development on an equitable and self-sustaining basis, in order to provide a unique visitor experience and an improved quality of life through partnerships among government, and private sector and communities" (CTO 1998b: 1). More and more hoteliers are also responding to advocacy of sound environmental principles after 1992 and the Rio Earth Summit; in 1997 the Caribbean Hotel Association agreed to introduce the Caribbean Alliance for Sustainable Tourism (CAST) which will enable it "to set, encourage and monitor environmental standards for the Caribbean" (CTO 1997b: 10). These environmental standards concern impacts on the islands; sustainable tourism of the Caribbean 'seascape', however, has yet to be established.

Seascapes

An increasingly distinctive aspect of Caribbean tourism is the touristed seascape, with the scattering of islands as mere background. The sea may be fast becoming the most touristed part of the region: the Caribbean Sea attracted almost half of world cruise capacity in 2000 (CLIA 2000) and provided over half of all scuba diving tours (CAST 2000). Cruise operations are one of the most rapidly growing sectors of international tourism and the Caribbean is the dominant world destination (Dwyer and Forsyth 1998, CTO 2000). The region's share of total world cruise capacity was 53.6 percent in 1995 (compared to 2.6 percent of total international tourists) (CTO 1996), falling to 45.9 percent in 2000—although the capacity in the Caribbean increased by 37 percent over the same period and by 130 percent between 1987 and 2000 (CLIA 2000). In 1999 there were 20.3 million stay-over tourist arrivals and 12.1 million cruise passenger arrivals in the Caribbean but only 10.3 percent of total tourist expenditures were by cruise passengers (US$825 per tourist versus US$159 per cruise passenger visit) (CTO 2000).

Many island governments have built special cruise passenger facilities in order to attract cruise ships. These consist largely of duty-free shopping outlets and differ little from port to port, encouraging cruise passengers to shop without moving more than a few yards from their floating hotel. Even islands such as Dominica, which is often considered the ecotourism island *par excellence*, have sent mixed messages by building a new cruise terminal while the local hotel industry was intensifying marketing of nature tourism (Sharkey and Momsen 1995). Dominica completed a cruise ship port in the Cabrits National Park in 1991, and between 1990 and 1994 cruise passenger arrivals rose from 6,800 to 125,541, when they made up 69 percent of all tourists (CTO 1994, Sharkey and Momsen 1995). Building such facilities is very costly for small island governments, and while numbers of cruise visitors have increased, cruise tourist receipts have not yet proved substantial. For example, the Pointe Seraphine cruise terminal in St. Lucia attracted 356 cruise ship calls in 1999 compared to 218 in 1988. The number of passengers visiting the island rose from 79,500 in 1988 to 351,233 in 1999 (57 percent of all visitors), but the expenditures of cruise passengers on the island amounted to only 5.5 percent of the total tourist expenditures (St. Lucia Tourist Board 2000). In order to improve their returns on their investment many island governments attempted to increase the per capita charges for cruise ships; the response of the cruise lines was to eliminate that port from their route. Furthermore, many cruise lines are now investing in private islands where passengers can play on the sand and snorkel without having to deal with customs and immigration or a non-American society. Increasingly passengers are staying on board during port stopovers or returning to the ship to eat rather than taking an island tour. The ship is the holiday, a place where tourists are coddled and protected; the specific geographical location of the cruise and the landscapes to be visited are of minor significance. The ship provides the view, and the islands of the Caribbean are the background, the focal points of the seascape.

Decline or rejuvenation?

Cruises may be seen as a way of avoiding the degraded landscapes of the Caribbean while retaining the heliotropic visions of escapism associated with the region. But cruise tourism brings few financial benefits to the countries of the region as the cruise lines are foreign-owned and tourists spend little ashore, if they even bother to set foot on land. As the ships increase in size and cruises become more frequent, even the seascape is being degraded. Several cruise lines have been prosecuted recently for dumping waste at sea but it is doubtful that this practice has completely stopped; tar balls from ballast washings at sea have been a plentiful component of marine debris on beaches in St. Lucia and Dominica (Corbin and Singh 1993). Smaller scale use of marine resources by tourists is less damaging on the whole, and independent sailing and bare boat rentals in the Virgin Islands and Grenadines especially are growing rapidly. In the western Caribbean, location of the second largest barrier reef in the world, snorkeling and scuba diving are a major attraction. The Caribbean region is the world's premier diving destination, supporting 57 percent of all international scuba diving tourism (Green and Donnelly 2003: 143). It might be assumed that those involved in this type of nature tourism would be more protective of the environment, but attractive pieces of coral—not a 'rock' but a live animal—are often broken off and taken home as souvenirs. Intensive diving pressure at some sites has led to increased numbers of physically damaged coral colonies (Hawkins and Roberts 1993). Thus the subsurface seascape is also being degraded.

In response to this, many Caribbean countries are looking for stability or rejuvenation. The island of Aruba, which has seen a very rapid growth of tourism over the last two decades, has declared a moratorium on the building of new tourist accommodations. Some of the best-known older hotels, such as the Sandy Lane in Barbados, are no longer leaning on their laurels but are closing for long periods in order to carry out major and very expensive renovations. Alternatively, exclusive resorts are opening on private islands in the Grenadines linked to major airport hubs by small planes, which connect directly with regular flights from New York and London. In this way the land holiday is copying the isolated position of the cruise ship, providing the elite tourist privacy, protection, and pampering. This type of tourism includes a fifth 'S'—for service and servility rather than sex or shopping. Exclusive resort tourism in the Caribbean is a direct attempt to copy the success of cruise ships' all-inclusive resort style, offering "many of the same product and marketing advantages that cruise ships have enjoyed in being able to deliver a product tailored to the purchase criteria of the target market segment" (CTO 1997a: 16).

Origins of the enclave

The all-inclusive resort was pioneered on the north coast of Jamaica as a way of protecting the tourist from what was perceived as a dangerous local environment. As early as the 1970s, Jamaica's enclave resorts had become well established:

> The traveler to Jamaica…not only gets away from it all but from the island as well. The apparent mission of the hotels is to insulate the tourist from the

political and economic realities as thoroughly as possible and create the illu-
sion of an island within an island.

(Loercher 1980)

But what is 'dangerous' for tourists has also been local protest by residents against
the state. In one well-known incident in 1976, opponents of government structural adjustment policies blockaded the road to the airport, unnerving tourists
(*The Washington Post* 1979). This event and many others fueled Jamaica's reputation
as a place both luxurious and dangerous. As one travel writer expressed it:

> Jamaica's highlife is notorious for its luxury, decadence and danger. The hills
> along the north coast heaving up from the cerulean sea between Port Antonio
> and Montego Bay are dotted with sumptuous villas for the international
> glamoratti, who, once safely out of Sangster Airport, seldom inflict them-
> selves on the general life of Jamaica.
>
> Their estates, often in enclaves such as Round Hill and Tryall, are self-
> contained and hermetically sealed with guard posts, security crews and
> buffer zones. Shopping is done by the cook, chores are done by the maid,
> transportation is the province of the driver. The only exertion expected of
> vacationers is to call for service. Otherwise, except for the occasional athletic
> enterprise or sexual escapade, life in the villas is a dreamy confection of indo-
> lence and self-indulgence.
>
> (Robinson 1991)

Columbus first landed on the north coast of Jamaica in 1494 between Ocho Rios
and Montego Bay, where today there is a concentrated zone of enclave resorts. In
postcolonial Jamaica, resort beaches have been treated as private property and
closed to local access, while others demand access fees unaffordable for most
locals. Not until 1998 was the issue of access to the beach publicly debated, in a
movement led by the Jamaica Conservation Development Trust, the country's
largest non-governmental environment organization (Pragg 1999). However, the
Jamaica Hotel and Tourist Association and the enclave resorts on the north coast
objected to the open access policy, which they viewed as a threat to the privacy and
safety of their guests. Legally, under the Beach Control Act of 1956, the public has
unrestricted access to only 'public recreational beaches,' but 'public' beaches are
generally poorly kept, with only 12 of 85 being safe to use, and even these are
controlled by commercial operators who charge for access (Res & Co. 1999).
Hoteliers pay an annual fee for exclusive use of their beachfronts. The coastal
access problem continues to generate public hostility toward tourism in Jamaica,
although the tourist penetration ratio is much lower than elsewhere in the region.

The all-inclusive resort concept has spread to many other islands, and has
attracted increasing numbers of tourists. Islands that encouraged the growth of

Club Med, Sandals and other types of all-inclusive resorts have seen dramatic increases in visitor volume but relatively small gains in their local economies. These resorts regularly import various tourist goods and products and do not form links with local suppliers; thus the bulk of profits 'leak' outside the country to international resort owners and airlines. Such resorts are self-contained transnational economic enclaves; everything from shopping and banking services to sailboating is provided at the site. Thus 'paradise' becomes a commercialized microcosm of the mythic island, maintained at significant cost and in diverse ways.

Conclusion

While the arrival of direct flights from North America and Europe to several Caribbean islands in the 1960s laid the foundation of mass tourism, and development of direct air access to various small islands stimulated the spread of vacation tourism (Mowforth and Munt 1998), some industry leaders now see accessibility as a negative externality because it encourages mass tourism. As David Harvey (1989) convincingly argues, time–space compression leads to heightened awareness of difference and an interest in local environments, as well as the global marketing of images as part of tourism. Now the most seductive landscapes of the Caribbean, desirable because they are still unpolluted by mass tourism, are those available only to visitors with the social contacts and the financial power sufficient to command access. These conditions remind us that nature historically became hegemonized as a social construction of scenery and views for the visual consumption and pleasure of the leisured class (Urry 1995). In these landscapes the poor were utterly absorbed into the aesthetics of the view. It was the elite who had the mobility and culture to enjoy different landscapes. A lifestyle of winter escapes to the Mediterranean in search of sunlight, warmth, and landscapes unpolluted by industrialization had been extended to the Caribbean by the 1930s. Today, the apparently pristine dreamscapes made so well known by such early Caribbean second-home owners (Momsen 1977) as Noel Coward and Ian Fleming, both in the Ocho Rios area of Jamaica, are once again offering refuges to the rich. The all-inclusive seascape from the ship, and the elite, relatively inaccessible island, offer vantages from which the local people have been removed except as smiling suppliers of the sybaritic desires of the holidaymaker.

14 The contested beach

Resistance and resort development in Antigua, West Indies

Jeff Baldwin

...as you get dressed, you look out the window. That water—have you ever seen anything like it? Far out, to the horizon, the colour of the water is navy-blue.... From there to the shore, the water is pale, silvery, clear, so that you can see its pinkish-white sand bottom. Oh, what beauty!... You see yourself taking a walk on that beach.... You must not wonder what exactly happened to the contents of your lavatory when you flushed it. You must not wonder where your bathwater went when you pulled out the stopper.... Oh, it might all end up in the water you are thinking of taking a swim in; the contents of your lavatory might, just might, graze gently against your ankle as you wade carefree in the water.

(Kincaid 1988: 12–14)

On Antigua, tourism is about beaches, fine white strands met by gentle blue warm seas; little in the way of nightlife, adventuring, wild areas, or historic sites draws tourists' attention away from the sensual and playful experience of the tropical beach. But as Jamaica Kincaid lets us know about the Caribbean landscape ideal, Antiguan beach landscapes, like other touristed landscapes, are characterized by a certain dissonance between the visible and the invisible, the unexamined and the manifest. Apparently natural, they are products of social relations at work in the context of ecological and physical processes—including a local political economy of sand mining. Antiguan beaches are also contested landscapes, especially sites along the island's Caribbean shore where domestic and international tourism development conflict with local ideas about land use and place. This chapter examines processes of tourism development on Antigua through conflicts over coastal land use and changing geographies of tourism, which have increasingly targeted mangrove wetlands and historic subsistence lands as sites for new resort development. The analysis begins by examining the wider social processes that have contributed to the Caribbean becoming a tourist destination, and then turns to focus on the particular spatial processes at work in Antigua and its contested landscapes, as produced by the state, local elites, foreign developers, and Antiguan residents. In the final section I examine the discursive and material practices of various actors in conflict over a nascent resort development, the Asian Village Resort, and how the conflict 'inter-acts' with cultural constructs of Antiguan 'selves' and the debate over appropriate use of Antiguan coasts.

The conceptual approach draws on the idea of "development landscapes" (Cartier 1997, 1999a, 2002), and concern for a social theory which encompasses agency of non-human or 'natural' processes (Wolch, West and Gaines 1995, Haraway 1991, FitzSimmons and Goodman 1998, N. Smith 1998, Wilbert 2000, Castree 2002). Along Antiguan coasts beach erosion, an ecological process, has become endemic and threatens the very foundation of tourism in Antigua; any analysis of Antiguan tourism development then must encompass these and other 'natural' processes. Toward this end I employ Doreen Massey's "progressive sense of place" (1993, 1994), a perspective that understands places to be produced by a unique set of interacting processes. The geographer's job lies in identifying all those processes, social and ecological, which come together to produce places and their transforming landscapes.

Constructing desire for sun, sand, and sea

Tourism researchers have identified numerous categories of tourism: heritage, eco, cruise, sports, adventure, hedonistic, amusement, and disaster among others. Each of these implies certain landscape ideals, and while local communities can modify places to fit the archetype of these touristic landscapes, few can significantly alter the archetypes themselves (Shields 1991: 90–1). The people who promote tourism in Antigua are but a single voice in a choir singing to the global circuit of sun, sand, and sea consumers. They rely upon dynamic cultures of meaning that have constructed tropical beaches as desirous, seductive places.

Beaches have represented different meanings in Western culture over the past 2,000 years. 'The beach' has been: a place of stoic contemplation for Classical Romans (Corbin 1994), the opaque reminder of God's wrathful flood (Tuan 1979: 13, Landon 1982), an eighteenth-century site for Romantic sublimity (Nash 1982), and a destination for healthful hydrotherapy at British and Baltic Sea coast resorts (Corbin 1994: 64). By the mid-1800s beaches were becoming places for cooling escape and amusement for industrial workers in Britain and the northeastern United States (Lencek and Bosker 1998, Towner 1996, Urry 1990). Tropical beaches became far more desirable places through the 1920s. Prior to that time, exposure to tropical sun was considered harmful and tanned skin marked one as a laborer. Following World War I, though, sanatoria on the Mediterranean coasts claimed to be effecting cures among wounded soldiers through heliotherapy, or sun treatment (Blume 1992). The redefinition of tanning as healthful was reinforced by the interiorization of factory workers, so that tans came to be recoded as 'leisure,' healthful, and upper class. European elites embraced and popularized tanning. In the 1920s on the French Côte d'Azure,

> The sun was *new*.... The sun covers wounds and provides a carapace against
> new ones: it makes one feel invulnerable.... [However] it took names to
> make a habit chic.... Then in 1923 Coco Chanel descended the gangway of
> the Duke of Westminster's yacht, brown as a cabin boy.
>
> (Blume 1992: 72)

Whether the result of heliotherapy or the embodiment of cultural and actual
capital in wearing an unseasonable tan from a trip to sunny shores, tropical
beaches became desirable places.

Desiring the Caribbean

Before Caribbean tourism operators could entice North American and European
tourists with their beaches they had to overcome significant obstacles posed by
image and distance. Though the new slogan of the Antiguan Department of
Tourism is "Antigua: the Caribbean as you've always imagined it" (ABDT 2002),
the North Atlantic imagination of the Caribbean has not always been paradisiacal.
Historic experiences of Americans, French, and British in the Caribbean and the
Panama Canal Zone were marked by malaria, yellow fever, and death. Only at the
end of the nineteenth century, after public health programs focused on eradicating
the mosquitoes *anophelinae* and *Aedes stegomyia* as the vectors of malaria and yellow
fever, could the travel industry successfully promote the healthful Caribbean environment (Taylor 1993: 4). By 1915 Havana hosted 72 hotels supported by a
growing American desire for sunny winter vacations (Pattullo 1996: 9).

The problem of accessibility was overcome largely by American operators.
Henry Flagler pioneered the integration of railroad and steamship companies
with tourist accommodations in Nassau, the Bahamas (Taylor 1993: 4). Early in
the twentieth century the United Fruit Company, the Imperial Direct Line, and
Elder, Dempster and Company each began to add passenger traffic to their
already established sea freight businesses (Taylor 1993: 84, 128). In the 1930s,
commercial air traffic further compressed the distance between these established
resort areas and eastern US population centers.

As a region, the Caribbean offers numerous distinctions that have set it apart
from sun, sand, and sea destinations in the Mediterranean, the Indian Ocean, and
the Pacific. The Caribbean comprises 26 states, overseas departments, and territories. The Caribbean is close to the primary sources of international tourism in
North America and Europe. Though exotic (particularly in its promised
encounter with Creole–African cultures), most of the region's states are English-
speaking as a result of its British colonialism. Perhaps the region's best known
cultural markers lay in its musical forms. Reggae (originally from Jamaica in the
northern Caribbean), soca (from Trinidad and Tobago in the southern Caribbean),
and steel pan all are closely associated with the Caribbean, and inform Northern

imaginations of the region. Calypso, first popularized in Western culture by Harry Belafonte's 1956 hit, "Day O," is a pan-Caribbean musical folk form. Most islands have annual calypso competitions as a central feature of their carnival, which state tourism departments throughout the Caribbean moved from their traditional pre-Lent time to the summer months, in an attempt to bolster tourism during that traditional low season.

Developing Antiguan tourism

Prior to the advent of Antiguan tourism, the colonial sugar economy dominated the island. Fueled by African slave labor, the sugar wealth accumulated in the British West Indies approximated the value of all gold and silver extracted from the Americas (Thomas *et al.* 1994: 29, Brown 1963: 40). In the first 80 years of plantation production (1665–1745) Antigua returned three times as much profit to Britain as did all its North American colonies combined (Smith and Smith 1986: 18).

However, by the 1930s, depressed sugar prices, unemployment among cane workers, and decreased remittances from Antiguans overseas created "near famine conditions" (Tunteng 1975: 37, Smith and Smith 1986: 18). In 1929, in the midst of political and economic turmoil, the Antigua Trades and Labour Union (ATLU) formed in order to increase the power of agricultural labor in wage negotiations with plantation owners. In 1943, Vere Bird, a 34-year-old rural organizer, wrested the ATLU from the control of a group of black professionals (Coram 1993, Midgett 1984: 37–8). As British colonial power waned in the Caribbean, Bird worked to assume state leadership; he "sought to dominate his island, and has been able to, despite major corruption scandals, because no Antiguan, even in the opposition, can forget how he got them their nationhood" (Kurlansky 1992: 292). The loyalty and political capital ascribed to Bird's legacy continues to legitimate domination of the state apparatus by the Antigua Labor Party; and Lester Bird, Vere's second of three sons, and served as Prime Minister from 1993 until 2004. The legitimacy of the national leadership, based in the independence movement, has fueled a political dynasty, whose entrenched leadership has guided the development of Antigua's postcolonial economy toward a new kind of dependency—on tourism.

As Bird gained control, the Antiguan government began to pursue tourism development as a way out of dependence upon the failing sugar industry. Government tourism operators emulated a set of relationships established by the successful Mill Reef Club. Founded in 1947 by a group of North American investors, the Mill Reef has always been an exclusive enclave (Midgett 1984: 35, Coram 1993: 36). As such, the Club has carefully regulated both access to and flows of wealth: working-class Antiguans are uniformly excluded from the resort, and people "can recite the name of the first Antiguan (black person) to eat a sandwich at the [Mill Reef] clubhouse and the day on which it happened" (Kincaid 1988: 27). The Antiguan government encouraged additional foreign investment and ownership in hotel development, and in 1951 the Legislative Council passed the first of many hotel aid acts promising tax holidays and free repatriation of profits to foreign investors (Midgett 1984: 46, CTRC 1978). Today, of the 2,630 beach-side rooms, 88.4 percent are foreign owned. This degree of foreign ownership is typical of mass tourism destinations throughout the insular tropics (Britton 1991, 1999), and its infrastructures, managed by culturally similar others, help tourists avoid unpleasant or dissonant encounters with exotic Others (Urry 1995: 135–8).

Antigua has successfully overcome accessibility constraints and has become a 'mature' tourism site. Tourism is the center of a national economy, directly accounting for about 60 percent of all economic activity. Prior to the World Trade Center attacks in 2001, Antigua annually hosted approximately 200,000 overnight guests and about a quarter-million day-trippers from cruise ships (ABDT 2000, CTO 1996). The Antiguan tourism industry could never have achieved this magnitude of flow were it not for its airport. Antigua's nascent tourism industry gained an early comparative advantage over its neighbors when, in 1951, the US Navy granted commercial airlines the use of a modern air field (Coram 1993: 133). By the end of the 1960s, only Jamaica and Puerto Rico were better connected than Antigua. By comparison, island states that had poor connections in 1968 are generally characterized today by less 'mature' tourism industries marked by small-scale resort accommodations and low magnitudes of tourist visits.

Antiguan operators have also worked to build distinction on the basis of popularity of local beaches and yachting activities, both of which have deep historic roots. The Antigua Hotel and Tourism Association has as its long-standing slogan: "365 beaches, one for every day of the year." Though Antiguan beaches have themselves become problematic, in ways which I discuss below, the island's geologic processes have in fact produced a plethora of the fine sand strands essential to sun, sand, and sea tourism.

Antigua is geologically older than its volcanically active neighbors to the west and south. As a result, the Antiguan coast is characterized by shallow seas and mature coral reefs which both protect beaches from wave erosion and supply them with the fine white sediment, its famous sands (Multer, Weiss and Nicholson 1986). Like many of its smaller neighbors to the north, Antigua is generally a low island. Lacking a high central mountain range, Antigua also lacks the orographic clouds and rain common to its younger neighbors. As a result, Antigua is less rainy and sunnier than many other islands and its resorts benefit from visitors' historic experience with regional weather, which leads to satisfaction and return visits.

In the late 1990s the Antigua Hotel and Tourism Association (AHTA) adopted a new marketing strategy, re-titling Antigua "the heart of the Caribbean" (AHTA 1997b). The resort landscape in Antigua fits especially well with romance tourism in two ways. First, Sandals, a luxurious "couples only," all-inclusive resort opened its newest addition in Antigua in 1995. The Jamaican-owned resort chain has successfully developed the "WeddingMoon(tm)," a week-long stay which rolls a tropical beach vacation, a wedding with all the trimmings (base price $1,500), and a honeymoon into a single package (Sandals 2002). The Antigua Sandals won the World Travel Award for Leading Honeymoon Resort in 1997, 1998, and 1999 (*ibid.*, *Daily Observer* 1997a). The second factor that characterizes Antigua as a quiet, romantic getaway is the lack of nightlife. Unlike Ocho Rios and Montego Bay in Jamaica, or the resorts in the US Virgin Islands, San Juan, Havana, and St. Martins, Antigua's resorts are spread out along its coasts, and each has remained a more or less quiet enclave. As a result Antigua has historically appealed to a clientele characterized by the President of the AHTA as "the newly wed and the nearly dead" (Ramrattan 1997).

Seeking to diversify its market niches, the Antiguan Tourism Department has re-focused promotional efforts on tourism related to yachting, a unique and long-standing asset in Antigua. On the south coast, English Harbour served the Royal Navy as one of the finest harbors in the Caribbean. In 1967 the Antigua Yacht Club formalized an annual regatta, bolstering Antigua's position in the Caribbean yachting circuit. With renewed government sponsorship, Antigua Sailing Week has become one of the top three regattas in the world (ABDT 2002). In 1996, Antiguan ports recorded 4,540 yacht and 96 windjammer visits (ABMI 1997a). Yachting draws both participants and spectators and conveys a certain elite distinction to Antigua.

Negotiating tourism development on Antigua

As suggested above, the state has been central to tourism development in Antigua: it assists foreign investors, participates directly in specific projects, and sets the regulatory tone for resort development. As a democratically elected government with historic roots in the labor movement, the Antigua Labour Party (ALP) has mediated the often conflicting interests of foreign investors, unionized resort laborers, and local citizens. While much of the literature on Third World and Caribbean tourism development represents sun, sand, and sea destinations as passive (Britton 1991, Mecker and Tisdel 1990) or even feminine (King 1997, Nair 1996), and North/South tourism as flows between states, organized by globally scaled firms (Britton 1999, 1982), the case of Antigua demonstrates how significant local agency characterizes tourism development.

As institutionalized leaders of the ALP, the Bird family have embodied the nexus of power in the Antiguan tourism industry. They have used the full extent of the resources available to them to seduce foreign capital. They have borrowed from various development agencies and individual investors and built roads, an airport, harbors, and electrical and communications networks—the local infrastructures upon which mass tourism depends. These two groups of specific actors, the Birds and foreign investors, have thus worked to constitute one another; each offering the other what they desire in exchange for what they already possess. By controlling access to their desired places, the Birds have employed the rich appeal of Antiguan beaches to satisfy labor and foreign investors and enrich themselves.

Labor and tourism

In Antigua, the origin of the political regime in organized labor presents a certain contradiction. The government has simultaneously addressed the interests of its union constituency and those of resort owners, two groups that typically vie against one another to raise wages and to cut labor costs. In the late 1960s, as union members became increasingly concerned about this conflict of interests, Vere Bird created the ALP separate from the ATLU (Tunteng 1975). Over the past 30 years, ALP governments headed by the Bird family have successfully negotiated union agreements for tourism workers, which raised Antiguan wages far above those paid in the sugar industry. Partially as a result of those negotiations, the standard of living now ranks 40th of 171 countries in the United Nations Human Development Index.

The ALP has equated its tourism development programs with more and better jobs for Antiguans in an effort to retain electoral support. Though Antigua has no official employment statistics, the government publishes quarterly statistics on visitor nights and cruise passenger disembarkations. The ALP widely uses these indicators of tourism growth as proxy for the number of jobs available and thus for the legitimization of the Bird administration. A declaration that "The People Want JOBS…Give The People What They Want" on the back cover of an 'educational' brochure produced in support of a proposed large resort demonstrates the equation of new resorts with new jobs and further support for the government (ABMI 1997b). This ideological conflation of tourism, jobs, and government legitimacy is endemic in the ALP discourse, as exemplified in this ALP editorial:

> As a matter of fact, the former Prime Minister and Father of the Nation, usually dismissed the destructive and delaying tactics of the elitist opportunists.… Lester Bird and his government would do well to remember they were elected to govern this country, to run things, to find work for people.… Nowhere in the civilised world would anyone demonstrate against the creation of jobs for people.
>
> (*National Informer* 1997: 9)

This passage reinforces Vere Bird's revered position in Antiguan nationalism, then appropriates both Bird's development program and the wider project of 'civilized' modernity to legitimate further tourism development. At the same time, this position dismisses any opposition to tourism development as elitist.

But the notion of continued tourism development for organized labor is not universally accepted. As tourism has matured on successive islands, each has become a regional destination for labor migration (McElroy and Albuquerque 1988, 1991). As many as 42,000 migrants passed through Antigua as guest workers between 1980 and 1996 (Dorsett 1997: 3)—a significant flow on an island of 70,000 people. Labor leaders estimate that the 1,970 guest workers registered in 1996 represent only about one-half the actual number, and are concerned that many are non-unionized (*ibid.*, Gomes 1997b). For unionized Antiguan workers, who have benefited from the regionally high wages which attract immigration, non-unionized workers threaten to both take away jobs and drive wage levels downward. Though labor leaders are beginning to question the wisdom of job expansion for non-Antiguans, the Bird administrations have successfully managed tourism development to serve their ends. They have maintained power while at the same time retained the support of foreign capital so necessary to further expansion.

Lester's alchemy: sand into dollars

Though Vere Bird did not remain a poor man, he maintained a modest lifestyle—unlike his sons. Lester Bird succeeded his father as Prime Minister in 1993 and was re-elected in 1999 to a second five-year term. Though the Birds' notorious self-service has been well documented (see Coram 1993), two particular projects illustrate how Lester Bird has personally benefited from partnerships with foreign investors in Antiguan tourism. In the late 1980s, as Minister of Tourism and Minister of Economic Development, Lester Bird established the Deep Bay Development Company (DBDC) to oversee the construction of the Royal Antiguan Hotel, a 278-room resort purportedly built under contract for ownership by the Antiguan government by an Italian developer. In 1990, Tim Hector, a local newspaper editor and Member of Parliament, revealed that as head of the DBDC, Lester Bird—rather than the Antiguan government—was named sole owner of the Royal Antiguan Hotel (Coram 1993: 143, *Outlet* 1997e: 12).

In 1997, Prime Minister Bird pre-empted debate over a proposed development project by presenting to Parliament a signed agreement to sell 800 acres (320 hectares) of Antiguan land (over 1 percent of the island) for the 1000-room Asian Village/Guana Island resort. PM Bird asserted that the project would add 4,100 jobs and US$200 million to national wealth (ABMI 1997b: 12). The PM's personal stake in the transaction never became apparent; however, in a personal interview a local business leader confided that Lester Bird is known as "Mr. Ten Percent...you may do whatever you want in Antigua so long as the PM gets ten percent." As Lester Bird began to personally benefit from resort development in the 1980s, the scale of resort development and related environmental degradation increased significantly (Table 14.1). With the state apparatus supporting new large-scale resorts, tourism development intensified on the Caribbean coast and became increasingly contested (Figure 14.1).

Table 14.1   Resort development along Atlantic and Caribbean coasts, 1947–97

Year	Number of resorts			Rooms in region			Avg. resort size		
	Atl.	Carib.	Total	Atl.	Carib.	Island	Atl.	Carib.	Island
1947	2	0	2	60	0	60	30	–	30
1960	7	1	8	173	33	236	24	33	30
1968	18	5	23	535	304	839	31	51	36
1980	15	13	28	580	690	1270	39	53	45
1988	18	15	33	710	930	1640	39	62	50
1997	12	15	27	610	2020	2630	51	135	97

Sources: Hector (1997a), ABDT (1996, 1997), AHTA (1997b), ABMI (1997b), Coram (1993), DOS (1977), Ullman and Dinhofer (1968)

Antiguan beaches as progressive places

By contrast to state and private tourism operators, many local people do not support continued tourism development. The power and usefulness of Massey's progressive sense of place becomes apparent as we try to encompass all of the disparate social, physical, and ecological processes that together produce Antiguan beaches. I begin this discussion with a brief description of some of the ecological processes now at work in Antigua. I then address some particular social conditions that have sought to influence specific beach landscapes. In the final section I comment more substantially on the interaction between place, culture, and identity manifested in the encounter with a large new resort proposal, the Asian Village.

Coastal ecologies as agents

The beaches that provide the literal foundation for resort landscapes are linked ecologically to the health of coastal wetlands. Across the insular tropics, naturally occurring beaches depend on coral reef communities for sand input and wave moderation. Reef communities are in turn dependent upon wetland mangroves, which sequester damaging terrestrial nutrients and sedimentation and serve as nurseries for fish vital to reef ecologies (Bossi and Cintron 1990, Ogden and Gladfelter 1983). Studies in Antigua indicate that mangrove destruction and dredging related to resort construction and beach nourishment have degraded reefs and contributed to beach erosion widely around the island (Figure 14.2) (CIDA 1988, COSALC 1996, AHTA 1997a: 3, UPR 1994b, Albuquerque 1991).

Beach erosion coupled with the construction of new beaches has further increased resort reliance upon 'beach nourishment,' or the importation of sand from reefs and undeveloped beaches (Baldwin 2000; Norse 1993: 110). Though the government passed two acts to regulate beach mining in 1957 and 1959, the Public

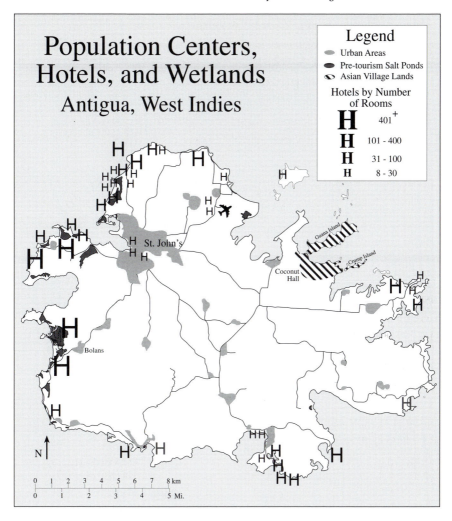

Figure 14.1 Map of Antigua showing relationship between population centers and the increased concentration of hotels along the Caribbean (eastern) coast. The extent and location of the Asian Village project near the North Sound is also indicated

Sources: ABLSO (1991), DOS (1962, 1977), Multer et al. (1986)

Works Department did not stop mining until the mid-1980s and small-scale mining continues clandestinely (Bunce 1997: 94, Baldwin 1998). With United Nations technical support, the DCA has twice generated comprehensive land use plans, once in 1975 and again in 1996, that would better regulate mining. The Antiguan Cabinet refused to approve either of the plans (Bunce 1997: 57). Antiguan development elites have relocated sand mining and its attendant problems to Antigua's sister island, Barbuda, which is nearly as large as Antigua but with a much smaller population (1,700) living primarily by subsistence (Berleant-Schiller 1991).

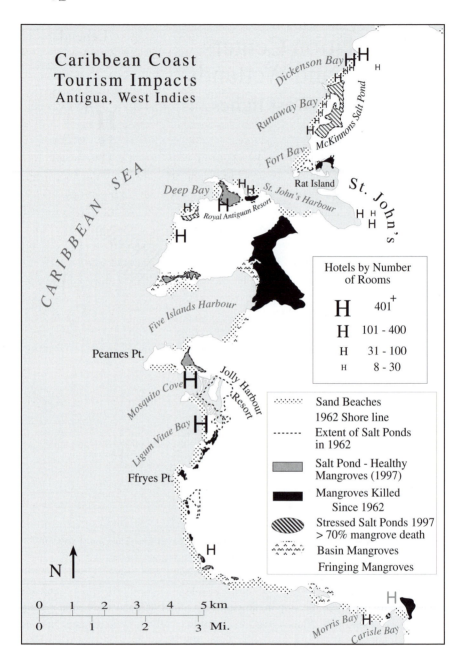

Figure 14.2 Caribbean coast tourism and resort landscape alterations
Sources: ABLSO (1991), DOS (1962, 1977), Multer et al. (1986)

Like Trinidad with Tobago and Nevis with St. Kitts, British administrators bound the sparsely populated island of Barbuda with its more populous neighbor, Antigua. Each of these islands resisted these constructed associations out of concern that they would be poorly represented in central governments and suffer exploitation (Knight 1990). Barbudans' fears have been realized in spades. In 1968 Vere Bird attempted to enter a contract with a Canadian development firm which would essentially give large tracts of Barbudan land in exchange for investment. Throughout the 1970s, for example, the Bird government attempted to sell Barbudan beaches to foreign investors, and in the 1980s the Bird government signed a contract to accept compressed sewage sludge from the United States to be dumped in Barbuda. These and other plagues failed; however, Barbudans have been unsuccessful in their efforts to stop industrial-scale sand mining. In 1979 the Antiguan Red Jacket Mines Company, controlled by Lester Bird, began mining sand from a dune field in southeastern Barbuda (Coram 1993: 105). By 1987 sufficient sand had been mined from the Palmetto Point Dune Field to diminish a rain-fed fresh-water aquifer and allow significant salt-water incursion, thus destroying the only reliable year-round source of fresh water on Barbuda, which is a notably flat, dry island (CCA 1991: 108). In 1990 two Antiguan companies, also controlled by Lester Bird and his associates, extracted 350,000 tons of sand from the dune field, and in 1994 Barbuda was identified as the single most significant source of sand in all of the eastern Caribbean (*ibid.*, UPR 1994a: 1). As sand mining has raised contention in Barbuda, resort construction and coastal alteration has made Antiguan resort beaches sites of contention.

Social resistance: economic forms

Assessing landscapes as "transformations of ideologies into concrete form" (Duncan and Duncan 1988b: 117) is a meaningful exercise in Antigua where shifts in the ideology of ALP leaders can be read in the landscape. Caribbean subsistence economies developed first under conditions of slavery and were maintained economically as well as culturally as a supplement to waged work. Faced with food insecurity in an environment of wage insufficiency and uncertainty, Antiguans and Caribbeans more generally developed 'moral economies,' informal supplemental production and reciprocal support networks (Watts 1987, Eltis 1995, Deere *et al.*, 1990, Smith and Smith 1986: 124, Besson 1987: 15–17). The landscapes where Antiguans engage in supplemental production, family provision grounds (garden plots) and wetlands, are also imbued with significant social value and meaning. These landscapes are also at stake in political debate. While union president, Vere Bird declared that until better wages come "we will eat cockles and the widdy widdy bush. We will drink pond water" (in Coram 1993: 17), i.e. Antiguans would sooner live off the land than work for near-starvation wages.

In contrast to his father's declaration 42 years before, in 1993 Lester Bird flatly stated that "Antiguans cannot eat mangroves" (Wilson 1993: 2). This statement reflects a modernist economic form: it negates the value of subsistence agency and assumes that scarcity is the result of under-production (rather than mal-distribution) and so advocates industrial production in the coastal zone, through tourism. In Antigua the tourism industry with government assistance has continued to transform these places of subsistence production into 'modern' industrial sites; into the archetypal sun, sand, and sea tourism landscape complete with imported palms and sand, overlooking dying reefs: artifice and illness largely invisible to mass tourists. As Lester Bird denies the social value of wetlands to Antiguans, the processes of resort development have destroyed places of meaning and security to many Antiguans.

These subsistence lands were especially hard hit by tourism development after 1968 (see Table 14.1). As the few pocket beaches on the remote Atlantic coast became fully occupied by small resorts in the 1970s, tourism developers increasingly focused on the more densely populated Caribbean coast. Caribbean coast resorts increased notably in terms of size, particularly in the period between 1988 and 1997. The increased proximity to Antiguan population centers and the ecological footprint of these developments is illustrated in Figures 14.1 and 14.2 respectively. As the tourism industry grew, government and resort developers increasingly focused on beaches associated with mangrove wetlands along the more heavily populated Caribbean coast of Antigua (see Figure 14.1). As a consequence between 1987 and 1994 the five largest wetlands on Antigua were destroyed, four as a direct result of government-led programs involving dredge spoil deposition, resort construction, and sewage dumping, all in support of tourism development.

Popular opposition

In its popular forms, local resistance to tourism development is poorly organized, but often well articulated and sometimes effectively manifested. "This Country," a calypso performed by Calypso Stumpy during the 1996 Carnival in St. Johns, illustrates the sophistication of local people critical of continued resort development:

> Casually we sittin' back
> An' watchin' dis drama unfold
> casually we so relax – as we selling we land and soul…
> look how we cover we white sand beaches
> wid den ugly concrete hotel
> So we fillin' up the swamp an' beachlands
> we doan give a damn bout environment
> let de fishes die by de millions
> de bottom line is dollar an' cents…
>
> (*Calypso Talk* 1996: 3)

This calypso and several others of similar stripe presented at the 1996 Carnival inspired the Bird government to pre-approve, i.e. censor, all calypsos to be presented in the 1997 Carnival. In 1997 this resistance was endemic in the Antiguan tourism discourse, expressed in other calypsos and in letters to the editor published in the two local opposition newspapers.

Antiguans have also sought to intervene in the development of coastal land-scapes through direct action. I present here two examples of such efforts in order to show the strategic nature of local agency in producing place. In 1993 Foster Derrick, an Antiguan environmental activist, organized protests against an attempted 86 acre (35 hectare) hotel, marina, and shopping complex at Coconut Hall, in the North Sound area (Pattullo 1996: 114–15). In the previous year, Italian developer Canzone del Mare presented the project to the Antiguan Development Control Agency (DCA). The DCA approved the one-page proposal without a public hearing and excavation began in late 1992. Derrick organized sit-ins to block heavy equipment and to draw public attention to the destruction of the mangroves along the coast. In response to the sit-ins, the head of the DCA eventually ordered that construction stopped (Pattullo 1996: 115). By placing their bodies around earth moving equipment, by acting strategically in place, a few Antiguans effectively stopped an international developer backed by the Antiguan government. In response to this anti-development move on the part of the DCA, Lester Bird fired the Agency's chief administrator and the DCA was transferred to the Ministry of State, directly under PM Bird (Bunce 1997: 58).

Another protest occurred in 2001 around the redevelopment of the Carlisle Bay resort on Antigua's southern coast. The Irish owner, Patrick Doherty, began construction of a new road through a mangrove, which would bypass Old Road, a nearby town. Local people appealed to the government demanding both that an Environmental Impact Assessment be conducted and that new concrete footings be removed as they were in violation of regulations (Cana Online 2002). On July 24 a group of about 100 Antiguans blocked three shipping container-loads of supplies outside the resort. Antiguan police eventually responded by firing tear gas and rubber bullets at the protesters (Anthonyson 2001). The conflict continued into 2002 as the foundations were first removed, then new concrete poured, instigating an arson at the resort, which was in turn followed by illegal searches of several of the protest leaders' homes. What is crucial in this action too is the thickness of presence each side is able to strategically manifest in place. Because the wetland is in the public domain, the government has legitimate control over the space. However, because the construction is illegal by Antiguan law, the courthouse in St. Johns may serve as the final strategic site where this conflict will be resolved.

Both of these brief cases suggest the importance of effective presence in producing place. In Oakes's terms (this volume), contestations of socially valued places are closely associated with the evolution of new forms of agency, or subject formation. Castells (1997: 8) also suggests this process of producing place and space plays a role in producing identities. He argues that people with an awareness of their subjugation, in this case of their personal losses from coastal development, may seek to redefine themselves as they transform both places and the processes by which meanings are ascribed to places. Through social movements organized around resistance identities and shared agency, people experience transformed subjectivities and create alternative power bases. In Antigua the interplay between place, meaning, identity, and tourism has reconstructed the meanings of each. This interplay is more clearly revealed in the protracted civil contestation over the Asian Village Resort project in Antigua's North Sound (see Figure 14.1).

The Asian Village: a cultural politics of place

The Asian Village resort project is now effectively defunct. It was to have been a prime example of tourism product differentiation and the production of place as simulacra (Baudrillard 1994). Malaysian developer Dato Tan Kay Hock explained that "the concept of The Asian Village is a heavily themed...integrated tourist destination, taking the best of architecture, culture, food and flavour from Malaysia, Bali in Indonesia, and Thailand" (ABMI 1997b: 22). The resort was to also feature an "Asian style water village" (*EAG'er* 1997: 1), consisting of 1,000 hotel rooms, a casino, two 18-hole golf courses, luxury homes, shopping and restaurant facilities, and housing and services for foreign laborers. Contractual provisions were also made to allow for the immigration, residency, and citizenship for approximately 2,000 Malaysian workers. In short, the Asian Village resort was to be an Asian theme park/resort located in the Caribbean. Though the development firm, Asian Village Limited, was a limited liability corporation registered in Tortola (*Daily Observer* 1997b: 2), it was a wholly owned division of *Johan Holdings Berhad* (*EAG'er* 1997: 1), a Malaysia conglomerate which was hit hard by the 1997 economic crisis in Pacific Asia.

In 1997 Lester Bird signed an agreement with Asian Village Limited that called for the sale of Guana Island, the largest island immediately off the shore of Antigua, and three smaller islands (Crump, Rabbit, and Hawes Islands) totaling 800 acres (320 hectares) of land. An additional 350 acres (140 hectares) was to be leased for golf courses on Antigua adjacent to the islands. The public discourse which ensued provides a contemporary opportunity to examine the evolving politics and ideology of tourism development in Antigua, processes integral to the social production of place. The case study is also illustrative of how place works to produce identity.

In the ways that 'truth' is as much discursively as materially constructed, in Antigua there is a "battle about the status of truth and the economic and political role it plays" (Foucault 1980: 132). In the struggle over the effective meaning of the

North Sound island lands, tourism development, and Antiguan belonging, there are two dominant points of contestation: what constitutes legitimate Antiguan national identity, i.e. whose voice is legitimate, and what is the meaning and most important use of Guana Island? In the following discussion I present representative examples of prevailing discourses from three primary groups: (1) the ALP, which seeks legitimacy in its heritage stemming from Vere Bird and job expansion, (2) the primary political opposition party, the United Progressive Party (UPP), which challenged the ALP's claim to Antiguanness for that party's dealings beneficial to non-Antiguans, and (3) the Environmental Awareness Group (EAG), which challenged continued tourism development generally by valorizing the resort-free environment around Guana Island. All sought to legitimize their 'truths' and 'meanings' on a state scale, the scale at which tourism policy and national identity is made. All three groups hoped to gain a mandate in support of their position through the 1999 general election. The ALP, led by Lester Bird, narrowly won the election; however, its legitimacy was called into question when the total number of ballots cast (52,348) nearly exceeded the entire population of the island (64,000), indicating a significant over-vote (Hector 1999). In conversations with Antiguans it became clear that the people were quite divided over the issue of continued tourism development. In 2004, UPP leader Baldwin Spencer defeated Lester Bird in a close contest.

To be Antiguan

In public discourse about the Antiguan self-image, Vere Bird serves as the exemplar of an Antiguan *ethnie* (Smith 1986), an actively constructed national identity. Like most of his generation, he was poor, dark skinned, and uneducated; he became the just and selfless father of the sovereign Antiguan state. The ALP and the UPP have both used the model of Vere Bird as a basis for debate, each seeking legitimation by representing the other as "un-(Vere) Bird-like" or as un-Antiguan. One ALP editorial (entitled "Are the elite fighting on in E.A.G. garb?") further distances both the EAG and the UPP from the exemplar by characterizing their members as "light skinned," "middle class," and "intellectual" (*National Informer* 1997: 9). The UPP has been more aggressive in representing the ALP, and Lester Bird in particular, as un-like Vere Bird. The tone and substance of these representations commonly appears in the popular press:

> Few people are more unrooted, more deracinated than Lester Bird. He studied English History at school, American history at University [of Michigan] and Caribbean History never. He studied English law in England he never acquired a sense of Caribbean jurisprudence.... He is Other-directed. The major change that has taken place in Lester Bird's life is that his local support group of his youth, has been replaced by a foreign support claque.
>
> (Hector 1997a)

Antiguans are aware that Lester Bird was not born in Antigua. On diverse occasions interviewees prefaced their response to my open-ended question "What do you think about Prime Minister Bird?" with exactly these words: "He's not Antiguan." Beyond charges of cronyism favoring non-Antiguans, i.e. non-citizens (*Outlet* 1997f), the UPP represents Lester Bird as privileging non-Antiguans to such an extent as to threaten Antiguan sovereignty (*Outlet* 1997c: 3, 1997e: 12, 1997f: 4, Hector 1997a). The Asian Village agreement served as especially fertile ground for this line of criticism, and the UPP discourse frequently linked the project to re-colonization (e.g. Hector 1997b, *Outlet* 1997g, Walter 1997b) and with white supremacy. Of the Prime Minister himself, an *Outlet* editor wrote:

> Lester Bird lived in the segregated Jim Crow United States in the late 1950's. He was no doubt infected by that deadly disease…it leaves the whole race inferior…. [into] the Asian Village, Antiguans and Barbudans, mainly black, but of whatever race, cannot go.
>
> (1997e: 8)

Through such statements the UPP constructs race, and privileges blackness as one marker of citizenship and non-blackness as signifying non-Antiguanness. To defend racially coded Antiguans against racially coded Others is thus emblematic of good Antiguan leadership.

This racial theme becomes central in UPP discourse in attempts to construct the Asian Village investors as threatening or inferior Others. In a speech delivered to a UPP street rally in 1997, Senator Lionel Gomes charged that Asian Village Agreement clauses 9.2.6 and 8.2.23, which grant unlimited work permits—and instant citizenship to all Malaysian workers (*Outlet* 1997b: 18)—would lead to Antiguans being electorally overwhelmed by Malaysian immigrants. The UPP attack often devolved into racial generalizations and the conflation of all Pacific Asians into a Chinese race. At a street rally attended by 250 Antiguans, Senator Gomes (1997a) made the racist comment "We all know da Chinyman breed like guinea pig." Within days this evolved into the slogan "Save our land from the Chinyman, dey breed like guinea pig," which was broadcast throughout St. Johns by car-top loudspeakers. Such generalized race positions continued to be articulated incorrectly in relation to China. Senator Baldwin Spencer stated:

> we want to send a national message to the people of Hong Kong and China that the Nation of Antigua and Barbuda is not for sale to anyone posing as a developer, who intends to subordinate this nation to their Asian whims and fancies.
>
> (*Outlet* 1997d: 3)

In this sometimes frenzied, discursive struggle to control what constitutes Antiguanness, development and environmental issues have been partly transformed into a discourse of racial politics and a struggle to control the definition of national citizenship, based in personal and embodied histories of particularly powerful individuals. The contest was not over tourism development per se. Senator Gomes, UPP leader and President of the United Workers Union, which represents most unionized tourism sector workers, explained that he favored the Asian Village proposal. He objected to Lester Bird unilaterally signing the agreement without consulting Parliament (1997b). In this racialized discourse what is contested is who will benefit from tourism development and who will be alienated. Mixed in with the issue of alienation remains a distinct anti-development, conservationist tone among Antiguans. The Environmental Awareness Group (EAG) is the only *organized* group which sought to define Guana Island (and other commonly held coastal environments) as *in*appropriate for tourism development.

Producing the Asian Village and Guana Island

The EAG sought to influence the disposition of Guana Island and nearby coastal communities by constituting the meaning of these places as valuable in their present state, as wildlife habitat and as potential lobster fisheries. (The EAG recognizes that no lobster fishery currently exists and advocates the sinking of wrecked cars to create reefs currently absent but vital to lobster habitat.) The EAG outlined its agenda for the North Sound, and Guana Island specifically:

> The coalition partners [local fishermen, members of the surrounding communities, and the EAG] are calling upon the government to reconsider this proposed project and to continue to work with the people of the nearby communities to develop this area in a sustainable manner. The coalition also calls on the Government to protect all offshore islands for posterity and future Antiguans and Barbudans.
>
> (*EAG'er* 1997: 3)

The EAG thus defines sustainability as conservation with increased economic production. The EAG legitimizes its call for conservation by asserting the importance of Guana Island and the North Sound for wildlife. The EAG notes that the northwest coast is the fourth largest mangrove community extant in the Lesser Antilles and hosts the "world's rarest snake, the Antiguan Racer Snake (*Alophis antiguae*)" along with numerous locally rare bird species (*EAG'er* 1997: 3, *Outlet* 1997a, *Daily Observer* 1997c). In a more instrumentalist tone, the EAG also notes "that over 80% of the commercial fish species spend part of their life cycle in the coastal mangrove areas and ... that Antigua and Barbuda sits on the largest fishing bank in the Eastern Caribbean" (*ibid.*). Thus the EAG invokes the social value of the North Sound, citing a history and a future of artisanal fishing in accord with widely valued, though increasingly nostalgic, subsistence economies. With the exception of anti-accumulationist Rastafarians, Antiguans have generally embraced 'modern' ideas of progress. Conservationists are left to compromise, then, between industrializing and preserving landscapes. Technical and mechanistic paradigms, antithetic to ecological and non-market functions, dominate and legitimate the ALP's response to the EAG. An editorial reflects the ALP's modernist privileging of jobs over environmental and cultural preservation as well as ecological integrity:

> When Labour [the ALP] decided to dredge the St. John's Harbour in order to convert it into a Deep Water Port, the then opposition...said that the project would destroy the mangrove and shrimp by the bridge at Perry Bay. The mangrove, shrimp and shallygo are all gone, but Deep Water Harbour stands as testimony to the vision of Labour and the thousands of people who make a livelihood from it.
>
> (*National Informer* 1997: 9)

The passage implies that environmentalists were correct in their assessment of environmental and cultural damage; however, these are unimportant issues when compared to the jobs created as a result of the dredging. The ALP has wrapped itself in the mantle of productionism, privileging the technical management of industrial, ordered landscapes and dismissing knowledge and consideration of the ecologies native to coastal environments.

For those people concerned about the ecologies (humans included) of Antiguan coasts, a far more powerful conservationist argument might be based in preserving beaches, the landscape which makes sun, sand, and sea tourism possible for Antigua. The disregard for the salt ponds and reefs, which ultimately produce beaches, has led to a cascading failure of ecological processes vital to the tourism industry (Davidson and Gjerde 1989). That collapse has in turn forced reliance upon expensive, short-term, and problematic technical solutions to vital environmental imbalances in the coastal zone (Zachariah 1997a, 1997b).

Conclusion

Antiguan beaches are complex places. They are the material expression of myriad social, ecological, and geological processes. In order to understand the beaches as places seductive to mass tourism we must examine variously scaled historic and contemporary processes, and the thickness or presence of process in specific places (Woods 1998). In Antigua, and across the tropics, local tourism operators have had little influence upon broader meanings that have constructed beaches, strong sunshine, and the tropics as desirable in Western culture. Antiguan leaders in the 1950s understood that Westerners did desire beaches, and that they had an ample supply. These leaders also understood that to be successful as a mass tourism destination Antigua must have both accessibility and amenities. The US military gifted the island with accessibility. Through support and cooperation the Antiguan state has successfully seduced foreign investors who have built the resorts.

As beach resort construction has both intensified and extensified, the state–capital alliance has also produced local opposition and resistance. Tourism infrastructure has incrementally destroyed salt pond and reef communities, and the industry has undermined the ecological processes that produce and maintain the beaches international tourists so desire. As the symbolic, material, and ecological wealth of these living coasts is diminished, large numbers of Antiguans have begun to question continued tourism expansion for the benefit of the global traveling class; Antiguans are ready to act both to protect and to 'modernize' their remaining coastlines. In this atmosphere of dissent, Antigua's coasts, as places of subsistence, tourist, and ecological production, are becoming sites of renewed identity formation, sites of political action producing possibilities for a progressive sense of place.

Part IV

'The Orient'

Why the Orient seems still to suggest not only fecundity but sexual promise (and threat), untiring sensuality, unlimited desire, deep generative energies, is something on which one could speculate.... [O]ne must acknowledge its importance as something eliciting complex responses, sometimes even a frightening self-discovery, in the Orientalists.

Edward W. Said, *Orientalism* (1978: 188)

The so-called European hegemony in the modern world system was very late in developing and was quite incomplete and never unipolar. In reality, during the period 1400–1800...the world economy was still very predominantly under Asian influences. The Chinese Ming/Qing, Turkish Ottoman, Indian Mughal, and Persian Safavid empires were economically and politically very powerful and only waned vis-à-vis the Europeans toward the end of this period and thereafter.... Asians were preponderant in the world economy and system not only in population and production, but also in productivity, competitiveness, trade, in a word, capital formation until 1750 or 1800....

The preponderance of Asian economic agents in Asia and of Asia itself in the world economy has been masked not only by the attention devoted to "the rise of the West" in the world, but also by the undue focus on European economic and political penetration of Asia.

Andre Gunder Frank, *ReOrient* (1998: 166)

15 Desiring Ashima

Sexing landscape in China's Stone Forest

Margaret Byrne Swain

Ashima is a hardworking, beautiful girl in Sani folk legend.... Because she would not submit to a despotic rich man, this man used his secret knowledge of the Rock Gods to cause a flood that drowned her. At her death she was transformed into a megalith towering in her hometown within Stone Forest. She is Stone Forest's most famous landscape.

> Translated Chinese inscription on a tourist souvenir plaque depicting a Sani woman's head, from Kunming, Yunnan, 1998

Scrutinizing the stone – Ashima seems to wear a head-wrapping towel, carry a large bamboo basket on the back with a lot of wild flowers in it, and wear an unlined long gown. Under the head-wrapping towel, there is also flowing hair that Sani girls have uniquely. She has experienced various vicissitudes of life, but still stands tall and upright firmly. She is a right and pretty, kind-hearted and lovely Sani girl – Ashima.

> English inscription on a Kodak photo sign at the entrance of the Ashima rock viewing area in the Stone Forest National Park, Yunnan, 1999

China's Stone Forest is a tourist site inscribed with the mythic Sani heroine 'Ashima.' This chapter asks about gender/sex/sexuality dimensions of touristed landscapes, such as the Stone Forest, that are intimately tied to human imagery. Ashima's story of an ideal ethnic minority girl defines, and is defined by, cultural and physical landscapes; Ashima is a sign of the Stone Forest's seductive topography. Her sexed allure, an embodiment of colonization and commoditization as well as ethnogenesis, is expressed in visual and written texts, behaviors, and expectations. The Stone Forest, dialectically formed and reworked by diverse people over time, attracts tourists by difference and extremes of uniqueness in natural karst topography and cultural characteristics. Desiring Ashima, as commodity, sex object, exotic other, or as an emblem of ethnic pride, homeland, and/or nationalism, drives production and consumption of this touristed place. It is also a site where hybrid notions of gender and political identity evolve and global/local economies connect and transform. As the Stone Forest becomes a cosmopolitan tourscape, we can see that its signifier of Ashima is also resituating into local Sani identity and place. In terms of local power relations, in the Stone Forest we see patriarchal, ethnic hierarchies of the Han intersect with more equity-based relations.

Touristed landscapes as sexed sites

Landscapes may be understood as sexed places in terms of gender, sex, and sexuality inscribed on the site and in the people who constitute the landscape over time. Gender relations in colonial regimes, Catherine Nash (1994: 235) has noted, shape convergence between appropriation of the landscape and production of the subject of woman coded as 'landscapes of desire' or primitive 'others' for the masculine colonizing state. Asymmetries of power relations within a state often determine the mapping of landscapes onto named identities, and the meanings of place (Blunt and Rose 1994: 8, Nash 1994: 241). These mappings demarcate supposed sexuality and seductive sensuality of the primitive, wild, natural, while also marking women to signify moral and cultural purity, sexual innocence, and need for protection. Tensions and asymmetries in these opposing images play out in claims of cultural identity and claims to place between the colonizing state and local subjects. We can see such convergence in China's internal colonization of ethnic minority territories along national borderlands, including Ashima's home.

The term touristed landscapes used here represents a range of experiences and goals acted out by diverse groups in locales subject to tourism, but which are also places of integral meaning. In treating the Stone Forest as this kind of landscape one may readily see the sexualized lure of novelty and exotica for tourists. For tourees, desires of identity, ideology, and economy shape presentations of self/collectivity in place. *Difference* is the hook for both the local touree, producing niche marketing, and the domestic or global tourist consuming the most unique.

When I began to think about these issues in China's Stone Forest, I framed my questions in terms of gender. In previous work, I have proposed a definition of gender as a site-specific analytical term based in diverse systems of culturally constructed identities. Gender is expressed in ideologies of masculinity and femininity, interacting with socially structured relationships in divisions of labor and leisure, sexuality and power, among women and men (Swain 1995a, 2002). How to sort out gender and sexuality is an interesting question raised by many researchers. Judith Butler (1990) has questioned if any distinction needs to be made between sex and gender. Thinking about sex in terms of 'social sexuality,' 'genders,' or a 'third sex' has also been pursued in efforts to break away from analytical dyads, based in biological sex norms and heterosexuality, that do not reflect real diversity (Connell 1987: 290).

Indeed, a number of feminist scholars have written about interconnections between gender and sexuality as a system or continuum of cultural meanings inscribed in and on people's bodies. How these connections are made is expertly negotiated by Margaret Jolly and Lenore Manderson (1997) in what they label a 'holy trinity' of sex, gender, and sexuality in Western feminist debate. They illustrate recurrent tensions among arguments for biological/natural essentialism or cultural construction of these analytical categories. Volatile connections between gender and sexuality, they suggest,

> can be understood through their…shared relation to that ambiguous third term, sex, which is used to mean either bodily desire undifferentiated by gender, or corporeal sexual difference…. In either case it denotes a refractory 'nature' which seems to resist or to escape cultural construction.
>
> (Jolly and Manderson 1997: 2)

However, the facts of sexed bodies (as female/male) and a range of homoerotic and heteroerotic desires themselves may be understood to be shaped, culturally constructed, as gender norms and identities into sexed bodies and sexualities. From this perspective ethnicity, race, and class are integral to the terrain of gender, sexuality, and reproduction. These sites of desire in Jolly and Manderson's terms (1997: 6) are located in the contexts of colonialism and other circuits of the local and the global through time such as tourism. Gendered and sexualized exchanges in the touristed landscape of China's Stone Forest are understandable within these contexts. Privileging hegemonic viewpoints, in this case Chinese over ethnic minority culture, will not adequately explain questions of power in these exchanges. Rather more complex scenarios that take into consideration historical depth of contact and othering need to be envisioned. In this study of Ashima and the Stone Forest, I explore some of the exchanges in which people construct and consume this site of desire over time.

Touristed landscapes are sexed by the playing out of gender and sexuality constructs based in multiple hierarchies of difference. Local folk, visiting tourists, state bureaucrats, tourism industry workers, advertisers, and media producers sex the landscape, while located in the place and processes of a tourist site. These sites vary widely, incorporating distinct types of human focus, such as ethnic, heritage, sex, indigenous, war, and 'other' kinds of tourism. Gender and sexuality branding of touristed landscapes has been well documented worldwide, from the hip-swaying Tahitian women (Kahn 2000) to the caber-tossing Scotsmen (Edensor and Kothari 1994). Similar processes occur in Stone Forest, where sexing the landscape and desiring Ashima draw from intersecting regimes of gender and sexuality in a changing political economy.

In this chapter I first frame the touristed landscape by describing the natural environment and associated cultural myths of Ashima's Stone Forest. I then examine gender and sexuality systems shaping exchanges among the diverse groups sexing the Stone Forest as a site of desire. We will see that these groups including Sani and Han locals, Chinese nationals, and global tourists and tourees form dialectical relationships that sex this place in popular imaginations. These exchanges and the hierarchies in which they are embedded are then explored in tourist representations and practices of written texts, performance, souvenirs, employment, development planning, tourist experience, and local community responses. In conclusion, I raise the question of theoretically locating analysis of sexed landscape in 'third space' (Kahn 2000) to accommodate these multiple inputs and timelines. Desiring Ashima drives production and consumption of the Stone Forest. Here hybrid notions of gender, sexuality, and political identity evolve and global/local market economies connect and transform into a cosmopolitan tourscape and locally constituted place.

Ashima's Stone Forest

Ashima and the Stone Forest became linked with the rise of state ethnography (*minzu xue*) and mass tourism after the founding of the People's Republic of China (PRC) in 1949. At that time the tightly compacted area of dramatic limestone karst was a scenic spot that drew visitors to wonder, but it was in the midst of Sani farmland. The Sani legend of Ashima had yet to be standardized and published in Chinese. This all changed in the early 1950s when the government constructed the first Stone Forest Hotel, and a government folklore production team collected some 20 'best versions' of the epic poem to be known as Ashima (Li 1985).

Ashima's homeland, the Stone Forest, is situated in eastern Yunnan Province. The administrative region encompassing Stone Forest was incorporated as Lunan Yi Nationality Autonomous County in 1956, which then became part of the Kunming municipality in 1983. The Stone Forest park itself became a national scenic protected area in 1982. A government master plan with tourism as the 'central pillar' was begun in 1986 for the county's local modernization and economic development. By the 1990s, with transport improvements, travel time to the Stone Forest National Park from Kunming, the provincial capital, decreased to less than two hours one way, via superhighway or deluxe train. By the mid-1990s more than 1.5 million visitors, including over 100,000 international tourists, visited the Stone Forest annually. The county was renamed 'Stone Forest' in 1998 in a direct move for tourist name recognition, despite considerable dissent from local residents (Swain 2001).

The karst topography of eastern Yunnan is difficult terrain for living and farming, but it is good for tourism, with its captivating and cruel topography. As Tim Oakes (1998: 7) remarks about conditions in neighboring Guizhou Province's ethnic/scenic tourist sites, it could also be said that the Stone Forest represents a region "steeped in desire" of the rural ethnic minority poor looking for affluence and freedom from laborious rural work. To the east of the Stone Forest park, in the county district of Guishan, lies territory so rocky that people build stone houses unique to the region. Here, near the picturesque stone village of Nuohei, local folk historians and PRC scholars have thought might be a possible home of historical Ashima, or at least of the myth itself. The Guishan area, a conservative bastion of Sani culture, was also the site of Sani Catholic conversions and communist resistance against the Guomindang (He 1994). It was a fitting place for PRC scholars to research Ashima's official story.

Ashima's people are of Tibeto-Burman origins, designated by the state as the Sani branch of the Yi minority nationality. Approximately 200,000 Sani live in southeast Yunnan. Their homeland is scattered with limestone formations that are highly concentrated in one county, site of the Stone Forest park. Sixty-seven thousand Sani live in Stone Forest County, constituting one-third of the total county population that is otherwise predominantly Han Chinese. Tourism-generated income has brought wealth to a few, while improved infrastructure for tourism has significantly affected the whole county, but especially formerly isolated rural Sani villages.

'Welcome to Ashima's Hometown' proclaim signs in English and Chinese along the superhighway completed in 1992 from Kunming to the national park. The signs denoted the purported 'native place' of the mythical Sani maid. The popular story of Ashima unfolds as follows. It begins with her birth and welcoming into Sani society. As she grows to be a young woman of renowned talent, strength, and beauty, she learns all of the skills needed to be an asset in her community. Ashima plans to choose her own spouse, but she is stolen away by an evil landlord. Ashima's companion Ahei, who in different versions is identified as either her brother or her lover, is an accomplished young man who rides to her rescue. After surmounting many trials, tricks, and tigers, Ashima and Ahei flee the landlord, only to be trapped in a raging flood that sweeps Ashima way from Ahei. She then returns as an echo heard in the stone forest.

Ashima's story has its origins in a long history of Sani oral tradition and written literature. But their folklore has been manipulated in the context of various colonial projects, including the French expansionists from Indochina, and the Chinese themselves through the state policy of "internal colonization" (Hechter 1975). In this kind of imperialism at home, a nation attempts to control its peripheral land and people through culture as well as resources. In China's case the focus has been on non-Chinese groups settled along national borders. During the 1950s, codification of folklore into the revolutionary canon was an important cultural task of the new PRC government. Subsequently, in the Cultural Revolution (1966–76), virtually anything not deemed 'Maoist' was suppressed, while the arrival of the Reform Era in the early 1980s ushered in a rush to commoditize culture in the new market economy.

French missionary Père Paul Vial made some of the first non-indigenous records of Sani folklore. Starting in 1888, for over a period of 40 years, he used his study of Sani culture to promote a civilizing project of global Catholicism in the Sani's idiom (Swain 1995b). We can see through Vial's writings pieces of Ashima's story, but no actual myth as told above. Vial (1898: 18–20) published a Sani wedding chant that matches almost verse for verse the later state-produced poems describing Ashima's character. The chant, sung by Sani male priests (*bimo*), who record and use such religious knowledge, tells about the birth, childhood, and courtship of 'Chema,' meaning golden girl.

In his research, Vial contrasted this literature with a genre of women's improvised work songs (*k'oure*). Vial (1898: 20) published an example of these 'Sani blues,' in a song "The Complaint of the Young Bride," which lists many of the same reasons against arranged marriages given in the contemporary state codification of Ashima. With respect to Ashima's fate of literal calcification in the landscape, where the stone pillars of the forest embody her essence, these story elements were possibly developed in a genre of Sani earth songs (*mifeke*) (Vial 1898: 22), which describe Sani migrations in relation to the origins of various geographic features.

State folklorists acknowledge such diverse Sani poetic genres (Yang 1981: 2), which they then 'discovered' to form one story. Official PRC state versions of the Ashima legend were constructed in 1954 by the Yunnan People's Guishan Cultural Workgroup (Li 1960). Based on published Sani folktales and a Kunming opera (*geju*) production of Ashima, government officials considered their possibilities for propaganda, and ordered a group of urban Han intellectuals out to collect Sani folklore, dances, and music (Yang 1980: 4). This common practice of the era can be seen as cultural imperialism, an example of Han chauvinism (which Mao himself warned against). State officials assumed from their privileged position and presumed cultural superiority that they should decide what is Ashima's authentic tale. Scholars who worked on the Ashima project documented original translations, and these have been analyzed and argued about over the years (Li 1960, Yang 1980, Li 1985, Bi 1986, Liu 1993).

In its revolutionary drive to modernize ideology, the state devalued the 'folklorized' Sani uses of Ashima-type narratives. Beginning in the Maoist era, Ashima served the state's cultural politics at local, provincial, and central levels by embodying shifting imaginaries of modernity. As rewritten by the state, Ashima's de-sexed sacrifice was an exemplar for the nation—dressed in exotic clothes. This Ashima narrative for the state socialist project promotes exemplary gender, class, and ethnicity narratives, and the Maoist model firmly wedded Ashima to the landscape.

During one of the brief periods of liberalization in the Mao years, Ashima gained sex appeal and cinema glitz with the filming of *Ashima*, the movie musical. It ultimately became a significant vehicle for her fame locally and nationally. This very popular musical was made in the early 1960s, then re-released in 1979. Over 20 years later, it is still readily available throughout China on video and VCD. The costumes are Sani, the staring actors are Han Chinese, and the musical delivery is pure global cinema. In many respects it is a Chinese cinema complement of the contemporary American film *Brigadoon* complete with romantic mist, doomed lovers, a lot of sheep, archaic material culture, and mock-ethnic song and dance routines. It uses dramatic Stone Forest scenery and follows the standardized epic poem's story line, with the major exception of transforming the character of Ahei from Ashima's brother into her lover. (This possibility was raised in the 1954 folk-lore data, but not used in the state's attempt to create an Ur-Ashima.) The movie's portrayal of Ahei makes it a universal love story, rather than a Sani story about sibling relationships. Ashima herself morphs from virtuous sister to lover in this movie, produced and then suppressed as dangerously bourgeois, on the eve of the Cultural Revolution. The sexual romance and glorification of minority culture was antithetical to the nation's revolution project, but other elements of the story line remained true to Maoist goals: the theme of class relations played well throughout the revolutionary era, and now has new meaning in the reform era as a market economy stimulates class divisions.

Tourism development history

Concurrent development of the touristed site (the Stone Forest) and the legend marked the production of an imagined landscape situating Ashima for national and international consumption. With the end of the Cultural Revolution, and the beginning of post-Mao reform, scenic areas and minority cultures became impor-tant exploitable resources. The 'civilizing' project shifted from consumption of feminized minorities by the state to consumption by tourists, driven by then new party chairman Deng Xiaoping's exhortation about 'the glories of becoming rich.' Tourism development of the Stone Forest region began in earnest. Park income grew from 5.25 million yuan in 1992 to 113 million yuan in 1998, and tourism bureau employees had grown to 508 from approximately 100 in 1992. Private sector employment in service and infrastructure continued to grow rapidly, and per capita income, though highly variable in the county, was increasing (Zhang Xiaoping, personal communication).

Ashima marks all of this development. In the Stone Forest park, the rock of Ashima, described in guidebooks and park signs as looking like a Sani maid with a basket on her back, is a distinctive formation. Colored lighting and landscaping designed to draw in tourists enhance the view. Tourists are encouraged by park guides to have their photo taken with the rock in the background, and perhaps try for an echo from Ashima. It has become impossible to walk around the Stone Forest without seeing many signs of Ashima on tourist goods, sights, and busi-nesses. Images of Ashima, as a young woman in Sani costume, and the written name of Ashima are everywhere. Photos or a drawn outline of Ashima's rock decorate the park admission ticket, postcards, and souvenirs (Figure 15.1). Outside the park's gates, restaurants and shops use Ashima's name and image, as either the rock or a Sani woman, in their advertising. News reports about economic development in the county draw on "Ashima's place" (*Ashima de difang*) as a land of minority nationality spirit and enterprise (Li *et al*. 1993). Her name is also used in various international joint ventures for tourism development and regional handicraft manufacturers.

Figure 15.1 Stone Forest park admission ticket

Ashima also marks the local built environment, and images of Stone Forest rocks and Sani women are intermixed with the name of Ashima to represent the larger region and province in the nation. In the county town of Stone Forest, the movie theater and a main street are named after Ashima, while a sculpture of the mythical maid has endured several decades of traffic at a crossroads. The Sani women's head-dress, which evokes images of Ashima, became the architectural design basis for public buildings in the town and the Stone Forest County Tourism Bureau (Figure 15.2). Ashima also evokes a larger territory. For a number of years during the 1980s and 1990s the logo of the Yunnan Provincial Tourism Bureau was an outline of the Ashima rock. Given the power of Ashima's image, it is not surprising that tobacco, Yunnan's premier global industry in the 1990s, also used these associations of place. Two prominent brands of Yunnan cigarettes sold nationally are 'Ashima' with a Sani woman's face trademark, and 'Stone Forest' with a karst rock logo. In such ways Ashima represented all of Yunnan as exotic, unique, feminized, accessible, and consumable. During this same period the Stone Forest was by far the most devel-oped and easiest to reach international scenic tourist attraction in the province. Chinese guidebooks about the Stone Forest will attest that this place is "the greatest wonder under heaven" and one of the best-known domestic tourist attractions. Ashima's place is national territory, marked by national pride. Scenic theme parks in major urban centers around China replicate small parts of the authentic, natural Stone Forest with concrete and landscaping. Almost always one of these sites is Ashima's rock, promoting nostalgia for invented tradition.

Sexing the Stone Forest

Multiple gender and sexuality systems—of local Sani, Han nationalism, and global tourism practice—sex the Stone Forest landscape in contradictory ways. It

Figure 15.2 Stone Forest County Tourism Board advertising brochure, featuring Ashima 'the rock' and young Sani women singing and dancing

is both a place of consumable, feminized 'others' and a place of ethnicity and equitable power relations. Ashima stories embody a range of potential exchanges and hierarchical arrangements enacted by the people who constitute Stone Forest.

Gender and sexuality systems

The state's standardized versions of Ashima's story provide a commentary on tension and desire in Sani gender and sexuality. As the story is told, Ashima's parents lament that every Sani girl must wed, even if she does not choose her spouse or agree with her parents' choice, and then must leave for her husband's village and household. But also woven through the story is an opposite ideal that every young person, female or male, should be able to choose their own mate. Our heroine, with the hero's help, actively resists male hegemony, bride price, and virilocal marriage practice; and the scenario plays out within a patriarchal class structure in which the common people's gender system, exemplified by the couple, has little in common with those of the state.

Contradicting patterns of gender relations are also evident in Vial's study of Sani society. Vial was puzzled by the existence of Chinese-like practices of bride price and arranged marriage which he found ideologically incompatible with Sani non-hierarchical gender values and a lack of patronymics. Vial recorded an era of change in Sani society, in which their own gender practices were changing in response to outside intervention, while also striving to maintain their own society. For example, while post-marital residence for most Sani had become virilocal, the Sani were not patriarchal like the Han majority society. As we have seen, male Sani sacred poetics stress the ideal of gender balance, while female Sani secular poetics voice discontent and resistance to the unequal fate of women due to marriage practices. Further information from Vial on Sani women's inheritance, property rights, divorce, and rights to children also reinforces an interpretation of continuing gender equity—substantiated by my fieldwork 100 years and several revolutions later.

One of the basic messages of Ashima's tale is about women's control of their own lives. This message was revolutionary in the mid-twentieth century for Han Chinese, while culturally expected among the Sani. Ashima may be seen as an ideal among Sani women, but Han Chinese also use her name as ethnic slang to label Sani people as 'those Ashimas.' This both acknowledges these women's independence, as in the story, and denigrates their identity. Chinese minority nationalities are often feminized in contemporary cultural politics, making minority people seem less threatening to the majority and under control (Schein 2000), but there is also ambivalence about minority power and difference. We should note that in the Stone Forest park there is no 'Ahei Rock,' no memorialization in the natural landscape of Sani masculinity.

In the national discourse, Ashima and the Sani people are remnants of ancient tribal peoples. Words used to describe them include 'hardworking,' 'hospitable,' 'simple,' 'colorful,' and 'traditional.' In a preface to the English version of the official state Ashima text the translator Gladys Yang (1981: 2) noted the poem's "simple unadorned language," the Sani people's "simple written script," and the youth's nightly gatherings for singing, dancing, and making love. Yang makes the point that sexuality had to be contained: "although they could love freely, they could not marry whom they pleased but had to abide by their parents' choice. This explains why, for many generations, the Sani people have expressed their longing for freedom and happiness" (1981: 1). In a 1993 interview with me, a member of the 1953 folklore production team recounted tales of Sani free marriage, babies who do not know their fathers, and other cultural details more mythical than real. The textual representations of Sani as simple, colorful, and lusty folk continue to do their work, now taken up in the touristed landscape. Much like popular tourist accounts of colonized aboriginal peoples all over the world, it is often noted that the Sani and Ashima like to "sing and dance" (*tiaowu changge*); these very words actually appear in contemporary Internet websites linked to Ashima. The presentation of minority folk as people in need of control and civilization continues.

As part of the touristed landscape to be consumed, Ashima is both an erotic minority and a transformed virtuous citizen. Likewise the Sani themselves, in the national discourse of assimilation, should be controlled and made one with China while retaining a few colorful exploitable characteristics. Ashima's landscape was inscribed by the nation's sexualized imagining of a lusty minority girl who turned into a piece of rock, signifying national patrimony. But paradoxically it also serves as a natural sign of territory and homeland for the Sani: woman as marker of an ethnic group boundary in the land.

Producing Ashima desire

Desiring Ashima—as commodity, sex object, exotic other, or as an emblem of ethnic pride, homeland, and/or nationalism—drives production and consumption of this touristed place. Intersecting and sometimes contradictory concepts of nation-state, romantic patriarchy, ethnic group, and gender equity shape Ashima desire. Desire in the touristed landscape is expressed through a variety of texts, including the landscape itself, advertisements, souvenirs, tourist services, state planning, local/national arts, travel writing, and globalized tourist expectations and experiences. Some texts explicitly link Ashima and the Stone Forest, while others focus primarily on the story or the place. We can see this by contrasting the packaging of two VCDs produced by different companies of the 1979 movie release of *Ashima*. One was purchased in Kunming and the other in Shanghai. Neither VCD cover names the Stone Forest, but the Kunming VCD shows the actress playing Ashima in the foreground, with a photograph of the Ashima rock in the background. A brief Chinese text ends with "Ashima turns herself into a mountain stone." The Shanghai VCD cover shows various photographs of Ashima and Ahei from the movie. The Chinese notes are quite extensive, ending with "out from the water a fossil resembling Ashima was revealed." While the Kunming story gives Ashima agency, the Shanghai story fossilizes her, a favorite phrase to describe backward minorities in state discourse.

While a dichotomy of local Sani vs. the Han nation does exist, local Sani government officials carry out the state project and many local Sani work in the tourism industry. Rural Sani in Stone Forest County have become increasingly involved in tourism by providing services and goods. They sweep the park, take tourist photos, work in restaurants and hotels, and sell souvenirs. They manufacture distinct types of handiwork: marketing simple bags embroidered with the name Ashima or Ahei to domestic tourists and more intricate expensive ethnic work based on Sani material culture to international tourists. Several generations of young Sani women and men have worked as government-trained ethnic song and dance entertainers and official guides to lead tourists through the Stone Forest park's labyrinths (Figure 15.3). Dressed in Sani women's costumes, usually Sani, but in the 1990s sometimes Han, these guides are called Ashimas, which they will say translates to 'a beautiful maiden.' In the process, guides' bodies also become Ashima's, which, like cigarettes, are mainly consumed, albeit differently, by Chinese men. The market economy has opened up a range of employment activities around the park site, including female prostitution by locals and migrants, which is typical of Yunnan's ethnic tourism sites as well as urban centers. In the late 1990s a few male tour guides, called 'Ahei' following the

Figure 15.3 Sani woman performing karaoke as Ashima, Stone Forest Park, 1993 (photograph by the author)

myth, began working for the Stone Forest Tourism Bureau. They appear to be moving into positions of authority within the guide group. This raises the possibility of masculinization of prestigious work, while women are doubly feminized by naturalized, gendered, sexualized work (as caretakers, domestics, prostitutes), having already been constructed as emblematic of a feminized minority ethnic 'other' in the national imagination.

We can see that these themes were evident early on in the reform era, after 1978. In a 1983 *Women of China* travelogue about the Stone Forest, written in English for an international audience, the author tells about "a legend that one of the tall stones is the incarnation of a beautiful Sani girl by the name of Ash[i]ma who paid with her life for loving the young man of her choice" (Zhong 1983: 44). Later a Sani woman is asked if "women here still suffer the way Ash[i]ma did in her marriage," and is answered "no that is all in the past now." The author then comments that Ashima "has become a symbol of the Sani women's past sufferings and their aspirations for freedom and happiness...[and] has become a popular legendary figure among Chinese people of all nationalities" (48). Another example of an un-Sani-tized Ashima is evident in the 1992 dance drama produced by the Yunnan Province Song and Dance Troupe with a multi-minority dance company, giving "a strong sense of authenticity and heritage" (Liu 1992: 39–40). In this drama, set in the Stone Forest, both lovers Ashima and Ahei drown in a flood caused by the evil rejected suitor. Liu's review notes that dance motifs are drawn from distinct nationalities, but also emphasizes production use of the Sani legend's themes of "Ashima drawing her love from the rocks and symbolizing the Sani people's spiritual sustenance found in the mountains and rivers...[and] the deeper leitmotif of Ashima coming from nature, growing in nature and returning to nature." He concludes that "Ashima is the pride of the Sani people; all Sani girls want to possess lofty characters like Ashima, a symbol of Yunnan Province" (*ibid.*).

Conclusion: Ashima's place

These claims of Ashima both being primordial Sani female and belonging to the whole Chinese nation apply as well to her homeland, the Stone Forest and its touristed landscape. There, hybrid notions of gender and political identity have evolved and global/local economies continue to connect and transform. What was once a place visited only by domestic tourists is now on the global tourism circuit. As Chinese communism gives way to a post-socialist society, ideas of gender, class, ethnic, and sexual identities are mixing up. The Stone Forest has been sexed in terms of gender, sex, and sexuality, inscribed on the site and in the people who constitute the landscape over time. Asymmetries of power relations are evident in the state's mapping of Ashima onto this landscape and the multiple meanings associated with her place. Neither spatial nor historical analysis alone captures the complexities of Ashima's Stone Forest.

As the two inscriptions in this chapter's introduction show us, Ashima is clearly part of the Stone Forest's landscape. She is emblematic of an ethnic folk legend, as well as national folklore. Her physical representation in a natural karst formation evokes embeddedness, endurance, and transformation. Her story tells of desire, virtue, sacrifice, survival, and belonging in an exoticized homeland. She is a girl, fresh and natural, and a gendered sexualized landscape. The first inscription is from a Chinese text aimed at the domestic tourist audience. This is a cosmopolitan souvenir, and so it also includes an English text. Both versions agree that Ashima was a legendary Sani girl who became a rock and is now a famous part of the Stone Forest landscape. The Chinese version, however, has much more detail about the exact location (Lunan/Stone Forest) and class and gender struggles (bad magical rich guy tries to possess her, then kills her); it also affirms that she is the most famous landscape feature in the Stone Forest. The second inscription is also bilingual with approximately the same information, in Chinese and English. It is from a series of signs by Kodak, marking scenic photography spots inside the Stone Forest park during the late 1990s. The signs are intended for an international audience, practicing global capitalist tourism.

In direct contrast to such consumption of the touristed landscape are local Sani who reclaim a distinct Ashima. During the fall of 2000, I had a conversation with several Sani farmers, as we ate dinner just a few kilometers away from the Stone Forest park. My friend said, "there are two Ashimas you know—the tourism one over in Stone Forest, and ours, Sani people's. It's all made up in the park—the Ashima rock, the Ashima girls, for the tourists." Yes, agreed the other man who, it turned out, is a Sani priest, "they are not the same, our Ashima's history is thousands of years old."

Culturally, Ashima's place is centered among Sani people. But in the mid-1980s, before the comprehensive tourism development plan was put in place, 'Ashima' was named in Stone Forest park signs while Sani people were not. Information about aspects of Sani life, including their writing system and native religious practices, was available in guidebooks, often authored by Sani intellectuals, but the casual tourist in the Park would have no idea who these people were living in and around the Stone Forest. During the 1990s, changes in the representation of Sani at the Stone Forest park increasingly featured information about their culture; they also normalized the legendary Ashima as a Sani Yi woman. Now she has local meaning as well as Chinese ethnic minority identity for domestic and international tourists. These opposite images of Ashima, and the Sani, as simple timeless natives and as competent modern locals, are found in national popular culture and global tourism. Ashima is a factor in this contradiction: for 50 years Sani people have reacted to Ashima as ethnic heritage and public culture, while the legend's links to modernity influence their increasingly cosmopolitan lives. By cosmopolitan (*shijie zhuyi*) I mean knowing about the world and being open to diverse ideas from modernity and heritage (Swain 2001). Sani people are cosmopolitan as they combine their distinct identities with other ways of thinking, at home or abroad. The problem is to keep cultural place and identity while being a hybrid citizen of the nation, and the world (Clifford 1998, Robbins 1998). Ashima shows the way, appearing in some very cosmopolitan places. One unusual case I found at a video rental store in my California town. A Japanese anime video cover illustration caught my attention because it was of the heroine wearing complete Sani women's dress. I watched the film, and discovered that this strange tale was about a three-eyed girl, the last of the "san-jiang" (or "Sani"?) race of immortals, who must find a way to become human. She starts her quest in Tibet and ends up in Hong Kong. Here we have a new, cosmopolitan version of Ashima's story, but in reverse, going from immortal to human. She is still traveling inside and out of the Stone Forest, located in the landscape and in people's imaginations.

16 'New Asia–Singapore'

A concoction of tourism, place and image

T.C. Chang and Shirlena Huang

Introduction

The world famous 'Singapore Sling' is a cocktail concoction unique to the grande-dame of Singapore's hotels—The Raffles. It has been said that a tourist's experience of the country is incomplete without sampling this cocktail, luxuriating in the breezy atmosphere of The long Bar, reliving the bygone era of colonial Singapore. Together with but even somewhat more than the 'Singapore Girl', the 'Instant Asia' image and the 'Merlion', the 'Singapore Sling' is a leading icon among a group that have come to epitomize Singapore's tourism image to the Asian region and world. As marketing tools, these icons have been fashioned by planners of the tourist industry and marketed as products unique to the country. In the context of tourism planning, Ashworth (1994: 17) tells us that 'producers' (planners, entrepreneurs, retailers and the like) often 'select', 'package' and 'interpret' whatever resources are at their disposal to create alluring tourism products and places. As such "the interpretation, not the resource, is literally the product" (Ashworth 1994: 17).

The creation of tourist icons illuminates an important characteristic of tourism development in Singapore. In the same way that icons have been carefully fashioned to portray a specific image to the world, tourism landscapes have also been developed with specific images in mind. In this chapter, we explore the strategies deployed by the Singapore Tourism Board (STB) to convey enticing images to global travellers as well as create landscapes that enchant and enthral. Specifically, we argue that these tourism development strategies are aimed at seducing tourists (and often residents) using a particularly potent concoction of image creation, touristic promotion and place-making.

Concocting tourism places: the seductive powers of 'interpretation' and 'images'

Before development commences, potential tourist sites are seldom 'attractive' for mass tourism in their own right. Tourist destinations are often the concocted products of slick publicity, trendy development plans and careful selection of resources. Here, we briefly discuss what we consider to be the seductive powers of tourism developers (policy makers and entrepreneurs alike) as they *interpret* places and conjure *images* attractive to visitors. The combined power of interpretation and imaging is evidenced in different ways. Urry (1995), for example, notes that the romantic allure of the English Lake District is a social construct rather than a natural given, engineered by works of poetry attesting to its beauty and skilful tourism marketing capitalizing on such poetry. With its rainy weather and rocky physical terrain, the district is anything but hospitable! Yet it is presented as a lovely place through a process which Shields (1991) describes as "social spatialisation;" the Lake District "had to be discovered; then it had to be interpreted as appropriately aesthetic; and then it had to be transformed into the managed scenery suitable for millions of visitors" (Urry 1995: 193).

In heritage tourism, the historical resources of a place are similarly interpreted before a suitable product can be created. According to Ashworth (1994: 17), "Interpretation involves a conscious series of choices about which history-derived products are to be produced, and conversely which are not." This interpretation process is almost always ideologically linked to the agenda of the ruling state, be it nation-building, social cohesion, cultural revival or economic enhancement. As such, the transition process from raw resources to final product is often subjected to "distortions, ambiguities, fragmentation, inaccuracies, editing and bias" (Goodall 1990: 273). Myths about places and people are often encoded in the interpretation process, resulting in images that are alluring although not always related to the realities of place. Many examples of partisan marketing abound in the tourism literature ranging from island resorts (Goss 1993b), Orientalist-themed parks (Teo and Yeoh 1997, Yeoh and Teo 1996), large metropolitan areas (Chang *et al.* 1996, Hall 1995, Richter 1989) to small heritage towns (McGinn 1986, Murphy 1985).

Sometimes, multiple images may even co-exist in the same time–space. In contemporary cities, Ashworth and Voogd (1990) explain that there exist at least three images: "entrepreneurial images" aimed at attracting investments; "residential images" aimed at retaining residents and promoting a sense of civic pride; and "tourist images" directed at foreign visitors. These images are usually conveyed through cultural codes such as slogans, advertisements, movies, novels and so forth. These images work seductively because their target audiences—locals, tourists, investors—seldom give pause to ponder the realities behind the image.

In his classic book *The Image*, first published in 1961, Daniel Boorstin is interested in the idea of the "public image" because "something can be done to it: the image can always be more or less successfully synthesized, doctored, repaired, refurbished, and improved, quite apart from (though not entirely independent of) the spontaneous original of which the image is a public portrait" (Boorstin 1992: 187). Applying this argument to places, we contend that the image of a tourist destination is likewise crafted to suit different circumstances and objectives. Destination areas may be depicted with a particular image at one point in time, and yet another image at a later point in time depending on the government's motives, political ethos or cultural objectives (Chang 1997a). In this way, place images are seductive because they can never be 'read' or 'understood' literally; instead they can only be interpreted contextually according to the socio-political realities and economic conditions of places in which they are embedded (Huang and Chang 2003).

This is not to say that all tourist images are false. Image-making provides a way for urban managers and the governments of countries to reinterpret their history and culture in a different light, sometimes creating new and equally legitimate forms of heritages (Ashworth and Larkham 1994). In the dawn of a 'New United Europe', for example, commentators have noted a fervent search by national governments and the European Community for new forms of identity, lifestyles, meanings and values that will unite the continent socially and culturally (Larkham 1994, Masser, Sviden and Wegener 1994). Heritage is therefore not a static concept, and place identities are likewise dynamic and ever changing.

Ultimately, whether elitist interpretations and images are seductive depends on public acceptance or rejection of them. It is thus crucial to explore the ground "where interpretation and participation meet" (Uzzell 1989: 10). All too often, a wide divide exists between planners and users of particular places because of the failure by "planning agencies to understand something of the sense of place felt by residents" (11). Quite apart from the touristic images imposed upon places, consumers thus create their own "unique representations or mental constructs" resulting in their own "personal images of place" (Bramwell and Rawding 1996: 202). Touristed places are therefore socially contested as much as they are socially constructed.

Background to Singapore: global tourism and landscape change

Tourism has grown phenomenally in Singapore since independence in 1965 when the country received about 100,000 foreign visitors. In 2002, Singapore attracted 7.5 million tourists, up from 7.3 million in 1996. Despite limited scenic sights, Singapore has remained one of the most attractive Asian cities because of a combination of factors: efficient airport facilities; a central location within Southeast Asia; and aggressive marketing efforts by the STB. Tourism plays a key role in Singapore's domestic economy. In 2000, it contributed S$10 billion (about US$6 billion) to the national economy, or 5 percent of GDP (STB 2001).

Despite the Asian economic slowdown in the late 1990s, the Singapore government remains committed to the tourism industry. Recently, the Committee on Singapore's Competitiveness identified tourism as one of five key 'hub services' to anchor the country's millennium goal to be a "premier services hub in Asia with a global orientation" (MTI 1998: 156). Other key services include transport/logistics, media/communications, trading and business/professional services. Further proof of the government's commitment to tourism is its constant (and critics would say obsessive) modification of landscapes and places to attract new waves of tourists and residents (Teo and Chang 2000). The resort island of Sentosa, Bugis Street and the Singapore River are all places that have been dramatically transformed over the years to cater to changing consumer demands and urban planning ethos. Our case study of Chinatown in this chapter similarly provides an example of the government's vigilant approach towards landscape modification.

We focus on Chinatown because it was the earliest of Singapore's designated conservation districts, and has undergone several phases of urban redevelopment since the 1960s. It has its origins in early Singapore, when the colonial powers marked out separate quarters for each of the 'native' communities, including the Chinese. For the next 150 years, the Chinese enclave was perceived as a landscape of moral and physical disorder and decay in the colonial imagination (Yeoh and Kong 1994) but to the Chinese who lived there, it was *Tua Po* (Greater Town), a vibrant albeit overcrowded centre brimming with street life and bustling with all kinds of activity—commercial, residential, cultural, social, educational, etc.— through most of the day and night. With political independence in the early 1960s and "a new government intent on improving the material conditions of the newly independent people" (URA 1988: 18), the landscape of Chinatown (alongside other slum areas within the central area of Singapore) was to be "dramatically redrawn along modernist lines informed by efficiency, discipline and rationality of landuse" so that Singapore could take "pride of place in becoming an integrated modern city centre" (Yeoh and Kong 1994: 20). However, despite being under the juggernaut of redevelopment, Chinatown managed to retain many elements of its historic landscapes, especially the Chinese shophouse, and lifestyles, as the centre of Chinese celebrations and festivals (Yeoh and Kong 1994: 21–2).

Ironically, urban conservation efforts from the mid-1980s to the mid-1990s had a much greater impact on the heart and soul of Chinatown's community. While conservation efforts conserved the shophouse as Chinatown's quintessential architectural form, studies of the various sub-sections of Chinatown (e.g. Yeoh and Kong 1994 on Kreta Ayer; Yeoh and Lau 1995 on Tanjong Pagar) reveal clearly that most of the residents, as well as Singaporeans in general, are of the opinion that conservation accelerated the passing of traditional ways of life and businesses in Chinatown. Long-term residents in particular felt the greatest sense of "personal loss and the dislocation of community life" brought about by the eclipse of traditional family-run Chinese businesses and trades by the numerous Western-style offices, pubs and lounges, which now dot the landscape (Yeoh and Lau 1995: 65). In addition, the celebration of key Chinese festivals such as Chinese New Year have been "re-defined by the institutionalized state organizations" and turned into essentially organized economic activities (Yeoh and Kong 1994: 24). Thus, to date, while Chinatown as a conservation area retains a sense of historical continuity in its Chinese identity, it may be said to do so much more in terms of its 'hardware' (a place that is Chinese in terms of its built environment and organized festivals and activities) than its 'heartware' (a local community with daily lived experiences in the area). Recent plans for the area suggest that conservation will continue to be used as an ideological tool to achieve state-defined goals, this time as part of the strategy of selling Singapore as the epitome of the New Asian landscape, an effective mix of modernity and tradition.

Marketing New Asia–Singapore: imaging or imagining identities?

Over the course of time, Singapore has been marketed as a tourist destination in different ways. In the 1960 and '70s, Singapore's exotic appeal was highlighted through the 'Instant Asia' slogan, portraying the country's diverse cultures and proclaiming its multicultural and multiethnic identity. With rampant modernization in the 1980s, the marketing slogan was changed to 'Surprising Singapore,' depicting the country as an intriguing mix of modernity and tradition, of old and new (Chang 1997a). In 1996, Singapore was re-imaged once again with the focus on New Asia–Singapore. These changing slogans, we argue, not only play a critical role in repositioning the country in a competitive global tourism marketplace, but also constitute what Ban (1992: 9) describes as "part of the inner narrative that forms the imaginative text of Singapore as she [*sic*] moves from colonialism to independence, from insecurity to a sense of her [*sic*] place in Southeast Asia." In this section, we discuss the ideological underpinnings of the New Asia slogan and show how it has been deployed by members both within and outside the tourism industry to promote their respective services and products.

The New Asia–Singapore slogan (together with its subtitle "So easy to enjoy") was developed by the STB in collaboration with Batey Ads (best known for its advertising campaigns for Singapore Airlines). After months of market testing, the slogan was announced in a worldwide advertising campaign in 1996, which featured television advertisements, posters, postcards, calendars and magazine layouts (Klyne 1996: 8). According to the STB (1996: 5), the New Asia slogan suggests "a Singapore which is progressive and sophisticated, yet still a unique expression of the Asian soul." The New Asia spirit complements Singapore's desire to "preserve and nurture its Asian heritage," expressed in the government's promotion of core 'Asian values' and the fervent conservation of old buildings, while harnessing "the marvels of high technology," epitomized by the country's vision to be an e-commerce hub (STB 1996: 25). Singapore is thus depicted as a place where "East meets West, old engages new, and tradition marries technology in harmonious co-existence" (STB 1998a: 3) (Figures 16.1 and 16.2).

Figure 16.1 Postcard from the 'New Asia–Singapore' campaign: images of cultural diversity superimposed on landmarks of Singapore's built environment

Source: Singapore Tourism Board

Figure 16.2 Postcard from the 'New Asia–Singapore' campaign: The Raffles Hotel at night

Source: Singapore Tourism Board

The STB unabashedly describes its New Asia publicity campaign as 'story-telling.' The STB's *Tourism 21* plan, a tourism blueprint for the twenty-first century, speaks of the need to adopt a new story to promote the tourism industry. "Not only must this story recognize the outstanding qualities of Singapore, it must also consider what it aspires to be" (STB 1996: 25). The importance of 'narratives' is further evident in the STB's plans to develop story-boards, plaques and brochures of significant tourist sites, prescribing exciting story lines to each place. Hence, information sites in the Museum Precinct convey the story of Singapore's journey into nationhood; those in the Singapore River depict the story of Singapore's colonial legacy and the entrepreneurship of its people past and present; while Chinatown and Little India tell the stories of a multicultural people living and working in harmony.

At the Singapore Tourism Conference 1998, the STB reassessed the efficacy of the New Asia–Singapore slogan: after two years, there was still disappointingly low public awareness. More needed to be done to indoctrinate the public about the phrase and its meanings: encourage retailers to manufacture products that "symbolize Singapore's 'New Asianess' in the form of clothes, souvenirs and other tourist icons;" provide artistic entertainment that publicizes the New Asia brand name through TV documentaries, drama and songs, public dances and festivals; collaborate with the Ministry of Education to spread the New Asia message to school children; and confer awards on individuals and corporations that champion the slogan (STB 1998a:2).

The STB has also hoped to infuse the spirit of New Asia into mainstream society. New Asia, the STB (1998a: 3) maintains, "presents a total picture of the way we live, work and think. It is dynamic and evolving." New Asia evokes the image of Singapore as "an exciting, energetic, and bold destination, whose people are forward looking, modern, yet clearly Asian" (4). The STB even goes so far as to proclaim the need for a mindset shift:

> Effectively, Singaporeans must metamorphose into "New Asian" Singaporeans. They must be called upon to be living representations of the "New Asia-Singapore" branding. This should come across in the way they carry themselves, and in their thoughts, actions and manner of speech. We should thus aim to encourage fellow Singaporeans to be ambassadors of "New Asia-Singapore." (5)

New Asia is therefore not just a tourist-attracting device but a social-engineering scheme with clearly ideological motives.

The notion of a New Asia is not peculiar to tourism or Singapore. Indeed the concept resonates in much of academic and popular Asian discourse on the emerging economies of Asia and the concurrent surge in Asian cultural pride. In his book *The Asian Renaissance*, for example, ex-Deputy Prime Minister of

Malaysia, Anwar Ibrahim (1996), spoke of an emerging "Asian esthetique" marked by a consciousness in the arts, architecture, music, fashion, religion and Asian values. This emergence springs from a growing confidence among Asians in representing their political views and asserting their cultural identities in the globalizing world order.

Within Singapore, the commercialization of the New Asia image is most evident in the tourism industry. A number of tour operators, for example, devised New Asia–Singapore tours that bring visitors to contrasting sites of exotic and modern appeal: "Flavours of New Asia-Singapore" introduced tourists to the country's multiethnic cuisines; "Heartlands of New Asia-Singapore" brought visitors to residential sites such as traditional *kampongs* (villages) and modern housing estates; the "Spirit of New Asia-Singapore" emphasized *fengshui* (Chinese geomancy) elements in modern urban landscapes; and "City Experience" focused on contrasting landscapes like Chinatown and the modern financial district. Of course such tour itineraries have long existed in the country. What is new is the way tour guides have been trained to emphasize the New Asia–Singapore elements to tourists, and the renewed emphasis by the state to conserve traditional activities and old buildings alongside new ones. In this light, New Asia tours are really nothing more than a case of the product remaining relatively unchanged but the packaging given a snazzy cosmetic touchup (Chang and Yeoh 1999).

Lifestyle entrepreneurs have also been incorporated into the New Asia–Singapore fold. In the restaurant business, for example, the global trend in 'fusion cuisine' spearheaded by Tex-Mex and Nuevo Latino cuisines has found an equivalent in the 'New Asia Cuisine' first introduced by Singaporean chefs at the World Gourmet Summit in 1997. New Asia cuisine has been described as a "branch of fusion [cuisine] involving a medley of ingredients and techniques unique to different Asian cuisines, but given a modernized, even Western slant in terms of presentation" (Teo 1999: 35). Local restaurants such as Club Chinois, House of Mao, Doc Cheng's and the aptly named Wok and Roll best epitomize this East-meets-West concept. Andrew Tjioe, the owner of House of Mao, a Chinese restaurant with a quirky Mao Zedong theme, realized the potential of developing this outlet right after the Singapore government lifted its ban on Mao's books in 1998. The London *Independent* even went so far as to interpret this lift as marking a new consciousness and maturity in Singapore—a symbolic move towards a 'New Singapore' (in Koh 1999: 92). A large number of restaurant operators have indeed openly declared the 'New Asia cuisine' to be STB's marketing brainchild (S.Y. Lim 2000: 65).

The image of New Asia has not escaped the attention of the media industry. The inaugural issue of the magazine *East* (first published in Singapore in 1999), for example, featured a cover story on "faces of new Asians." Described as people with "a finger on the pulse of the modern [and] with an ear to the past" (*East* 1999: 19), personalities like Chinese actress Joan Chen and Japanese author

Banana Yoshimoto were highlighted as Asians who have made their mark on the global scene while remaining firmly planted in Asian traditions. According to its editor, *East*'s target audience is new Asians "who are constantly seeking out new experiences and new information about the region," people who are forging their identities in an era of globalization (8).

The commercialization of New Asia–Singapore has its difficulties and problems. As the manager of one local company promoting New Asia–Singapore tour packages explained, the New Asia theme is difficult to market because Singapore is more 'Westernized' than 'Asian.' Few companies are willing to develop such tours because of the fear of limiting themselves to a niche market, which is not unfounded because Western tourists are more likely to patronize New Asia tours than Asians—although the latter constitute about 75 percent of Singapore's total tourist market.

Another challenge is 'depth.' Some entrepreneurs feel that New Asia–Singapore is merely a marketing slogan created to sell tourist kitsch, and that it says nothing about Singaporean society and culture. In the same vein, we interpret Singapore's insistence on being a bridge between east and west as masking its insecurity of being neither east nor west. Ang and Stratten (1995: 71) neatly express Singapore's quandary on the Western-dominated international stage as "neither in the West, nor properly in the Asia constructed by the West." Commenting on its tourism landscapes, Urry (1990: 63) similarly remarks that Singapore is more western than eastern: it is " 'in the east' but not really any more 'of the east.' It is almost the ultimate modern city and does not construct itself as 'exotic/erotic' for visitors."

To be truly effective, 'New Asia–Singapore' would need to be accepted by Singaporeans as a way of life and self-image rather than strictly an advertising tool. Tourism images can serve as mission statements to which a place and its people can aspire rather than just a statement on what the place can offer visitors (Boorstin 1992). Put another way, tourism advertisements create expectancies on the part of tourists which local societies must fulfil if they wish to continue attracting tourists (Hall 1995). Tourism images are therefore a 'means' and as such represent a "search for self-fulfilling prophecies" (Boorstin 1992: 198). However, as we shall see next, the New Asia campaign, while attractive as a tourism marketing image, is not universally accepted by Singaporeans as a way of life and an image for their country and its landscape.

Creating and contesting New Asian landscapes

As part of the quest to translate the New Asia–Singapore branding to the urban landscape, the STB launched the Thematic Development Programme in 1997. The underlying aim of this programme has been to create "thematic zones with a unifying character or theme" to enhance Singapore's tourism product. Eleven areas were targeted, each focused around a "unifying character" to create areas of interest to tourists and residents. Of the eleven thematic zones, four were identified as priority areas: Ethnic Singapore (comprising Chinatown, Little India and Kampong Glam), Orchard Road, the Singapore River and the Entertainment District. Of these, Chinatown was selected as the "test bed for the STB's plans for the other zones" (STB 1998b). We shall thus focus on Chinatown in our discussion.

The use of thematic zones to re-engineer urban landscapes is not new to Singapore. Rather it is a refinement of policies that have guided urban redevelopment in Singapore since the mid-1970s when the Urban Redevelopment Authority (URA) initiated studies of conservation opportunities in historic districts. Such studies "signified the first steps towards conceptualising a conserved area to retain [the country's] distinct identity and character" (Perry, Kong and Yeoh 1997: 254). By the 1980s, however, state institutions began to recognize that apart from identity, the conservation of buildings, structures and districts also provided "the sign posts from the past to the present" (Committee on Heritage 1988: 29), and hence was integral to supplying "the substance of social and psychological defence" against the influences of Westernization among young Singaporeans (Perry, Kong and Yeoh 1997: 255). The official discourse against the infiltration of Western values centred on the need to preserve *Asian* values, and more significantly, the search for and the maintenance of *local* cultural identities as inscribed in, *inter alia*, the built environment.

The decline in Singapore's tourist arrival growth rates in 1983 (the first time since independence in 1965), followed by a recession in the mid-1980s, added further impetus to urban conservation. The Tourism Task Force, set up in 1984 to make recommendations to rejuvenate the tourist industry, counselled for the need to redevelop key ethnic enclaves and historical districts as tourist attractions, areas embodying the "Oriental mystique and charm" found in the "old buildings, traditional activities and bustling road activities" (MTI 1984: 60). By the late 1980s, heritage conservation of the built environment had become "intimately connected with redevelopment strategies designed to cater to tourist demands for uniqueness" (Perry, Kong and Yeoh 1997: 257).

Deploying heritage as a tourist-attracting device also characterizes the STB's current thematic development. However, in at least two respects, thematic development differs from past conservation efforts: first, the objective to transform these sites into more attractive places is targeted mainly at tourists; and second, the scale at which these plans are being carried out is much larger. Nowhere in the *Tourism 21* report is there any reflection of earlier concerns for these areas as repositories of the nation's past and keepers of its unique identity. This inordinate focus on tourism was new given the many heated debates over the state's efforts to benefit the local population through heritage conservation (see Chang 1997b, Huang *et al.* 1995, Kong and Yeoh 1994, Lee 1991, Teo and Huang 1995, Yeoh and Kong 1994, Yeoh and Lau 1995). Furthermore, the URA, as the main state agency overseeing Singapore's conservation efforts, had always maintained that while the tourism dollar is crucial, "we conserve primarily for Singaporeans. Tourism is secondary" (*The Straits Times*, 29 July 1990). With past assurances waning, the fear that local sites would be adulterated and reinvented for tourism would now be transposed to the entire district of Chinatown.

The STB explained it took "extremely thorough and painstaking" efforts to "produce a set of proposals which capture not only the imagination but also the heart and soul of Chinatown" (STB 1998b). However, its plans for Chinatown met with intense public concern. First, it was feared that the STB's plans would reposition Chinatown from a place that once served the economic, social and spiritual needs of its residents and Singaporeans to one serving tourism. Second, the public was also concerned that the conscious development of Chinatown along the New Asia–Singapore theme would create a place with little to do with the Chinese in Singapore. In other words, thematic development would create places that are 'new', 'Asian' and 'tourist-friendly,' but neither distinctive for Singapore nor beneficial for Singaporeans.

The STB's overall strategy for Chinatown is simple: provide visitors with sufficient "visual clues" to "give the feeling that they are in a special place" (*The Straits Times*, 22 November 1998). To achieve this, the plan called for the whole of Chinatown (85 hectares) to be divided into three zones with color-coded infrastructure, flags, signs and street furniture: "Greater Town" with a red theme (distinguished by a new theatre, museums and theme streets), "Historic District" in gold (with original temples and clan associations), and "Hilltown" in green (a hilly area with boutique hotels, pubs, cafés and gardens) (Figure 16.3). Four themed streets distinguish the zones: "Bazaar Street" (a shopping strip featuring Chinese crafts), "Food Street" (food outlets and *al fresco* dining), "Tradition Street" (traditional craftspeople and merchants) and Market Square (where fresh produce can be bought).

In addition to discrete zones and themed streets, the revamped Chinatown plan has several focal points. These include an Interpretative Centre (housing artefacts and dioramas of indigenous life) to serve as a first-stop and 'gateway' to Chinatown, as well as a Village Theatre to serve as a hub for Chinese cultural activities such as opera performances, poetry readings, calligraphy and traditional exercises. Even the five parks proposed for the area (including a water garden, fire garden and metal-sculpture park) are themed according to the five-element correlations of traditional Chinese cosmology. The entire package aimed not so much at recreating the place but "recalling and revitalising the Chinatown spirit so fondly remembered by visitors and Singaporeans alike" (*The Straits Times*, 26 September 1998). Given the grassroots furore that erupted in reaction to the plans, we must question whose memories were being recalled—or what memories were being constructed.

A key concern that emerged from letters to the local press, discussions on the Internet and opinions expressed in public meetings was that theming Chinatown to

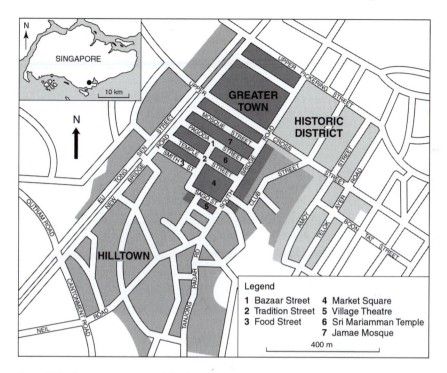

Figure 16.3 Chinatown as a themed district

capture the tourist dollar would destroy the essence of the place and create a "tourist trap with no cultural authenticity" (*The Straits Times*, 5 and 22 December 1998). What Singapore does not need is a would-be fossil of a bygone era serving as a "money-churning tourist experience" at worst, and a yuppie hangout for locals at best (*The Straits Times*, 5 December 1998). Many of the views ventilated in the public arena also referred to Chinatown as a "living organism"—not to be "drain[ed of] life and vitality" but instead to be "nurtured and cared for"—and whose "heartbeat" would be prematurely terminated and "soul" lost if it was displaced from its life-source, the local people. Ultimately, while many acknowledged that Singaporeans were practical enough not to want the "dirty and unhygienic" Chinatown of old, it was agreed that freezing it as "a prettified theme park" (*The Straits Times*, 18 December 1998) would destroy its charm. Some even went so far as to argue that if the idea of an "authentic Chinatown remains but an ideal for purists" (*The Straits Times*, 1 January 1999), it is preferable for it to be left to die a natural death and be replaced by a new offspring than re-engineered into "a caricature of its original self," a place that is no longer relevant to the lives and times of the local community (*The Straits Times*, 18 December 1998) (Figures 16.4 and 16.5).

A second line of debate centred on what was seen as the STB's plans to turn Chinatown "into a place more Chinese than it ever was" (*The Straits Times*, 22

Figure 16.4 Untouched shophouses from early twentieth-century Chinatown

Figure 16.5 Renovated shophouses occupied by new pubs and restaurants

November 1998). The thrust of these criticisms can be divided along two related lines. First, it was claimed "the STB proposal contains no attempt to incorporate cultural heterogeneity into the Chinatown story it weaves" (*The Straits Times*, 22 December 1998). Although carved out by early colonial planners to house mainly the successive waves of Chinese immigrants, Chinatown was never solely a Chinese quarter. The Singapore Heritage Society (SHS) charges that the STB was taking "liberties with the truth" and culturally re-engineering Chinatown as a supremely Chinese landscape in its New Asian representation (*The Straits Times*, 22 November 1998). This ignores the fact that the Malays and Indians also resided and located their businesses and schools in Chinatown.

A related issue is that in the STB's eagerness to emphasize the Chinese-ness of the area, the revamped Chinatown would be little more than what prominent local architect Tay Kheng Soon has labelled "a kitsch representation of mainland Chinese icons," which depletes and distorts local history (*The Straits Times*, 2 February 1999). On one level, the STB's version of 'Chinese' highlights the 'high culture' of ancient Chinese civilization rather than the daily life culture of Singapore's working classes. On another level, the STB's conception of Chinese-ness also simplifies the complexity of the Chinese community, which was divided spatially and culturally in Chinatown according to dialect groups, mainly Hokkiens, Teochews and Cantonese. Hence, for example, the proposed adoption of Mandarin for the naming of Chinatown as *Niu Che Shui* (literally bullock cart water, reflecting how fresh water was distributed in the Chinese quarter), to replace its original dialect names, *Ngau Chea Su* in Cantonese, *Gu Chia Chui* in Hokkien and *Kreta Ayer* for the Malays, overlooks each group's special association with the area (*The Straits Times*, 22 November 1998).

The public's contestation of the STB's Chinatown plan reveals a degree of 'romanticizing,' in both popular and state versions of how Chinatown should be remembered. Chase and Shaw (1989) explain that nostalgia is most likely to surface in a society undergoing rapid changes, where reminders of the past and vastly differing conditions of the present co-exist in the same lifetime. In such situations, "elements of the past tend to be treated with sanctity and characterized as constituting 'loss', worthy of recapturing.... Yet, in harking back to a lost place and time, a selective amnesia and concomitant mythologizing is at work" (Kong and Yeoh 1995: 18). What the debate surrounding place-imaging clearly shows in this case is a disjuncture in what the public and the state view as worthy elements to be preserved. To the public, the plans for Chinatown are, ironically, too "Asian/Chinese" to be Singaporean, and many of the elements too "new/exotic" to belong to Singapore. For the STB, an important part of achieving the New Asia identity is the requirement that local landscapes take on a broader cloak of "regional/pan-Asian" characteristics. Hence while heritage and tradition are important, being part of New Asia also means updating traditional activities, landscapes and architectural styles with modernistic flourishes.

Conclusion

In this chapter we have argued that the STB has fixated on a particular image for the tourism industry in the 1990s and this image, for better or worse, has served as a vision-statement for tourism and lifestyle entrepreneurs, as well as urban developers in the planning and marketing of city landscapes. New Asia–Singapore 'imagines' (and 'images') Singapore as a city-state equally proud of its modern sophistication as it is of its Asian cultures and traditions. The New Asia–Singapore identity is essentially a 'top-down,' state-led initiative conceived in the hope of repositioning the tourism industry for the new millennium, creating a new identity for the Singaporean community, and providing an avenue to rejuvenate urban places. How seductive the New Asia–Singapore campaign ultimately is depends on its reception by the public. In this regard, we have shown that while the image has been well internalized by entrepreneurs hoping to capitalize on the New Asia brand name, it has been far less endorsed by the average Singaporean. Using the case of the STB's thematic enhancement of Chinatown, we interpret the vociferous public reactions as an outright rejection of the state's imaging strategies and its attempts at imposing what are perceived as discordant identities upon people and places.

The differing degrees of success in implementing New Asia–Singapore reveal much about the processes of tourism marketing. Two points are worthy of mention. First, we see that while it is possible to market tourism products and services using 'brand names,' slogans and catchy titles, the same is certainly not true in the marketing of places and communities. To reduce Chinatown's essence to a formula, which can then be replicated in other thematic zones, like Little India or the Singapore River, diminishes the uniqueness of each place while ignoring the complexities of the lives of people in each area. The public furore over Chinatown sends a clear message that people and places are not to be readily 'commoditized' or 'objectified.'

Second, the global success of the New Asia–Singapore campaign contrasts markedly with its lukewarm local reception. The STB's New Asia–Singapore advertising campaign won in the "Best Recreation/Tourism category" at the 1997 Mobius Advertising Awards competition held in Chicago, in a field of 5,000 entries from 30 countries (STB 1998b). The judging panel comprising leading advertising agencies across the United States was impressed not only with the visuals, but with the entire concept of New Asia–Singapore—its story line, captions and promotional materials (see Figures 16.1 and 16.2).

Contrasting with this global endorsement is the general apathy of Singaporeans towards the campaign and their derision of the Chinatown plan. A survey by S.Y. Lim (2000) revealed that out of 118 Singaporeans surveyed, 81 percent had never heard of the phrase 'New Asia–Singapore.' How then can this global–local divide be explained? We argue that what is being endorsed (by international advertising agencies) is purely at the level of abstraction, removed from its context. The originality of the New Asia concept, the visual appeal of publicity material, the slick photographs and catchy wordings are thus highly praised, regardless of whether they actually tell the viewer anything concrete about the country. On the other hand, what is being derided by the lay Singaporean is the STB's imposition of order and precision on a society and its landscapes when clearly their identities are fluid, contingent and slippery (Massey 1993). In the thematic development of Chinatown, therefore, we see there exists "no common language between public and planners" (Uzzell 1989: 11). Likewise, in the conception of the New Asia–Singapore campaign, the opinions of Singaporeans were neither sought nor considered.

At the start of this chapter, we identified the Singapore Sling as a cocktail inextricably linked to the country's tourism appeal and colonial heritage. Today the world-famous Singapore Sling has been repackaged to be more tourist-friendly, available as a "nifty" gift set of "six pre-mixes" that can be purchased at the Raffles Hotel Shop along with an "authentic Singapore Sling glass" (*The Sunday Times*, 8 August 1999). The Singapore Sling epitomizes much of what we have discussed about the marketing and packaging of New Asia–Singapore. Like the cocktail, the New Asia campaign is tourist-friendly—skilfully packaged to appeal to visitors seeking more than just a "holiday site" but also a "complete Singaporean experience". The Singapore Sling has become a trademark of The Raffles, a 'must' for all visitors to the hotel. Similarly New Asia–Singapore is the self-proclaimed brand name for the country, serving as a rallying call to all players and planners in the tourism industry. Like the cocktail with its heady concoction of different spirits and fruit juices, the New Asia–Singapore campaign is also a seductive concoction born from a mix of influences—tourism marketing, place-imaging, identity creation and ideological intentions. By proclaiming itself as New Asia–Singapore, the STB hopes that Singapore, like the world-famous Raffles Hotel, will become internationally recognized as a 'tourism capital' appealing to global and regional visitors looking for both modernity and tradition.

17 Is the Great Wall of China the Great Wall of China?

Trevor H.B. Sofield and Fung Mei Sarah Li

> A monument of antiquity is never seen with indifference.
>
> Thomas Whatley (1770)

> If you have never seen the Great Wall your life has not been fulfilled.
>
> Attributed to unknown Chinese sage

Postmodernism's challenge of positivist research as the basis for establishing 'truth' has placed emphasis on relativist and multiple ways of seeing to reveal alternative truths. Our exploration of images and myths about the Great Wall of China depends on the intellectual space opened up by the postmodern project and is also inspired by "post-disciplinary" thinking (MacCannell and MacCannell 1982: 1). We are interested in a process of "trying to make what we all know we know seem less familiar" (Clunas 1996: 14) by de-familiarizing the single greatest touristed landscape in China, the Great Wall.

The Great Wall of China dominates a touristed landscape, which fundamentally symbolizes 'Chineseness' and far beyond the geographical boundaries of China. Modern tourism has contributed significantly to the current image of the Great Wall, motivating mass visitation by both domestic and international travellers. This chapter explores the historic range of people, institutions and events that have contributed to the process of shaping this iconic touristed landscape and its complex meanings, in order to reveal how what we know as the Great Wall is not the Great Wall.

Heritage tourism

Since the end of the Second World War heritage tourism has proliferated. Modern society has turned back to the past to understand how it has arrived at the present. For highly modernized societies, the breach with the past has meant that history, devoid of its role as an active constituent of the present, becomes a matter for nostalgia and curiosity. These issues underlie the difficulty of defining 'heritage' because, from a postmodern perspective, heritage can have multiple meanings, which are socially constructed. The social and political role of heritage helps individuals, communities and nations define who they are, both to themselves and to outsiders (Kymlicka 1995), providing a sense of place belonging. In this context, heritage may be used to sustain or demolish a particular version of history or promote certain political or social values. Each generation redefines its heritage in response to new understandings and new experiences. Interpretation may change to suit or satisfy particular needs because heritage, its ownership and its presentation necessarily involve changing values, power structures and politics. Heritage is thus not a 'thing,' static in time, but, through continuous interpretation, is better understood as a process. Invention, substitution, replication, reinterpretation and simulation, all of which will employ symbolism to a greater or lesser extent, are accepted by many tourists as 'the real thing' and play havoc with the concept of authenticity. Symbols become the reality: 'authenticity' is manufactured for consumption. The paradox is that over time the originally inauthentic may become authentic.

The Great Wall of China

Tourism utilizes symbolism in many different ways, especially iconography (famous sites, buildings, people or places), to represent places or countries. In terms of imaging, the Great Wall of China is one of the most universally recognized icons of national identity in the world, ranking with the pyramids of Egypt and the Eiffel Tower of Paris. For many tourists those five words (*the-Great-Wall-of-China*) work as a synecdoche, capturing all that is Chinese. Globally, the Great Wall symbolizes human heritage, the one man-made object that is visible, it is erroneously claimed, from space. Touristically, it is a compelling symbol for both Chinese (akin to a pilgrimage) and international tourists. Symbolically, it combines elements of both the sacred and the secular. However, the idea of the Great Wall bears numerous images conveyed at different times by different actors with different histories. Each interpretation presents its own story of myth and legend, fact and fantasy. Tourism intervenes to take symbolism beyond the duality of signification and representation because of the insertion of a third factor—the object itself (MacCannell 1994). Is the Great Wall of China really the Great Wall of China?

The historical record

Searching for the origins of the Great Wall compels analysis of both Chinese and Western discourses. Chinese histories have recorded the construction of many walls in different parts of China—city walls, frontier fortress walls, earthen rampart walls, defence walls built around provinces, some of them more than 1,000 *li* in length (1 *li* is about a third of a mile or half a kilometre). And then there is the 'long wall' of the Qin emperor, Shihuang, who is credited with unifying China about 2,200 years ago, and who is recorded as having built a northern border defense around 220–205 BCE. While fortresses and city walls were built of stone, the long wall and similar constructions were all built of earth and mud brick. Not until the Ming dynasty of the fifteenth and sixteenth century was a long wall built of stone. There is, however, no historical record of a single 'great wall' forming the northern border of China. The wall built by Qin Shihuang was believed to be 10,000 *li* long, but it was not regarded as and was never called "the Great Wall" by the Chinese. The wall now presented to the world as *the* Great Wall was constructed only in the fifteenth and sixteenth centuries during the Ming dynasty and was probably about 7,000 *li* (2,300 miles or 3,700 km) in length. Reassertions of the Great Wall as a continuous barrier, thousands of kilometres long, first built 2,200 years ago and periodically repaired and reconstructed by successive dynasties, were socially constructed in the early Western accounts of the wall, and they continue to influence both popular imagery and academic discourse.

A discourse of European romanticism about ruins in the seventeenth, eighteenth and early nineteenth centuries glorified images of the Great Wall (Aldridge 1972, Mason 1939, Spence 1992). The Great Wall of China has been known in the West for centuries, and many European travellers recorded apparently accurate descriptions of it, one after the other, for more than 300 years. One of the earliest accounts is from a Dutch missionary, Father Athanasius Kircher (1617, in Jenner 1992), whose map and vivid images of the Wall had a significant impact on European thinking about Chinese civilization. The Jesuit, Martino Martini (1665 in Waldron 1990), provided perhaps the most influential of the early accounts in his *Atlas Sinensis*, which described the Wall as extending for thousands of kilometres and forming the entire northern border of China. The French traveller Ferdinand Verbiest (1685 in Jenner 1992) described the Wall as greater than the other seven wonders of the world combined. Du Halde (1741, in Spence 1992) asserted seemingly authoritatively that it had been built 215 years before the coming of Christ with the most massive labour force ever assembled by a ruler. John Barrow (1805), who was a member of the first British ambassadorial mission, led by Lord Macartney, to China in 1793–4, and who later founded the Royal Geographical Society, calculated that the stone in the Wall would suffice to construct two smaller walls around the world at the equator. Illustrations of the Wall prepared by Lieutenant Henry Parish, who accompanied the mission, provided 'proof' of its immensity as it wound up and down over the hills to the far horizons, an image which is retained to this day. All of these descriptions were of the Ming wall from the vicinity of Beijing (Waldron 1990: 208–9).

These images reflect the prevailing fondness of the European Regency period for ruins. Travellers of the era scoured the world for ruins, while the poetry and art of English Regency and Victorian romanticism captured their imaginations. This allure was translated into landscape gardening. The landed gentry would erect artificial ruins of famous buildings as elements of their estate gardens, covering them with rambling roses and ivy to make them appear genuine, authentic. As Macaulay (1984: 24) notes:

> the fashion of building artificial ruins raged over Europe throughout the eighteenth and well into the nineteenth centuries.... We all know how ruins, classical Gothic, and even Chinese, sprang up in every fashionable gentleman's grounds, in Great Britain, France, Germany, Austria and the Netherlands.

Western worldviews have held the Wall to symbolize traditional Chinese attitudes towards outsiders as a protective barrier between civilization and barbarism; it was thus more than a defence line (Lattimore 1962, Wakeman 1975). The Wall "demarcated the entire Chinese culture from an ecological, social and cultural region of foreignness. Behind the Great Wall, the processes of Chinese history could work themselves out" (Mancall 1984: 11). This thesis contended that the Wall provided the perennial security that allowed China to develop as a unified, continuous civilization over several millennia. It was designed to keep Chineseness in as much as to keep others out (Lattimore 1962), a symbol of a country turned in on itself (Bauer 1980, in Waldron 1990).

Chinese scholars too have echoed these perspectives. After the dynastic era, Sun Yat Sen (1922), the 'Father of modern China,' wrote that the Wall had preserved the Chinese race and prevented it from being conquered by the nomads. In *The Great Wall in Ruins*, Chu and Ju (1993: 3) assert the existence of a single Great Wall:

> In northern China, starting at the ancient battleground near Fort Shan-hai and stretching across wind-swept expanses of virgin mountains and empty deserts nearly all the way to Mongolia, lies the Great Wall. Or more accurately, what is left of that mammoth stone structure put together by Emperor Qin. For more than 2000 years the Great Wall proudly stood guard protecting the Han Chinese against the nomadic horsemen of the north, and silently watching the rise and fall of many a dynasty.
>
> For centuries Chinese were protected by another Great Wall...a wall of symbols and ideas, of traditional values and beliefs. This cultural bulwark held Chinese society together for millenniums, shielding Chinese life from external encroachment and internal erosion. Just as the stone Great Wall reflected an inward looking mentality of China's ruling elite, the cultural Great Wall became a mantle that protected the Chinese from innovation and change.

In a meticulously researched examination of both Chinese and Western historical records, Waldron (1990) debunks the foundations of much Western scholarship related to the Wall. His deconstructivist approach differentiates myth from legend, fact from fiction and foreign policy objectives from imperial court power struggles in which the Wall played a pivotal role. His research demonstrates that there never has been a single continuous Wall demarcating China's northern border. Nowhere in the Chinese historical records could he find reference to a Great Wall. Rather, there have been many walls constructed in numerous different places over the centuries, not only in northern areas, and perhaps totalling as much as 55,000 kilometres of walls. Numerous terms exist to describe the different walls constructed over the centuries but there is no single, fixed name. Only in twentieth-century Chinese does he find contextually a phrase which equates to the English-language phrase "the Great Wall" (*wanli changcheng*, literally 10,000 *li* wall). There was never a concerted, continuous policy of repairing and maintaining a single northern wall as the essence of a single continuous foreign policy. In ancient times, no conception of anything like our modern acceptance of a "Great Wall" existed in China.

Alternative Chinese views of the Great Wall

China's myths and legends, songs, poems and plays yield further alternative histories of the Great Wall. If we return to the first long wall of Emperor Qin:

> Though that wall disappeared in fact, it lived on in histories and in the popular imagination.... Ch'in walls exemplified the futility of Ch'in Shih-huang's military policies and his tyranny toward the people, while among the people the memory of forced labor and death on the frontier led to the development of a tradition of legend and song.
>
> (Waldron 1990: 195)

The failures of the Qin dynasty demonstrated that force could never take the place of virtue and this morality entered the canons of Confucian thought. "Thereafter the Wall of Ch'in became for the literati an unambiguously negative symbol" (Waldron 1990: 196), and this view has remained influential in Chinese attitudes down to the present. Walls in general were criticized as failures—in terms of military effectiveness, as counter-productive diplomatically, as wrong morally, and as symbols of repression, used metaphorically when rigid controls were imposed on free thought (Chou Shu-jen 1925: 59, in Waldron 1990: 215).

A number of poems and folk-tales originated from the Qin Shihuang long wall. One is the story of Meng Qiangnu, which is widely recognized by Chinese around the world. Meng Qiangnu's husband had been conscripted to work on the wall and, worried about him with the approach of winter, she set out to take him warm clothes. She travelled for 1,000 *li* along the wall looking for him, only to find that he had already died. Overcome by grief she cried with such anguish that the wall cracked open to reveal her husband's remains. These she took back to their home for an appropriate burial. This Chinese morality tale is representative of the genre called "Songs of the Long Wall" which depict the wall as the work of a tyrant, costing more than 100,000 lives. Another tale concerns the general in charge of wall construction, Meng Tian, who was ordered by Qin Shihuang's successor to commit suicide, ostensibly because of an unexpiated crime but in reality because he was an alternative power and therefore a threat. In bemoaning his fate, Meng was recorded as saying that his only crime was "to cut the veins of the earth" (a reference to *fengshui* or geomancy); but historians rejected this, asserting that Meng's true crime was the immensity of the bloodshed and suffering he had caused as the builder of the long wall, and for which he deserved to die (Fryer 1975). Such popular railing against the wall construction is far removed from the Western myth of a noble Great Wall.

It would appear that the elaborate stone walls constructed by the Ming dynasty lent concreteness to the legends and traditions of previous walls, and gave them a spurious genealogy. This was the *mirage* seized upon by the first European travellers to China. They embellished and expanded their singular view until there arose only one Great Wall, which absorbed the identities of all previous walls. Its consolidation into one wall even paved the way for Western analysis of Chinese civilization and foreign policy. The notion of one Great Wall is widely accepted even among the Chinese both at home and abroad, and any attempt to refute it is likely to be met with the allegations of attempting to deny China's greatness.

The twentieth-century Great Wall

How then have we arrived at this situation where the Great Wall is not the Great Wall but is of course the Great Wall, as any tourist will happily tell you? How has myth replaced history, and how has tourism utilized its imagery to persuade visitors to the Wall that they are touching and seeing the very essence of China? The Wall's vast imagery and iconography, portrayed worldwide for generations, has persuaded visitors in advance that when they step onto the stone ramparts they are touching and seeing the very essence of China. The effectiveness of the Wall's sustained twentieth-century imagery may be found in the need by the new China to recreate its national image and identity. Influential figures, beginning with Sun Yat Sen, accepted and used the Wall to generate patriotic feelings as a necessary component of developing modern national identity and culture. The Great Wall of China as mythologized in the West had taken on a life of its own; absorbed back into Chinese society on Chinese terms, it has been adopted as a national symbol.

Mao Zedong and the Communist Party played an equivocal and ambiguous role in developing this new symbolism. In 1935 near the end of the Long March, Mao wrote a poem about reaching the Great Wall as the portent of Red Army victory over the Japanese. In 1936, another famous Mao poem, "Snow," treated the Great Wall as a symbol of China's magnificent heritage. A 1935 film, *Children of the Storm*, included a song, "March of the Volunteers," exhorting: "Arise, ye who refuse to be slaves; With our very flesh and blood; Let us build our new Great Wall!" In 1949 the song was adopted as the official national anthem of the People's Republic of China. In 1952 the new communist government began to restore parts of the Wall. It was to the Great Wall of China that Mao took Richard Nixon in 1972. Yet Mao Zedong thought urged destruction of 'the four olds'— tradition, religions, ancient philosophies ('feudal superstitions') and reverence for past imperial dynasties (Lin 1979). As an outcome of this policy, the Cultural Revolution (1966–76) witnessed the destruction of hundreds of kilometres of the so-called Great Wall by the Red Guards and its stone bricks were used for roads, levees, dams, factories and houses.

In 1984, Deng Xiaoping issued his clarion call to the nation: "Let us love our country and restore our Great Wall!" Whether or not there was widespread pride in the Wall as the epitome of China, thousands of Chinese responded to Deng's call, and gave both time and money to support its reconstruction. This resurrection of the Great Wall reveals how its symbolic power continues to be used for national purposes. One of the problems for post-Deng China was that "holes had to be knocked through the mental wall" (Jenner 1992: 95) constructed by Mao in order to implement the 'open policy' of 1978. At such policy junctures, the state has elevated the Great Wall as a symbol of unity and strength of purpose, reconstructed on the pride of a glorious past: the Great Wall "and everything it stands for" (98) is the sign above all others of Chinese greatness.

The Great Wall and tourism

While the Great Wall was being elevated to its current position of paramount icon, tourism simultaneously became an acceptable sector of economic development. Under Mao, the only visitors to China were 'special friends of the Revolution'; fewer than 250,000 were permitted visas between 1949 and 1972 (Chow 1988). In China under reform after 1978, tourism began to emerge as a major economic force. At the same time, the regime embarked on a revival of traditional Chinese culture with tourism often used as the vehicle for its development (Sofield and Li 1998).

The Great Wall is a central feature of these activities; many kilometres have been restored, roads and cable cars enable several million visitors a year to reach its heights, and hotels have opened adjacent to the most visited sites. It is understood as a mandatory stop in the schedule of all visitors to Beijing. The Great Wall's touristed landscape of most common experience for both domestic and foreign visitors is the Badaling section, which is now easily accessible via expressway from Beijing. In 1987 it was listed as a UNESCO World Heritage Site, followed by development of high-quality tourist facilities. Around the base of the site, the Great Wall of China Museum and the Great Wall National Theatre provide opportunities for seeing the Great Wall without climbing the Great Wall. These cultural historical centres wrap the history of the Wall into the twentieth-century nationalist project, providing visitors with in-depth Chinese perspectives and making the landscape of the Great Wall into what Duncan (1990: 17) termed "a signifying system through which a social system is communicated," a site which is not 'just a tourist attraction' but a representation of Chinese values about great-ness and modernity. Of course consumption is possible too: craft shopping lines the roadway through the Badaling site, and a local ATM machine dispenses cash with your global bankcard.

When Western tourists look at the attenuated stone ramparts disappearing over distant hills they see the Great Wall of Western myth: the sign is the reality and the view confirms their preconceptions, while their cameras, videos and tee-shirts authenticate the Wall for the 'folks back home.' The Chinese see poems replete with philosophical ideals, a folk tale of morality, a history of many walls, the record of a past tyrant, the pain and memory of the blood of the thousands who died in its construction, a chronicle of battles fought and lost against the nomads of the steppes, a tale of past imperial intrigues and divisive foreign policy, all this and yet a national symbol affirming Chinese greatness in the eyes of the world, the positive displacing the negative, the quintessential image of the country, an icon of enormous pride. The seduction of place, generated through the romanticism of the Great Wall, is only too apparent in its power to attract travellers both domestic and international, from the four corners of the world; regardless of culture, ethnicity and language, all who visit the Wall are able to 'read' it. The images of the Great Wall as a national symbol and a global heritage site are two more layers of representation that now embrace and encompass the Wall.

Places form part of the 'common knowledge' of Chineseness: picturesque tower karst of Guilin, the sea of clouds around Huangshan (Yellow Mountain), the Three Gorges of the Yangzi River and the Yuntai Arch of the Great Wall. Images of these places bring spiritual unity even if the people have never visited them; but when they do visit the importance of these images is reinforced (Sofield and Li 1996). Ying Yang Petersen (1995: 150) has suggested that Chinese domestic tourism to such places constitutes "a voluntary cultural decision more akin to a pilgrimage" made in order "to validate the poetic knowledge of places." This might be read as a shift from a 'naturalistic' to a 'symbolic' concept of tradi-tion (Handler and Linnekin 1984: 273). The former assumes that tradition "is an objective entity, a core of inherited culture traits whose continuity and bounded-ness are analogous to that of a natural object" but in reality, as conveyed by the latter, there is an "ongoing reconstruction of tradition in the present...which is not natural but symbolically constituted." Without having been nurtured in the Chinese cultural milieu it is difficult for visitors to enter Chinese places with the same experiential understanding. Nevertheless, both the Chinese visitor and the foreign tourist to the Great Wall seek out images conveyed by photographs, folk tales, Confucian poets or travel brochures and, having experienced the reality beneath their feet, can return home with their preconceptions virtually intact.

In theorizing how attractions work, MacCannell (1989) utilized a semiotic treatment, in which the attraction is a sign.

> It represents (marker) something (sight) to someone (tourist). The marker or signifier provides information (e.g. name, picture) about the sight, and as a representation of the sight is usually the first contact that the sightseer (tourist) has with the sight. Markers are either off-sight (e.g. travel books, stories of people who have previously visited the sight) or on-sight (e.g. notices). Since off-sight markers anticipate the sight they are often superior to the sight.... Off-sight markers can also stereotype a destination by high-lighting certain 'must see' features.
>
> (Dann 1996: 9)

So the constant inclusion of the Great Wall in most of the tours offered by the China Travel Service fortifies markers as symbols, standing in for the represented object.

For Western tourists to the Wall, most of their off-sight markers will have been formulated by outsiders (the perpetuation of myths about the Great Wall is but one example, the ubiquitous guidebook another), while for mainland Chinese their off-sight markers will be local, incorporating the politics of history, the ballads and poems of past ages, Mao Zedong's literary efforts and Deng's more recent patriotic call. The 'off-sight markers' tend to be formulated by outsiders, but vis-à-vis the Great Wall the Chinese voice (as exemplified by the China Travel Service) is as vigorous as any outside organization in expounding and perpetu-ating the myth of the Great Wall. The language of touristic images of the Wall promotes and reinforces its authenticity. For MacCannell, the quest for the authentic is the crucial characteristic of tourist motivation, by contrast with Urry's (1990) view that in a postmodern world, authenticity often resides in an interpreted representation of reality rather than in the reality itself.

Off-site markers are also perpetuated by discourses of orientalism, which treat 'the Orient' as a stage for consumption of its mythical settings, such as Cleopatra, the sphinx, Babylon and more. Contending histories of the Great Wall affirm its orientalist origins in 'ruins' and ideational notions embedded in Western romanti-cism. Reflected back on itself, that is, Western narratives of the Great Wall have been consumed in China, which have bled into China's twentieth-century nation-alist ideology, even as the modern state's embrace of the Wall's powerful symbolism contends with historic domestic Chinese interpretations of its 'construction tyranny.'

As a touristed landscape, the Great Wall has multiple images and realities. In 1989 the China Travel Service opened a theme park, the Splendid China Miniature Scenic Spots, in the city of Shenzhen adjacent to the border with Hong Kong. It features miniatures of China's best known scenic spots, distinguished by "the world's longest rampart," the Great Wall of China, and "the world's biggest palace", the Imperial Palace in Beijing. Tourists pose for photographs in front of this theme-Wall, which provides a simultaneous fusion of and alternative authentic and the marker for it, consuming the reproduction through the visitor's interpretation as if it were the original (Sofield and Li 1998). The result is a post-modern expression of the original, providing a would-be authentic experience. As Urry (1990: 85) interprets: "what we increasingly consume are signs or represen-tations.... This world of sign and spectacle is one in which there is no originality, only what Eco terms 'travels in hyper-reality'." This compels him to conclude that sites which have been made into attractive spectacles are not necessarily inau-thentic, but rather that there is no one simple "authentic reconstruction of history" (156); instead there are various kinds of interpretation and reinterpreta-tion with their own validity.

A visit to the Great Wall, in whatever form, is an example of Nuryanti's (1996) "individual journey of self-discovery" coupled with, for the Chinese, a reaffirma-tion of the Chineseness of the participant observer. Where MacCannell (1976: 44) considered the distinction between authentic original and reproduction as essential to the marking of an authentic tourism object or sight (its 'sacraliza-tion'), for the Chinese visitor to the Splendid China, the "aura of the original" (Harkin 1995: 653) may imbue the reproduction and thus produce a tourism experience of satisfaction. So, "What is considered the authentic sight, versus the tourist marker, is not necessarily fixed" (*ibid.*). In this touristed place, the Great Wall, the global and the local come together, drawing upon both Chinese and Western histories, upon ancient and modern political power, upon cultural and social values from both the 'East' and the 'West' to fuse many interpretations in a process that creates a single new reality. The Great Wall of China is not the Great Wall of China but it is the Great Wall of China.

18 Existential tourism and the homeland

The overseas Chinese experience

Alan A. Lew and Alan Wong

Introduction

'Home' is a relative concept. For many people, the physical places that they call 'home' will change over the course of their lives. The reasons for changing one's home are diverse, though geographers have traditionally viewed them as 'push' factors and 'pull' factors. Following a change in home, old and new places often vie for 'homeness,' and as each new place becomes more of a home, each old home comes to hold a sense of 'past homeness.' For many people, these past homes form important parts of their life experiences and personal identities. For some, these past homes can hold a greater sense of belonging and 'at home' feelings than their current physical home. In addition to past homes, there are also many 'other homes' that shape individual and group identities, often consisting of places that have never been physically lived in for any significant period or even visited. Although the origin of such ties to other homes in distant locations can be many, cultural ethnicity (including religion) and ancestry (including race) are probably the two most common.

The idea of the home, as a place of security, refuge, comfort, family and simple shelter, is one of the most enduring, pan-cultural desires. But what if 'home' becomes diverse places? Past homes and other homes are significant generators of travel and tourism, as people are drawn to and seek out places that hold special, personal, and often more 'authentic,' meaning (cf. Oakes, this volume). These types of tourists have been considered distinctly from the majority of tourists who are motivated to travel more by recreation, education, general curiosity, and escape than by an existential calling. Erik Cohen (1979a) suggested that the majority of mass tourists are either fully centered (in a psychological sense) in their physical home, and therefore travel mostly for recreation and curiosity; or they are largely decentered, and travel because it serves as a temporary distraction from an already alienated home existence.

Cohen proposed one other form of highly centered tourist; one whose center was not in the physical home, but in another home: the existential tourist. The existential tourist is not decentered, but also is not centered wholly in their place of residence. Instead the existential tourist negotiates space to return to a home away from home to past homes and other homes. With the increasing ease of travel and migration that has accompanied the end of the Cold War and the rise of economic globalization in the late twentieth century, ever larger numbers of people are living far from their places of birth and ancestry (Ma 2003). In a time when many nationalities (defined here in cultural terms) can claim to have their diasporic populations scattered across the world, the potential for existential tourism is greater than ever.

This chapter explores conditions of existential tourism among overseas Chinese, focusing on relations with their ancestral homeland areas in China. Like other diasporic ethnic groups, overseas Chinese migrants, in both historic and contemporary times, have followed long-established paths, bound by 'networks of ethnicity,' which "extend the group's identity spatially, and are an important facet of social and economic organization, particularly within migrant communities" (K. Mitchell 2000: 392). Highly structured ethnic networks support existential tourism to China and several major fields of influence shape this structuration process, overlapping in different ways (Figure 18.1). Overseas Chinese institutional structures support ideas about traditional Chinese values, thereby working to enable and maintain a sense of 'Chineseness.' 'Traditional values,' however, have also adapted to meet the special conditions of the migrant/diasporic community, as migration creates both 'outsider' and 'home out there' experiences, the evolution of multiple homes, and the need for mechanisms to overcome geographic

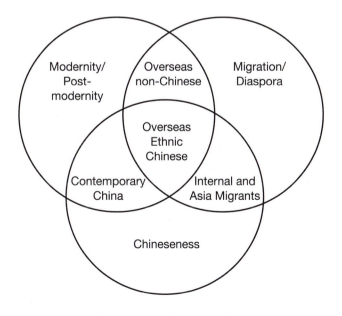

Figure 18.1 Influences shaping the overseas Chinese experience

spaces between old, new, and transitory homes (Leung 2003). The influence of space-shrinking technologies and globalizing modernity provide further realms of influence, shaping the form and experience of both migration and 'Chineseness,' by, for example, enabling closer relationships and easing the strain of return visits.

For overseas Chinese, modernity as a social and historical phenomenon has shaped migration from China and existential tourist travel back to China. Economic migration, which has characterized most overseas Chinese migration, has been an integral part of modernity's global project (Giddens 1990). Individuals who participate in the migration process have come to epitomize modernity through their transnational/postnational and transborder organizational structures, multiple and hybrid cultures, and cosmopolitan openness (Anthia 1999, Bhabha 1994). Indeed, the overseas Chinese, probably more so than those who stayed behind, are modern subjects, "cut loose from [their] moorings in the reassurance of tradition" (Giddens 1990: 176). Yet at the same time, they are often more bound to the *idea* of China than those who never left (Sun 2002). For them, existential tourism, and the social institutions that have been built to support it, is a modernist system that holds the promise of transcending both geographic and social space, and overcoming the risks of an uncertain future.

What are the existential tourist's landscapes of experience? Among overseas Chinese tied to ethnicity networks, touristed landscapes are sites where ties in the network touch down and locate, where 'flows' of 'ethnoscapes' (Appadurai 1990) become emplaced. These sites may as easily be 'Chinatowns' or affluent suburbs of world cities, immigration lines or airplanes, and, ultimately, the hometown of the patriline, the ancestral home of the family. For overseas Chinese who themselves or whose forebears left rural China, the ultimate destination is the ancestral village, the ancestral home and the gravesites (if they survived grave removal in the name of recovering arable land during the Maoist era) of one's ancestors. For those who left towns and cities, attempting to locate the past may result in another kind of alienation: the discovery that the family home has been torn down to make way for new high-rise and commercial development. This contemporary existential tourist undergoes an even more extreme 'search' for 'home,' one in which all homes are displaced.

These are some of the general conditions, social and spatial, in which overseas Chinese have constructed their relationships to China, and their existential tourist trips to home provinces, towns, and ancestral villages. Some aspects of these conditions are universal, such as modern alienation and the postmodern search for a center(s) 'out there.' Other aspects are situated in the realm of diaspora and migration writ large, including displacement, hybrid identities, and existential tourism. And then there is the realm of 'Chineseness,' with a complex set of arrangements centered on extended family relations and obligations.

Overseas 'Chineseness'

Chinese diasporas have largely been a modernist and capitalist phenomenon. Although migrations from China to Southeast Asia are recorded as far back as the third century BCE, the earliest significant Chinese settlement in the Philippines, the Malacca Straits area, and the islands of Indonesia occurred during the Ming dynasty (1368–1644) (Poston and Yu 1990, G. Wang 2000). Most of the merchant traders of that era were sojourners, though some migrated permanently. They occupied a kind of 'third space' (Bhabha 1990, 1994): for example, between the Malay and Chinese cultural worlds, by establishing families with local, non-Chinese women, while successfully maintaining a high degree of Chineseness through centuries of descendants. The result was a hybrid culture, known as *Peranakan* in Malaysia and Indonesia. Distinctive forms of Peranakan culture still exist in Malacca and Penang in Malaysia, as well as in Singapore (Cartier 1996, 1998).

Early Chinese migrants undertook their quasi-modernist project, driven by economic imperatives in a process of regionalization (if not globalization) rather than nationalization, forging new transborder hybrid identities that maintained aspects of traditional ethnicity while adjusting to the politics of 'outsideness' and the economics of regional relationships (Cartier 2003, Pieterse 1995, Wong 1997). By the mid-sixteenth century, European traders found ethnic Chinese merchants and communities well established throughout Southeast Asia and occasionally beyond. In 1712 the Manchurian Qing dynasty, fearing revolutionary influences, made the return of Chinese from abroad punishable by death. Chinese were not allowed to legally leave China and emigrants were accused of having had "deserted their ancestors' graves to seek profits abroad" (Duara 1997: 42). This edict had greater moral than practical meaning for the Chinese maritime communities of the South China coast for whom long-distance trade was a way of life; they largely continued active trading across the East and Southeast Asian seas through the Qing dynasty (Cartier 2001).

But the conditions for migration changed appreciably after the British forced China to open its southern ports following the first Opium War in 1841. Through the turn of the century, impoverished Chinese laborers emigrated in droves to Southeast Asia and beyond (Pan 1990). These 'coolie' laborers differed from their merchant forebears in that most, though not all, were of rural peasant origin and intended to return to China after they made their fortune overseas; they were classical 'sojourners.' In an effort to capture the growing wealth and political clout of overseas Chinese, the Republic of China (founded in 1910) made all Chinese throughout the world citizens of China; the People's Republic of China (PRC) discontinued this practice in 1949. In the early 1950s, China was again closed to out-migration, the Southeast Asian nations gained independence, and many sojourners became permanent residents in these countries and others. The

PRC at times embraced the overseas Chinese and at other times rejected them, while all the while attempting to manipulate them in its ideological war with the Republic of China on Taiwan (Tu 1994). During windows of acceptance, some overseas Chinese returned to China to settle, mostly due to turmoil in their adopted lands, such as in Indonesia in the 1960s and in Vietnam in the 1970s.

After the end of the Maoist era, in 1976, China reopened its doors beginning around 1980. The newest and continuing wave of Chinese migrants have more diverse backgrounds and motivations than earlier generations (Wong 1997). Some are sojourners, but most are permanent immigrants. They come from both rural and urban areas, and they work as laborers and professionals. Aihwa Ong describes this new wave of migrants as exerting 'flexible citizenship,' with different locations and countries selected for family, work, and investments: "Flexibility, migration, and relocation, instead of being coerced and resisted, became practices to strive for" (Ong 1999: 19). What all of the more than 60 million ethnic Chinese living outside of China proper have in common are historical, cultural, and racial ties to the homeland of China (Lew and Wong 2002). For some, this relationship becomes manifest through travel and existential tourism. Such travel could be considered a postmodern form of sojourning—one that is more flexible and less essential than the permanent return of the sojourner.

Motivations for maintaining ties to the homeland are many. For all Chinese there is a racial identity that ties them to China and separates them from other racial groups in their adopted lands. China itself, especially in the Han areas of eastern China, is among the most racially and ethnically homogeneous countries of the world. A trip to China allows immersion in racial (if not fully ethnic) sameness that is possible in only a few places outside of China (such as in Hong Kong, Taiwan, Singapore, and the larger Chinatowns around the world). At the same time, race alone provides a somewhat superficial basis for establishing identity (Chang 1997). In the Chinese case, however, Han Chinese ethnic culture is so closely tied to the Chinese race that it is able to overcome a good portion of the great diversity of origin and migration experiences (Wong 1997). Sub-ethnic language and home region affiliations (at provincial, county, or better yet, township and village levels) allow for even greater degrees of kinship identity formation among overseas Chinese traveling back to China. The Chinese practice of genealogy is another important force structuring family ties. Most villages in China maintained detailed genealogical records that followed the male line of descent back to the legendary periods of Chinese history over 3,000 years ago (even as some were fictive in part). Many of these original genealogy records were destroyed during China's Cultural Revolution (1966–76), but many also survived and in alternative forms (P.P.H. Lim 2000).

For overseas Chinese, their Chineseness is a centering force that has a clearly defined 'homeland' in the country of China, even if the individual's home village has been long lost and the home of residence is thousands of miles away from China

(Ma 2003). While Chineseness can represent a minority status in a non-Chinese place (Chang 1997), structures that support the centering quality of a Chinese homeland can help to overcome this 'othering' experience, which can otherwise lead to a questioning of personal identity and values. Prasenjit Duara (1995), for example, identifies the sharing of historical memories (through the development of both formal and informal institutions), along with the experience of both 'othering' and being 'othered,' as crucial elements in the community building process among diasporic populations. Genealogical records are a form of group memory that can transcend space and offer new opportunities for community-building. Meeting a distant or long-lost relative reaffirms one's human and place relations in the world and offers new possibilities to transcend differences and share common stories. Visiting a homeland offers this same experience, in addition to new memories that can be shared at future opportunities. On the geographic space that separates diasporic populations from their homelands, Sun (2002: 131) says, "This geographic lack will continue to be compensated for through the growing importance of memories—memories of the past that are portable, potent, patriotic, and available for constant repetitious retrieval." In this way, memories, and traveling as a way of creating new memories and rebuilding genealogical ties, affirm ideas of well-being about one's place in relation to distant relations and homelands.

Sojourner migration

The Chinese sojourner tradition, from the middle of the eighteenth to the middle of the nineteenth centuries, represents the clearest example of the adaptation of traditional 'Chineseness' in the diasporic community. Several conditions recommended the sojourner lifestyle over permanent migration (Woon 1989). The leading historical condition was the importance of the extended family and insecurity in a place lacking the extended family reference group. Overseas Chinese adopted pseudo-extended family experiences through overseas voluntary associations in their countries of residence, based on lineage/surname, province, town or village, spoken dialect, trade/profession, and other special interests (Brogger 2000, Lew and Wong 2004, P.P.H. Lim 2000). A second condition derived from the classic notions of filial piety, which pressured migrants to return home. Chinese traditions and holidays have drawn on obligations of filial piety, as expressed in terms of caring for aged parents and for maintaining graves of the ancestors (Pan 1990: 55). In southern China, despite many decades of communist rule, most of the Chinese who left before 1949 maintained ownership of their village homes, or passed the deeds on to their foreign-born children (Lew and Wong 2003). A third condition has been the presence of a relatively open class society in South China—a 'cultural borderland,' where the Confucian land-owning class was less dominant—allowing greater openness and upward socioeconomic mobility (Pan 1990: 13). The idea of leaving China to improve one's position in life

was simply more accepted in southern China, with its maritime ports and traditions of overseas trade, than in the north. A final condition was the social prestige
conveyed to expatriate sojourners through home visits and donations to hometown
causes and infrastructure. Making more trips home before retirement increased
one's prestige among fellow sojourners. So did the amount of money sent home to
relatives, and donated for public works projects for the villages.

The sojourner experience transformed following the establishment of the PRC
in 1949, but these values remained strong among Chinese who left China in the
first half of the twentieth century. Within China itself, Overseas Chinese Affairs
Offices became established at national and local levels immediately after the
People's Republic was founded. These offices continue to assist overseas Chinese
in finding their ancestral roots and finding ways to contribute to their home
villages (Lew and Wong 2003). They regularly publish magazines for the overseas
Chinese community containing news of root-finding visits, homeland donations
and economic investments. And many now provide this information on the World
Wide Web, for example the Guangdong Overseas Chinese website, which has
links to similar sites in other provinces (GOCW 2002). Many overseas Chinese
villages and towns in South China are full of monuments to the sojourning and
overseas Chinese experience. Temples have been rebuilt with donations from
visiting local descendants, in addition to other kinds of public infrastructure,
including roads and bridges. Local schools are especially favored for donations,
and visible subscription boards honor leading donors, thereby continuing to
construct social prestige for the overseas Chinese visitors.

Traditional sojourner values are less prominent among the more heterogeneous migrant populations that have emerged since the 1960s (Wong 1997). The
shrinking time–space economy of the 1990s, with economic globalization, the
expansion of the Internet, and cheaper and more accessible travel, have made
traditional sojourning less necessary, though still an option. Recent migrants,
however, still share some common characteristics with earlier generations. In
addition to shared racial and cultural characteristics, these can include a sense of
existential 'outsideness,' common among migrants and expatriates, a desire to
build communities with shared values and experiences, and a goal of improving
one's situation in life. The long history of Chinese filial piety, extended family, and
genealogical traditions may be less apparent among younger overseas Chinese,
but these values and conditions are likely to exert themselves over time as this
cohort matures and comes to recognize the 'center out there' as offering a unique
opportunity for self-identity formation and ethnic solidarity.

Modern motivations

The field of modernity/postmodernity (see Figure 18.1) has influenced the
migration/diaspora realm by structuring both the motivation and form of

Chinese diasporic movement. The global time–space convergence brought about by transportation and telecommunication advances in the twentieth century has further intensified at the beginning of the twenty-first century, helping to alleviate the strains of displacement, while also encouraging more people to risk migration, enabling new levels of both permanent and temporary migration, easier communication among dispersed family members, and more frequent home visits for larger numbers of migrants. Prosperity has increased for many, access to international travel and the Internet has become more widespread, rates for international telephone services have plummeted, and economic and cultural globalization has made transnational lifestyles more commonplace. One way that these new levels of 'hypermodernity' have changed the overseas Chinese migration equation has been to reenable the sojourner model, through frequent home visits and continuous access to news from the homeland via telephone, email, and the World Wide Web.

In the contemporary world of 'ultramodernity,' 'flexibility' prevails far and wide: governments deregulate, economies are irrational, social relativism prevails, and personal relations are impermanent (Harvey 1989, N. Wang 2000, Wong 1997). Cultural and moral relativism in social relationships, the pressure of global competition in the workplace, and the uncertainty wrought by constantly shifting economic and political fortunes (including the unpredictability of terrorism) are some of the major sources of 'postmodern' stress. This new world of uncertainty, fragmentation, displacement, lack of center, and constant change requires a high degree of flexibility in response, which can also generate anxiety and doubt about one's identity (Ong 1999). At the same time, rapidly developing technologies offer continually emerging opportunities to transcend time–space barriers, transform traditional structures and forms of social relationships, and strengthen the existential homeland tie for today's sojourner (Ohmae 2000). People are finding new ways to participate in communities that are scattered across the globe, as well as new ways to express meaningful self-identities.

Just as the sojourner's strengthening of existential homeland ties served to accommodate the stress of modernity, so too might existential tourism contribute to addressing the needs of a postmodern world. An existential homeland is particularly relevant to issues related to a sense of placelessness and alienation. Through detached and objective modernity, and deconstructed and centerless postmodernity, people have "suffered from a deepening condition of *homelessness*.... It goes without saying that this condition is psychologically hard to bear. It has therefore engendered…*nostalgias*…for a condition of 'being at home' in society, with oneself and, ultimately, in the universe" (Berger, Berger and Kellner 1974: 82, italics added)

There are many ways that these nostalgias emerge in our daily lives, ranging from the more authentic to simulcra and kitsch. Many are part of the tourism economy, which since at least the 1970s has been viewed as a means of addressing alienation and the social and psychological needs of modernity. Dean MacCannell (1976), for example, argued that through the tourist experience, individuals sought to overcome the alienating forces of the modern world by gazing upon, and being situated in, places of shared social and symbolic significance. While MacCannell saw the tourist as a decentered modern subject negotiating space in search of a group or social center, Cohen's existential tourist was one who was individually centered, but with a center or home 'out there' in a place different from the place of residence. The home out there addresses the condition of feeling homelessness, and partially overcomes the tensions of displacement. This is the situation of many ethnic Chinese residing outside of mainland China, though in other ways their tourism experience is similar to MacCannell's search for a center. The ancestral homeland in China also satisfies the nostalgia for an earlier time when purpose, authority, and ethics were more clearly defined.

For the overseas Chinese visitor, both the idea of China and the touristed landscape are commodified and shaped through the act of being a tourist. Unlike the traditional sojourner for whom the return home was an important life cycle event, the temporary and non-committal nature of the existential tourism visit has the paradoxical affect of othering the home place even as the visitor seeks union through this place encounter. Despite the expected rights (and sometimes unrecognized responsibilities) that accompany a blood relationship to a place, the commodity nature of the existential tourist trip divides the buyer and the bought, as the existential tourist is visiting a place but moreover creating a nostalgia. The overarching narrative is the genealogical past, which makes the geographic present a scene for negotiating the historic bloodline. John Goss (this volume) goes even further, pointing out that

> landscapes of tourist consumption are structured by allegory: saturated in the pathos of material loss, represented in verbal narratives, visual images, and objects in various stages of decay and obsolescence, and in the spatial displacement evoked in themes of departure and arrival.

The scenes of existential tourism too are subsumed by historical narratives, and, together with the consumptive nature of the tourism phenomenon, can have the effect of sidelining the lived lives and contemporary landscapes of the homeland.

At the same time, forces of modernity/postmodernity are rapidly changing the meaning of Chineseness within China itself, reshaping nostalgic landscapes and the relationships of both residents and visitors to them. This is probably best displayed in the urban landscapes of contemporary China. Postmodern-style façade treatments are rife in the new residential high rises and recreational shopping streets that are now found in every large city in the country. Beyond the surface, China is changing in many other ways, as well. In less than a decade China will have more Internet users than any other country (in 2002 it was second to the United States) (Greenspan 2002) and in less than a couple of decades it will generate more international leisure tourists than any other country (WTO 2001c). This is the reality of contemporary China—especially in its urban areas. Although rural China is also undergoing change, the pace has been much slower (especially in the central and western parts of the country) and for those who have not migrated to cities, premodern Chinese values and traditions remain relatively strong. For the existential traveler, the China visited today will rarely match the image of traditional China, though for a more pragmatic (and less nostalgic) visitor it may offer opportunities for the development of more contemporary networks of ethnicity than were possible in the past. In essence, the modernization of China in the twentieth century and the postmodern economic and social globalization worldwide in the twenty-first century have reshaped the Chinese sojourner experience into a modified existential visiting-friends-and-relatives and quasi-business tourism and travel experience. In this era of rapidly changing geographic space and place, the existential trip to the homeland will often fall short of its identity-affirming goals (Oxfeld 2001).

Travel behavior

At the center of this whirlwind of tradition, change, and physical movement through geographic space is the overseas ethnic Chinese subject, balancing competing influences and negotiating a path toward wholeness of culture, history, and identity. Existential tourism is one of the most effective ways in which this goal can be achieved. A survey of overseas Chinese visitors to China (Table 18.1) provides an empirical basis for understanding some of the characteristics of existential tourism and its broader manifestations in China. The survey was part of a larger sample of residents from selected countries and administered at Hong Kong's Chek Lap Kok International Airport (cf. Hui and McKercher 2001). Residents of China ($n = 218$) and non-ethnic Chinese visitors ($n = 312$) were excluded from the overseas Chinese questions. The total valid responses from overseas Chinese was 350, which was a fairly high percentage (52.9) of all of the valid non-mainland Chinese respondents ($n = 662$). Of these, 6.3 percent of the respondents were part Chinese.

The results of the survey indicated that although Chinese residing within Asia visited China in larger total numbers, those living in non-Chinese cultures far from their ancestral homeland demonstrated the strongest propensity to reconnect and reaffirm their Chineseness. This was seen in many ways. The visitation rates to Hong Kong among the more spatially displaced traveler groups were extraordinarily high, with Chinese residing in the United States visiting at rates over 30 times their actual proportion of US citizenry. Furthermore, over half of these long-haul ethnic Chinese visitors have visited their ancestral village area in China and they are more likely to have relatives and friends in the home village to whom they continue to send remittances (Table 18.2).

Table 18.1 Ethnic Chinese among visitors to Hong Kong by country of residence

Residence	Taiwan	Singapore	Malaysia	USA	Australia	W. Europe	Mean
No. of All Respondents, including non-ethnic Chinese	136	96	82	123	88	137	
% ethnic Chinese [a]	95.3	88.5	100	26.0	15.9	5.1	52.9
% ethnic Chinese citizens in country's total population	98.0	76.7	27.0	0.85	1.58	0.40[b]	34.1
Visitation rate [c] s	0.97	1.2	3.7	30.6	10.1	12.8	1.6
% ethnic Chinese who visited their home village area	39.6	36.8	28.8	50.0	66.7	100.0	53.7

Sources: Singapore, Malaysia and Taiwan: CIA (2002); USA: Kim (2001); Western Europe: Huaren.org (2002).

[a] "Chinese" includes the survey category of "part Chinese," which had an overall mean of 6.3 % of all ethnic Chinese respondents.

[b] Western Europe's "% ethnic Chinese citizens" is a mean based on census rates in the UK, France and Spain.

[c] Visitation rate = "% ethnic Chinese respondents" divided by "% ethnic Chinese citizens" in each country. Higher rates indicate higher than expected proportions of ethnic Chine se among visitors from that country.

Table 18.2 Respondent's connection (%) to home village by country of residence

Residence	Taiwan	Singapore	Malaysia	USA	Australia	W. Europe	Mean
Are you connected to your home village?							
Yes	38.2	42.1	39.0	59.3	75.0	85.7	56.5
Types of connections:							
Relatives	30.4	35.1	30.5	59.3	66.7	71.4	48.9
Friends	6.9	12.3	6.8	11.1	25.0	57.1	19.9
Remittances to relatives	10.8	10.5	15.3	18.5	16.7	28.6	16.7
Investments/work[a]	9.8	5.3	3.4	3.7	8.3	14.3	7.5
School donations	0	3.5	3.4	0	0	0	1.1
Other donations[b]	1.0	3.5	5.1	0	0	0	1.6
No. of respondents	102	57	59	27	12	7	

[a]Investments: car repair, consultant company, information technology, real estate investment (2), seafood, textile industry, tourism, toy manufacturing, retail trade. Work: electronic goods, medical, training.

[b]Ancestral hall, temples, roads, bridges, hospi tal/clinic, and unspecified.

For the ethnic Chinese who resided in countries in which they were a substantial minority, existential travel to their ancestral homeland, whether a specific village or China as a whole, provided a promise of overcoming the stresses of modernity, displacement, and minority status. This is accomplished through the continuing influence of traditional sojourner values, the continuing patronage of voluntary associations in overseas Chinese communities, and the modern transformation of communication and transportation that support and enhance continued communication between immigrants and their relations in China. Their existential travel reaffirms their Chineseness, though few would probably recognize it in these terms.

While these same patterns emerged for ethnic Chinese residing in places closer to China, the degree of intensity was somewhat lower. Minority status is less of an issue for these Asian overseas Chinese, and thus the need for reaffirmation is less, though it still exists. At the same time, Asian overseas Chinese have easier and cheaper access to a wider range of Chinese identity-forming places within Asia where they can actually collect a more complex range of experiences through which to identify for themselves what it means to be Chinese (Lew and McKercher 2002).

The survey found one indicator that overseas Chinese in Asia are more closely tied to traditional Chineseness. This was in the likelihood of making donations to, and investments in, the ancestral village area in China (see Table 18.2). Asian-based overseas Chinese were somewhat more likely to participate in these activities, possibly indicating stronger sojourner values in this population than among those who are further removed from China. In addition, because of geographic propinquity, investment and tourism promotion trips by Overseas Chinese Affairs Officers in China were more common to Southeast Asia than elsewhere (Lew and Wong 2003).

The difference between Asian and non-Asian overseas Chinese relates somewhat to the economic and political differences between 'Greater China I' (Guangdong, Fujian, Hainan, Hong Kong, Macau, and Taiwan) and 'Greater China II' (ethnic Chinese beyond the immediate vicinity of China) (Chang 1995). Greater China I has been the core focus of overseas Chinese-driven economic development in China. It is the region of origin of most overseas Chinese, it is the destination of most of their trips to China, and it is where they have reestablished their familial and economic relationships (Ma 2003). It is no exaggeration to say that Greater China I is a driving force behind much of China's phenomenal economic growth in the 1980s and 1990s.

Greater China II is a more difficult concept, both economically and politically, due to its dispersed and heterogeneous nature (Lew and Wong 2002). What ties these dispersed overseas Chinese together is the race-based cultural notion of Chineseness and the seduction of China that Chineseness tends to generate (Ang 2001, Chan 1999). The results of this study, however, indicate that there is a need to distinguish a 'Greater China III' which would consist of ethnic Chinese residing outside of East and Southeast Asia, a group that Laurence Ma (2003) estimated to number about 7.5 million. Though the results of the survey cited here are cursory and exploratory in nature, they seem to indicate that these co-ethnics residing outside Asia share a significantly stronger likelihood of maintaining social ties to China, and of participating in existential tourism experiences, than do those residing in Asian countries.

As suggested above, the desire to establish and maintain existential ties to a homeland in China is generally strongest among older generation Chinese migrants who were born in China. The survey, for example, found that respondents over age 45 were far more likely to have visited their home villages (56 percent) than those age 45 and under (34 percent). In addition, 86 percent of those born in China and 63 percent of those born in Hong Kong have made existential trips to their ancestral village. The influence of Chineseness and migration would be somewhat weaker, or perhaps just different, for the overseas-born descendants of Chinese immigrants. The survey results supported this contention as the proportion of those having visited their home village dropped to a mean of 31 percent for those born in other Asian countries (including Taiwan), and near 0 percent for the small number included in the survey who were born elsewhere. For overseas Chinese, being born in China appears to serve as a major divide in the relationship that overseas Chinese have to their ancestral villages. Those born in China have the strongest existential ties to the homeland, while those born overseas, and especially outside of Asia, have the weakest.

Conclusions

Figure 18.1 defined three major social structures or forces that have shaped and influenced the identities of overseas Chinese: Chineseness, migration/diaspora, and modernity/postmodernity. The survey data indicated that these social structures can explain the phenomena of existential tourism of overseas ethnic Chinese back to China, though residents of different countries tend to occupy different locations in the overlapping arena of conditions, depending on the relative importance of each realm.

Stephen Chan (1999) argued that what we see today as the overseas Chinese community has been created by a series of smaller 'geographical diasporas,' each of which has its own distinct causal structure and social organization. For him, they are tied together only by the centrality of 'Chineseness,' which is essentially a racial category that cannot be escaped, but which also carries with it powerful cultural significances. Ien Ang (2001: 51), for example, stated "if I am inescapably Chinese by *descent*, I am only sometimes Chinese by *consent*," and then questioned whether one can, "when called for, say no to Chineseness?" She could not definitively answer this question. Similarly, Chen (2002: 1) reflected on his inability to speak Chinese fluently while living in Hong Kong, "Perhaps more so than any other race, being Chinese carries with it expectations beyond the physical. It's a complete package: linguistic, historical, psychological as well as physical. To be Chinese and not speak the language fluently, well, the mind boggles." Can one have 'partial Chineseness'?

We have contended here that in addition to Chineseness, the migration/diaspora experience and modernity/postmodernity conditions have also been significant influences in defining the overseas Chinese subject, and need to be considered equally. The networks of ethnicity that these three influences engender become strongly articulated in the phenomenon of overseas Chinese traveling back to China. The existential tourist experience is not necessarily an easy one, though it is most assuredly reflexive and has the promise to be identity affirming—if only so in a temporary and perhaps illusory manner. Just how long does the existential relationship to an ancestral place last? Given the changes that China has undergone in recent decades, for some it may not even survive the initial arrival. But still the seduction of the homeland remains.

People participate in existential tourism because sojourning is not an option for most immigrants today. We have become too entrenched and attached in myriads of ways to our new, physical homes—even if they are lacking an existential essentialness. At the same time, the ancestral homeland is a real place that has also undergone modernization and change at a pace and purpose separate from our overseas home and life. While existential homeland ties are also real, the interactions that surround them are largely structured in a traditional and historical narrative which may often be poorly suited to contemporary diaspora needs, especially among second-generation overseas Chinese (Lew and Wong 2003). Further complicating the situation are individual social relationships between homeland and overseas Chinese that emerge as part of the existential travel process. These can be complex, ambivalent, superficial, and, at worst, exploitative (Oxfeld 2001). In *Imaginary Homelands*, Salman Rushdie (1991) claimed that we can never go 'home' again, no matter how many visits we make, because the homeland and the home village are as much imaginary places as real ones. We construct our ethnicity, and our own authenticity, from what Mathews (2000) has called the postmodern 'cultural supermarket,' in which all identities are impermanent and the only real home is the entire world. Once one enters the cultural supermarket, which is both reflexive and fun, there is no turning back, there is no going home again. In this context, the existential tourist is a socially constructed role, though one which has been clearly informed by traditional Chineseness, shaped to the migration/diaspora experience, enabled by global modernization, and seduced by the promise of a sense of home in China.

19 Conclusion

Centering tourism geography

Alan A. Lew and Carolyn Cartier

We first started talking about this book in the canteens of the National University of Singapore, the scene of outdoor food stalls where the mix of smells and spices from Malay, Chinese, and Indian dishes hangs heavy in the constant humidity. The food vendors continue the tradition of hawkers who historically worked streetside; the Singaporean government organized them into food centers during the modernization drive, separating functional spaces and ordering public landscapes. As visiting academics, we were there for research, driven by understanding such landscape transformations as well as interested to explore the city's diversity of places. We were, in essence, tourists. And like many tourists, we were motivated by Singapore's cacophony of cultures, its juxtaposition of modernity and tradition, and the social contradictions that result from being an economically open yet politically circumscribed society. As tourists we were seduced by the promise of new sites/sights, foods to savor, the sounds and pace of the urban scene, and full body immersion in the equatorial climate. Being in Singapore (like many of the most seductive tourist destinations) was a continual intellectual and sensory experience. It is often out of such complex milieux that new ideas emerge to challenge conventions, and that was the context in which we forged the collaboration for this book.

In particular, our conversations converged on the limitations of tourism literature in geography, and mainstream human geography's relative lack of engagement with tourism subjects. We were motivated by several concerns. The first concern was recognition of the everyday nature of tourism in contemporary society—the touristed landscaping of virtually every corner and every aspect of the world—and how to better represent this reality in our work. The second was the idea of places as having a seductive quality—more than 'sense of place'—and the special insights that the idea of place seduction might provide in understanding touristed landscapes. The final challenge that we wanted to address was the gap between contemporary cultural geography and the type of tourism research undertaken by many tourism geographers, and how to address the divide between two fields that should be so obviously more conjoined.

We see tourism as a phenomenon and set of processes that has increasingly become embedded, whether intentionally or unintentionally, in the relationship between modernity and place (both contemporary and historic), in how places are created, and how they are experienced. Touristed landscapes, as tourism's emplaced material conditions, span the world, and it becomes increasingly difficult to delineate the tourism realm from places of non-tourism. By touristed landscapes we are referring to places that are leisure-oriented, places that offer promise of escape from daily life—for a week, a day, or even an hour—as they exist in our areas of residence and our regions of work, as well as more distant destinations. So the touristed landscape exists on a continuum, as it often seems that tourism processes and forms are increasingly defining traditionally non-tourism processes and forms. We also see how images and ideas about distant places are mapped onto touristed landscapes, as in Claudio Minca's relational study of the 'two Bellagios,' and how, in Anne-Marie d'Hauteserre's chapter, the tourism industry deploys images of 'Tahiti' to conjure imaginations about paradise worldwide. State interest in the promotion of urban economies also relies on place re-imaging, as in the case of Sydney and Melbourne, discussed by Michael Hall, and in T.C. Chang and Shirlena Huang's analysis of Singapore. In these places, the state works to transform and transmit place images, in effect sending them out into the global cultural economy in order to attract resources. Trevor Sofield and Sarah Fung Mei Li demonstrate how historical Western ideas about monuments and imperial China constructed a singular Great Wall of global imagination that is not much at all like the real Great Wall. Such global interrelating of place images lies at the heart of cultural geography's project— but only if it would take a distinctively global turn, away from its historic Anglo-American bias. Thus insights of several chapters point to how global tourism studies offer new lines of inquiry for established theoretical fields like the new cultural geography. While the new cultural geography recognizes multiple histories and cultures of place, analysis of the touristed landscape and place desire suggests another lens through which to view cross-cutting issues of culture and economy, experiential questions about mobility, embodiment and desire, and in the context of scale-transcending processes from local to global, innovative perspectives on multi-sited geographies.

So while our ideas about the relationship between the fields of tourism geography and cultural geography initially viewed the latter as the main arena of theoretical innovation, we began to see how research concerned with perspectives on tourism itself led to opening up innovative lines of inquiry. This situation is especially evident in the work of David Crouch, whose perspectives on embodiment and 'doing tourism' have yielded alternative ways of thinking about the tourist as agent, in terms of sensory mobility, gendered awareness of bodily emplacement, and the idea of doing tourism as an intermittent yet ongoing aspect of daily life–leisure. Such a perspective arguably better captures the empirical

reality of embodied leisure/tourism practice. Jon Goss and Tim Oakes also inno-
vate theoretical ideas about some of the most fundamental aspects of modernity
via complex understandings of tourism. Tim Oakes succeeds in providing a
nuanced definition of place seduction, and Jon Goss's chapter realizes a stunning
portrayal of the 'lifepath' of landscape, in ideas about how (im)mortality becomes
inscribed in touristed landscapes and their representative objects.

Perspectives on leisure/tourism economy also suggest alternative ways of
analyzing the globalization of service industries and their relationship to leisure.
Ginger White's presentation of the world tourism economy provides a global
economic perspective on the pervasiveness of tourism and how it is tied closely to
the globalization of service industry technologies. The relationship between
tourism as a set of service industries and the idea of the increasing personalization
of services in contemporary consumer society, discussed in the introduction,
points to how tourism as a set of embodied practices, as in the rise of 'health
tourism,' is on a course of increasing diversification—which suggests that the
evolution of leisure/tourism 'products' will continue to diversify. Dean
MacCannell brings the theme of global economic and technological restructuring
to the local landscape of individual work and leisure experience, showing how
personalized and embodied diversification in leisure is also emerging in 'silicon
landscapes.' His analysis of work-related values in Silicon Valley suggests how the
global computer revolution and the 'get rich quick' economy it engendered have
created new forms of leisure, exemplified by X-games and digital pop stars, even
prompting instant and fleeting dimensions of human relations. Dean MacCannell
argues that the edge of the global techno-economy seems to be creating an
emerging tourism/leisure/recreation landscape that is better understood in
terms of speed and rhythm ('leisure Taylorism' for the global era?) than from
traditional geographic perspectives on place—since Silicon Valley's wealth has not
generated remarkable public places but instead more personalized ones, i.e.
monster houses. Both of these cases underscore the significance of embodied
practices of leisure/tourism emerging in this era of increasing personalization.

Travel and tourism are also avenues along which we may address the complex
contradictions and perplexities of modern life, especially in transitions between
places and explorations of places with different cultural–economic conditions and
values. Tim Oakes finds a discerning definition for place seduction in located
cross-cultural experience, the encounter with a cultural other whose emplaced
conditions constitute the moment of seduction. Here too, place seduction is a
thoroughly embodied experience. Our authors also find different subjects of
seduction. Jon Goss sees the relationship of seduction to the touristed landscape
through the power of things—the souvenir—to represent allegories of historic
demise and resurrection, and, in turn, how the seduction of consuming the
souvenir places tourists in the drama of landscape history. In the case of domestic
tourism to Guangzhou, Ning Wang finds questions about negotiating modernity

in relation to industrial modernization in the form and symbolism of the contemporary built environment, which itself draws large numbers of tourists. Guangzhou's new built environment has all the appearances of a service sector post-industrial city being shaped by images of leisure values and facades of affluence. It is no wonder that the first signs of post-communist modernity in the Chinese city were found in the touristed landscapes of international hotels. Though perhaps less apparent than in many developing economies, modernity and leisure/tourism similarly come together in the image-making of developed world cities. Michael Hall, in describing the competitive nature of tourism development between Sydney and Melbourne, argues that redefining a city's leisure persona is an increasingly important tool in the civic booster's arsenal. T.C. Chang and Shirlena Huang's study of Singapore examines a similar state logic but with a different outcome, underscoring struggles between state and society over local meanings of place, reflecting ways that people understand their histories and will not readily allow their local leisure landscapes to be themed-over for globalizing tourist consumption.

Tourism, as a complex landscape and social phenomenon, is so deeply intertwined with both the leisure and working worlds of the modern city that extracting its impacts and implications from the landscape requires uncovering a city's larger history of place. Carolyn Cartier's historical examination of the seduction of place in San Francisco uncovers how it selectively builds upon a complex assemblage of multisensory human–environment relations and local political economic achievements, the results of which are portrayed in positive place-imaging by diverse media and institutions. In San Francisco, a major global destination, the complex texture of the city's landscape history allows almost every site of distinction to become a touristed landscape, reflecting both resident and non-resident values, imaginaries and desires. In San Francisco we also see the full, embodied, and direct experience of place that David Crouch theorizes in doing tourism. Such embodied experiences are part of daily life in San Francisco.

Many landscape studies in the new cultural geography concern place contestation, and tourism studies of contested landscapes have typically focused on how tourism may threaten local meanings of place. By contrast, all of the studies presented here are dealing with well-developed tourism economies, and so questions about the dynamic role of tourism in the landscape shift—to more nuanced and complex concerns about the contradictions of culture and economy embedded in highly touristed places. Mary McDonald's analysis of wedding tourism in Hawai'i especially captures such issues by demonstrating ways that different people and groups perceive, experience, and relate to touristed landscapes. Here we are dealing with much more than whether people want or resist tourism, since Hawai'i is already intensively touristed. In her study, McDonald shows how communities in Hawai'i must cope with highly constructed (both socially and materially) matrimonial landscapes that are in some places overwriting meanings of Oʻahu's most historic, indigenous landscapes—and tourist sites. That these weddings of Japanese citizens have no legal meaning makes them all the more culturally interesting. Jeff Baldwin's study of Antigua further demonstrates the limits to long-established island tourism economies, in how local meanings of coastal environments become a basis of resistance to alliances of the local state and transnational capital, revealing how the state's discourse of 'localness' is a front for an elite accumulation project. The problems of over-touristed landscapes in the Caribbean seem even to be compelling tourists to shift perspective gaze to 'seascapes.' Janet Momsen's study of Caribbean tourism reveals how intensified focus on the beach is leading to further development of already developed coastal beaches, creating more enclave resorts and pushing the place of tourism out to sea, to the cruise ship where people are swaddled in creature comforts. In the process, islands of destination are reduced to viewscapes and imaginaries, neatly distanced from the luxury bubble of the ship—its own form of personalized comfort and seduction.

The concept of place seduction plays out in every chapter, though not always explicitly. Together, the chapters demonstrate a diversity of seductive conditions and environments. Some are highly managed and marketed; others are a happenstance of history, location, birth, and life choices, such as migrants who continue to negotiate and re-negotiate attachments to their ancestral homelands, as Alan Lew described in the case of overseas Chinese. Some seductions are crude and thoughtless, others appear refined and self-reflective. All involve encounters between tourists and place, perhaps resulting in place transformation, and possibly a change in the tourist. In the context of circuits of travel, both destinations and homes have strong seductive qualities; so place seduction can be thought of as a fulcrum that binds and supports the relationship between people and places—both home places and non-home/tourism places. Desire to travel may be motivated in part through the tourism industry's semiotic system, through advertising themes including sun, sand, and sea, adventure and culture, by the allure of information gleaned through myriad bits of news and other media, as well as by stories and photos of returned tourists. People drawn to such representations are arguably looking for something, yet even together, such representations create partial imaginaries. So seduction, as Tim Oakes urges, is a kind of psychological operative; we believe that certain places and emplaced encounters have the potential to satisfy deeply held needs to know and connect with other people and places in the world.

Among the chapters, the more highly constructed and managed seductive encounters include those discussed by Claudio Minca, as above, and Margaret Swain. Both authors creatively account for the multiple interpretations and manipulations of imaging tourist places, as well as the relationship of the image to more 'authentic' realities of place. In the case of China's Sani people, the state re-images the legendary Ashima for national and international visitor markets, transforming the very display of Sani culture within the tourist settings of the

Stone Forest. The tourist Ashima is not the Sani Ashima, it is both less and more: it is less in its redefinition of Sani-ness; yet it is more in its evolution into a global icon. In the process, the state genders female 'minority peoples' as an element of borderland control. Perhaps no less contrived, but certainly structured and sold in a different manner and at different scales, are the seductions of mythical and exotic places through transhistorical reiterations in literature and the arts. Tahiti exemplifies these conditions, historically framed by the allure of sex, sun, and sand, which Janet Momsen also portrays in the Caribbean. In all of these instances, the landscape promises a sensory experience to entice a visitor. People visit places in the process, but extracting the image from the real in the visitor experience is as problematic as finding the real in the imaged place.

While scholarly perceptions have been changing, geographers (and other social scientists) who study tourism have historically been perceived in part to be more interested in personal travel experiences than in serious place study. Perhaps there was some truth in this, as much tourism-related research was highly descriptive and overly empirical, lacking both compelling methodology and theoretical relevance. What is probably more the case, however, is that tourism geographers have been drawn to applied work, which requires research that is oriented to public policy and economic profit, and which is relatively disconnected from core disciplinary debates. But this situation is also changing. More theoretically oriented geographers become interested in tourism when we cast its questions in complex ways, located at the intersection of the local and the global, the cultural and the economic, the psychological and the physical, dialog between residents/visitors, the worlds of work and play, and the 'real' and the 'make-believe.' The geographer's desire to understand the formative contexts of places and landscapes makes tourism a compelling platform from which to study such issues. Rather than a travel opportunity, study of tourism provides perspectives from which to understand the complexities of the modern human condition and contemporary landscapes; and because tourism is as pervasive as it is, it provides a central vehicle for understanding contemporary place processes, prevailing ideologies, and what becomes acceptable and contestable in landscape change.

Bibliography

ABDT (Antigua and Barbuda Department of Tourism) (1996, 1997) *Air Passenger Movement Statistics*, Unpublished.

——(2000) *Statistical Report 1999*, St. Johns, Antigua: ABDT.

——(2002) Online. Available <http://www.antigua-barbuda.org> (accessed 31 July 2002).

Abdullah, D.O. (2002) "M'sia gears up for health tourism blitz: ministries and private hospitals team up to woo patients," *The Business Times* (Singapore), 25 November.

Abedor, C. (1998) "Wedding chapel precedent threatens neighborhoods," letter to the editor, *Honolulu Star Bulletin,* 2 January.

ABLSO (Antigua and Barbuda Land Survey Office) (1991) *Aerial Survey of Antigua, 1:10,000.*

ABMI (Antigua and Barbuda Ministry of Information) (1997a) *A Glance at Statistics*, St. Johns, Antigua: Statistics Division.

——(1997b) *Let's Take Our Country Forward with the Guiana Island Project*, Barbados: Antiguan Ministry of Information.

Abram, S. (1997) "Performing for tourists in rural France," in S. Abram and J. Waldren (eds) *Tourists and Tourism: identifying with people and places*, Oxford: Berg.

Adamski, M. (1999a) "Neighbors want no liquor at chapel," *Honolulu Star Bulletin*, 29 July.

——(1999b) "Wedding company's bid for champagne license still on ice," *Honolulu Star Bulletin*, 29 October.

——(2000) "Wedding chapel won't pursue liquor sales," *Honolulu Star Bulletin*, 20 December.

Adorno, T. (1987) *Minima Moralia: reflections from damaged life,* trans. E.F.N. Jephcott, New York: Verso.

Agnew, J. (1987) *Place and Politics: the geographical mediation of state and society*, Boston: Allen & Unwin.

——(1989) "The devaluation of place in social science," in J. Agnew and J. Duncan (eds) *The Power of Place: bringing together sociological and geographical imaginations*, Boston: Unwin Hyman.

Aguiar, E. (2000a) "Waimānalo panel opposes park chapel liquor license," *Honolulu Advertiser*, 15 November.

——(2000b) "Kahalu'u chapel license irks board," *Honolulu Advertiser*, 28 December.

——(2002) "Liquor license foes seek full probe," *Honolulu Advertiser,* 13 June.

Ahmad, R. (2002) "Check-in for a holiday, check-up at the hotel, check-out in good health," *The Straits Times* (Singapore), 3 January.

AHTA (Antigua Hotels and Tourist Association) (1997a) *Tourism Vibes*, May.

——(1997b) *Heart of the Caribbean*, St. Johns, Antigua: ATHA.

Albuquerque, K. (1991) "Conflicting claims on the Antigua coastal resources: the case of the McKinnon and Jolly Hill Salt Ponds," in N.P. Grivan and D. Simmons (eds) *Caribbean Ecology and Economics*, St. Michael, Barbados: Caribbean Conservation Association.

Aldrich, R. (1990) *The French Presence in the South Pacific 1842–1940*, Honolulu: University of Hawai‘i Press.

Aldridge, A.O. (1972) "Voltaire and the cult of China," *Tamkang Review*, 2(2–3): 25–9.

Alexander, K. (2002) "Delta ending many travel agent commissions," *Washington Post*, 15 March.

——(2003) "Fliers wonder how it can be worse in 2003," *The Washington Post*, 1 January.

Altrocchi, J.C. (1949) *The Spectacular San Franciscans*, New York: E. P. Dutton.

Amendola, G. (1997) *La Città Postmoderna: magie e paure della metropoli contemporanea*, Roma: Laterza.

Amin, A. and Thrift, N. (2002) *Cities: reimagining the urban*, Cambridge: Polity Press.

Anderson, I. (1898) *Yellow Fever in the West Indies*, London: H.K. Lewis.

Anderson, K. and Gale, F. (eds) (1992) *Inventing Place: studies in cultural geography*, Melbourne: Longman Cheshire.

Ang, I. (2001) *On Not Speaking Chinese: living between Asia and the West*, London: Routledge.

Ang, I. and Stratten, J. (1995) "The Singapore way of multiculturalism: Western concepts/Asian cultures," *Sojourn*, 10(1): 65–89.

Anthia, F. (1999) "Theorising identity, difference and social divisions," in M. O'Brien, S. Penna, and C. Hay (eds) *Theorising Modernity: reflexivity, environment and identity in Giddens' social theory*, London: Longman.

Anthonyson, D. (2001) "Community update," *The EAG'er*, St. Johns, Antigua: The Environmental Awareness Group, 31: 3.

Anwar, Y. (2001) "Wedding dispute splinters Kawaiaha‘o Church," *Honolulu Advertiser*, 7 January.

Anzaldúa, G. (1987) *Borderlands: la frontera = the new mestiza*, San Francisco: Aunt Lute Books.

Appadurai, A. (1986) "Introduction: commodities and the politics of value," in A. Appadurai (ed.) *The Social Life of Things*, Cambridge: Cambridge University Press.

——(1990) "Disjuncture and difference in the global cultural economy," in M. Featherstone (ed.) *Global Culture: nationalism, globalization and modernity*, London: Sage.

——(1996) *Modernity at Large: cultural dimensions of globalisation,* Minneapolis: University of Minnesota Press.

Apple, R.W. (1999) "On the road: even through the fog, that city on a hill dazzles," *The New York Times*, 25 June.

Ashworth, G.J. (1990) *The Tourist-Historic City*, London: Bellhaven.

——(1994) "From history to heritage: from heritage to identity: in search of concepts and models," in G.J. Ashworth and L. Larkham (eds) *Building a New Heritage: tourism, culture and identity in the New Europe*, London: Routledge.

Ashworth, G.J. and Larkham, L. (eds) (1994) *Building a New Heritage: tourism, culture and identity in the New Europe*, London: Routledge.

Ashworth, G.J. and Voogd, H. (1990) *Selling the City: marketing approaches in public sector urban planning*, London: Belhaven Press.

Associated Press (2002) "FAA: air travel to rebound big in 2003." Online. Available <http://www.CNN.com/2002/TRAVEL/NEWS/03/3/12/rec.airline.forecast.ap/index.html> (accessed May 2002).

Augé, M. (1995) *Non-Places, Introduction to an Anthropology of Supermodernity*, trans. J. Howe, London: Verso.

——(1997) *L'Impossible Voyage: le tourisme et ses images*, Paris: Editions Fayot.

Austin, J.L. (1962) *How to Do Things with Words,* London: Oxford University Press.

Averini, S. (1968) *Karl Marx on Colonialism and Modernization*, New York: Doubleday.

Bachelard, G. (1994) *The Poetics of Space*, Boston: Beacon Press.

Baldwin, J. (1998) "Tourism, Development, and Environmental Alteration in Antigua, West Indies: wetland reclamation and changing views of coastal ecologies," Unpublished MA thesis, University of Oregon.

——(2000) "Tourism development, wetland degradation, and beach erosion in Antigua, West Indies," *Tourism Geographies*, 2(2): 193–218.

Ban, K.C. (1992) "Narrating imagination," in K.C. Ban, A. Pakir, and C.K. Tong (eds) *Imagining Singapore*, Singapore: Times Academic Press.

Bangkok Post (1999) "Two approaches to hotels and health," 4 February.

Barbados Ministry of Tourism (1997) "Barbados stayover visitor survey, Oct-Dec, 1996," Prepared by CTO, Barbados: Ministry of Tourism.

Baré, J.F. (1985) *Le Malentendu Pacifique*, Paris: Hachette.

Barnes, T. and Duncan, J. (1992) *Writing Worlds: discourse, text, and metaphor in the representation of landscape*, London: Routledge.

Barnett, C. (1999) "Deconstructing context: exposing Derrida," *Transactions of the Institute of British Geographers*, 24(3): 277–94.

Barrow, J. (1805) *Travels in China*, Philadelphia: M'Laughlin.

Barthes, R. (1986) *Mythologies*, trans. A. Lavers, New York: Hill & Wang.

Bataille, G. (1991) *The Accursed Shared: an essay on the general economy, Vol. 1: Consumption*, trans. R. Hurley, New York: Zone Books.

Baudelaire, C. (1964) *The Painter of Modern Life, and other essays*, trans. J. Mayne, London: Phaidon.

Baudrillard, J. (1981) *For a Critique of the Political Economy of the Sign*, trans. C. Levin, St. Louis: Telos Press.

——(1983) *Simulations*, trans. P. Foss, P. Patton, and P. Beitchman, New York: Semiotext(e).

——(1988a) *America*, trans. C. Turner, London: Verso.

——(1988b) *The Ecstasy of Communication*, trans. B. and C. Schutze, Brooklyn, NY: Autonomedia.

——(1988c) "On seduction," in M. Poster (ed.) *Selected Writings: Jean Baudrillard*, Stanford, CA: Stanford University Press.

——(1988d) *Selected Writings*, ed. W. Poster, Cambridge, MA: Polity Press.

——(1990) *Seduction*, trans. B. Singer, New York: St. Martin's Press.

——(1994) [1981] *Simulacra and Simulation*, trans. S. Glaser, Ann Arbor: University of Michigan Press.

——(1996) *The System of Objects,* trans. J. Benedict, New York: Verso.

——(1998) *The Consumer Society: myths and structures*, London: Sage.

Bauer, W. (ed.) (1980) *China und die Fremden: 3000 Jahre Auseinandersetzung in Krieg und Frieden*, Munich: C.H. Beck.

Bauman, Z. (1993) *Postmodern Ethics*, Oxford: Blackwell.

——(1998) *Globalization: the human consequences*, New York: Columbia University Press.

Bean, W. (1968) *California: an interpretive history*, New York: McGraw-Hill.

Behdad, A. (1994) *Belated Travellers: orientalism in the age of colonial dissolution*, Durham, NC: Duke University Press.

Belk, R. (1991) "The ineluctable mystery of possessions," *Journal of Social Behavior and Personality*, 6: 17–55.

Bell, D. and Valentine, G. (1995) "Introduction: orientations," in D. Bell and G. Valentine (eds) *Mapping Desire: geographies of sexualities*, London: Routledge.

——(1997) *Consuming Geographies: we are where we eat*, London: Routledge.

Bellagio Dove (2000) Associazione Operatori Turistici e Economici "PromoBellagio."

Benjamin, W. (1977) *The Origin of German Tragic Drama*, trans. J. Osborne, London: New Left Books.

——(1983) *Charles Baudelaire: a lyric poet in the era of high capitalism*, trans. H. Zohn, London: Verso.

——(1985) "Central Park," *New German Critique*, 34: 32–58.

Berger, J. (1972) *Ways of Seeing*, London: Penguin Books.

——(1992) *Ways of Seeing,* Harmondsworth: Penguin.

Berger, P., Berger, B., and Kellner, H. (1974) *The Homeless Mind: modernization and consciousness*, New York: Vintage Books.

Berhad, J. (2002) *Home Page*, December 28, 2002. Online. Available <http://www.johanholdings.com/financial.html> (accessed 12 June, 2002).

Berleant-Schiller, R. (1991) "Statehood, the commons, and the landscape in Barbuda," *Caribbean Geography*, 3(1): 43–52.

Berman, M. (1970) *The Politics of Authenticity: radical individualism and the emergence of modern society*, New York: Atheneum.

——(1982) *All That Is Solid Melts Into Air: the experience of modernity*, New York: Simon and Schuster.

Besson, J. (1987) "A paradox in Caribbean attitudes to land," in J. Besson and J. Momsen (eds) *Land and Development in the Caribbean*, London: Macmillan.

Bhabha, H. (1990) "The third space," in J. Rutherford (ed.) *Identity, Community, Culture and Difference*, London: Lawrence & Wishart.

——(1994) *The Location of Culture*, London: Routledge.

——(1996) *Locations of Culture: discussing post-colonial culture*, London: Routledge.

Bi, Z. (1986) "Ashima: women minzu de ge (Ashima: our nationality's song)," *Lunan Wenshi Ziliao*, 1: 176–96.

Biddle, G. (1968) *Tahitian Journal*, Minneapolis: University of Minnesota Press.

Billig, M. (1991) *Ideology and Opinions: studies in rhetorical psychology*, London: Sage.

Birkeland, I. (1999) "The mytho-poetic in northern travel," in D. Crouch (ed.) *Leisure/Tourism Geographies: practice and geographical knowledge*, London: Routledge.

Black, D., Gary, G., Seth, S., and Lowell, T. (2002) "Why do gay men live in San Francisco?", *Journal of Urban Economics*, 51: 54–76.

Blainey, G. (1983) *The Tyranny of Distance: how distance shaped Australia's history*, revised edn, South Melbourne: Sun Books.

Blume, M. (1992) *Côte d'Azur: inventing the French Riviera*, Slovenia: Thames and Hudson.

Blunt, A. and Rose, G. (1994) "Introduction: women's colonial and postcolonial geographies," in A. Blunt and G. Rose (eds) *Writing Women and Space: colonial and postcolonial geographies*, New York: Guilford Press.

Bocock, R. (1993) *Consumption*, London: Routledge.

Bolz, N. (1998) "The user-illusion of the world (1): on the meaning of design," *Mediamatic*. Online. Available <http://www.mediamatic.net> (accessed 15 June 2002).

Boorstin, D.J. (1964) *The Image: a guide to pseudo-events in America*, New York: Harper & Row.

——(1992) "*The Image: a guide to pseudo-events in America*," original publication 1961, New York: Vintage.

Bossi, R. and Cintron, G. (1990) *Mangroves of the Wider Caribbean: toward sustainable management*, Washington, DC: United Nations Environment Program.

Bougainville, L.A.D. (1771) *Voyage Autour du Monde*, Paris: Saillant & Nyon.

Bourdieu, P. (1984) *Distinction: a social critique of the judgment of taste*, trans. R. Nice, Cambridge, MA: Harvard University Press.

——(1986) *The Production of Belief: contribution to an economy of symbolic goods*, Thousand Oaks, CA: Sage.

Bowles, P. (1949) *The Sheltering Sky*, New York: Vintage.

Bradbury, M. (1976) "The cities of modernism," in M. Bradbury and J. McFarlane (eds) *Modernism, 1890–1930*, Harmondsworth: Penguin.

Bramwell, B. and Rawdding, L. (1996) "Tourism marketing image of industrial cities," *Annals of Tourism Research*, 23(1): 201–21.

Bratton, D. (2002) Interview, San Francisco Convention and Visitors Bureau, Research Manager, 15 June.

Briguglio, L., Butler, R., Harrison, D., and Filko, W.L. (eds) (1996) *Sustainable Tourism in Small Island States: case studies*, London: Pinter.

Britton, S. (1982) "The political economy of tourism in the Third World," *Annals of Tourism Research*, 9: 331–58.

——(1991) "Tourism, capital and place: towards a critical geography of tourism," *Environment and Planning D: Society and Space*, 9: 451–78.

——(1999) "Tourism, dependency and development," in D. Pearce and R. Butler (eds) *Contemporary Issues in Tourism Development*, New York: Routledge.

Brogger, B. (2000) "Singapore *huiguan* members' donations and investments in *qiaoxiang* areas— reasons, problems and rewards," in C. Huang, G. Zhuang, and T. Kyoko (eds) *New Studies on Chinese Overseas and China*, Leiden: International Institute for Asian Studies.

Broussard, A.S. (1993) *Black San Francisco: the struggle for racial equality in the West, 1900–1954*, Lawrence, KS: University of Kansas Press.

Brown, B. (1963) *After Imperialism*, London: Heinemann.

——(1998) "How to do things with things (a toy story)," *Critical Inquiry*, 24(4): 935–64.

Brown, M. (1998) *The Spiritual Tourist*, London: Bloomsbury.

Bruner, E. (1994) "Abraham Lincoln as authentic reproduction: a critique of postmodernism," *American Anthropologist*, 96(2): 397–415.

Bucher, G. (2000) Personal interview.

Buci-Glucksmann, C. (1994) *Baroque Reason: the aesthetics of modernity*, Thousand Oaks, CA: Sage.

Buck, E. (1993) *Paradise Remade*, Philadelphia: Temple University Press.

Buck-Morss, S. (1989) *The Dialectics of Seeing: Walter Benjamin and the arcades project*, Cambridge, MA: MIT Press.

Bulcroft, K., Smeins, L., and Bulcroft, R. (1999) *Romancing the Honeymoon: consummating marriage in modern society*, Thousand Oaks, CA: Sage.

Bunce, L.L.M. (1997) "Integrated coastal zone management of common pool resources: a case study of coral reef management in Antigua," West Indies, Unpublished PhD dissertation, Duke University.

Bushnell, O.A. (1993) *The Gifts of Civilization: germs and genocide in Hawai'i*, Honolulu: University of Hawai'i Press.

Butler, J. (1990) *Gender Trouble: feminism and the subversion of identity*, New York: Routledge.

——(1993) *Bodies that Matter: on the discursive limits of "sex"*, London: Routledge.

——(1997) *The Psychic Life of Power: theories in subjection*, Stanford, CA: Stanford University Press.

Butler, R.W. (1980) "The concept of a tourist area cycle of evolution: implications for management of resources," *Canadian Geographer*, 24(2): 5–12.

——(1993) "Tourism development in small islands: past influences and future directions," in D.G. Lockhart, D. Drakakis-Smith, and J. Schembri (eds) *The Development Process in Small Island States*, London: Routledge.

Cabezas, A.I. (1999) "Women's work is never done: sex tourism in Sosúa, the Dominican Republic," in K. Kempadoo *Sun, Sex and Gold: tourism and sex work in the Caribbean*, Lanham, MD: Rowman and Littlefield.

Caen, H. (1994) "Open other end," *The San Francisco Chronicle*, 14 April.

Calvino, I. (1974) *Invisible Cities*, trans. W. Weaver, New York: Harcourt Brace.

Calypso Talk '96 (1996) St. John's, Antigua: Wadadli Publications.

Camp, W.M. (1947) *San Francisco: port of gold*, Garden City, NY: Doubleday.

Campbell, C. (1987) *The Romantic Ethic and the Spirit of Modern Consumerism*, London: Oxford University Press.

Campbell, I.C. (1989) *A History of the Pacific Islands*, Berkeley, CA: University of California Press.

Campos, F. (1997) "Waikīkī Lutheran wants piece of wedding cake," *Pacific Business News*, 17 January.

Cana Online (2002) "Antigua and Barbuda: Premier says civil unrest will not be tolerated," Bridgetown, Barbados: Caribbean Media Corporation. Available <http://cananews.com> (accessed 28 July 2002).

Carrier, J.G. (1994) *Gifts and Commodities: exchange and western capitalism since 1700*, New York: Routledge.

Cartier, C. (1996) "Conserving the built environment and generating heritage tourism in West Malaysia," *Tourism Recreation Research*, 21(1): 45–53.

——(1997) "The dead, place/space, and social activism: constructing the nationscape in historic Melaka," *Environment and Planning D: Society and Space*, 15(5): 555–86.

——(1998) "Megadevelopment in Malaysia: from heritage landscapes to 'leisurescapes' in Melaka's tourism sector," *Singapore Journal of Tropical Geography*, 19(2): 151–76.

——(1999a) "Cosmopolitics and the maritime world city," *The Geographical Review*, 89(2): 278–89.

——(1999b) "The state, property development, and symbolic landscape in high rise Hong Kong," *Landscape Research*, 24(2): 185–207.

——(2001) *Globalizing South China*, Oxford: Blackwell.

——(2002) "Transnational urbanism in the reform era Chinese city: landscapes from Shenzhen," *Urban Studies*, 39(9): 1513–32.

——(2003) "Conclusions: regions of diaspora," in *The Chinese Diaspora: place, space, mobility and identity*, L.J.C. Ma and C. Cartier (eds) Landham, MD: Rowman and Littlefield.

Casey, E.S. (1996) "How to get from space to place in a fairly short stretch of time: phenomenological prolegomena," in S. Feld and K. H. Basso (eds), *Senses of Place*, Sante Fe, NM: School of American Research Press.

——(1997) *The Fate of Place: a philosophical history*, Berkeley, CA: University of California Press.

——(2002) *Representing Place: landscape painting and maps*, Minneapolis: University of Minnesota Press.

Castells, M. (1997) *The Power of Identity*, Malden, MA: Blackwell.

Castles, S. and Miller, M. (1998) *The Age of Migration*, New York: Guilford Press.

Castree, N. (2002) "False antitheses? Marxism, nature and actor-networks," *Antipode*, 34(1): 111–46.

Catholic Bishops' Conference of Japan (1992) "Concerning church weddings in which both bride and groom are non-Christians," *Japan Missionary Bulletin*, 46(3): 256–60.

Caygill, H. (1998) *Walter Benjamin: the colour of experience,* New York: Routledge.

CCA (Caribbean Conservation Association) (1980) *Antigua: preliminary data analysis*, Washington, DC: CCA.

——(1991) *Antigua and Barbuda: country environmental profile*, Bridgetown, Barbados: USAID.

CIA (Central Intelligence Agency) (2002) *CIA World Factbook*. Online. Available <http://www.cia.gov/cia/publications/factbook/index.html> (accessed 23 May 2002).

Chakrabarty, D. (1992) "Postcoloniality and the artifice of history: who speaks for 'Indian' pasts?", *Representations*, 37: 1–26.

Chambon, C. (2003) "Paris sets fashion for sand and the city as Europe falls in love with the urban beach," *Independent* (London), 21 July.

Chan, S. (1999) "What is this thing called a Chinese diaspora?," *Contemporary Review*, 274(February): 81–3.

Chang, M.H. (1995) "Greater China and the Chinese 'global tribe'," *Asian Survey*, 35: 955–67.

Chang, S. and Liu, J.K.C. (2000) "Re-visioning housing in the information age: a comparative study of high-tech housing developments across the Pacific," Unpublished manuscript, School of Architecture and Environmental Planning, University of California, Berkeley.

Chang, T.C. (1997a) "From 'Instant Asia' to 'Multi-faceted Jewel': urban imaging strategy and tourism development in Singapore," *Urban Geography*, 18(6): 542–62.

——(1997b) "Heritage as a tourism commodity: traversing the global-local divide," *Singapore Journal of Tropical Geography*, 18(1): 46–68.

Chang, T.C., Milne, S., Fallon D., and Pohlmann, C. (1996) "Urban heritage tourism: global-local nexus," *Annals of Tourism Research*, 23(2): 284–305.

Chang, T.C. and Yeoh, B.S.A. (1999) " 'New Asia-Singapore': communicating local cultures through global tourism," *Geoforum*, 30(2): 101–15.

Chang, W.C. (1997) "Ethnic identity of Overseas Chinese," in G. Zhang (ed.) *Ethnic Chinese at the Turn of the Centuries*, vol. 2, Fuzhou, China: Fujian People's Publishing Co.

Chase, M. and Shaw, C. (1989) "The dimensions of nostalgia," in M. Chase and C. Shaw (eds) *Imagined Past: history and nostalgia*, Manchester: Manchester University Press.

Chen, E.W. (2002) "No place like home," *South China Morning Post*, 24 June.

Chesnaux, J. (1991) "The function of the Pacific in the French Fifth Republic's Grand Design," *Journal of Pacific History*, 26(2): 256–72.

China Travel & Tourism Press (1998) *Travel in China: Guangzhou*, Beijing: China Travel & Tourism Press.

China Travel Service (1997) *"China, the Middle Kingdom, a world apart," 1997 Brochure*, Hong Kong: CTS.

Chow, W.S. (1988) "Open policy and tourism between Guangdong and Hong Kong," *Annals of Tourism Research*, 15(2): 205–18.

Chu, G.C. and Ju, Y. (1993) *The Great Wall in Ruins: communication and cultural change in China*, Albany, NY: State University of New York Press.

CIDA (Canadian International Development Agency) (1988) "Caribbean environmental programming strategy," *Final Report: volume 3 background information*, Ottawa: CIDA.

City and County of Honolulu (2000) "Ordinance No. 00–69, Section 15–16.6. Storage parking of commercial vehicles prohibited," *Revised Ordinances of the City and County of Honolulu 1990*, Honolulu: City and County of Honolulu.

Clastres, P. (1994) *Archeology of Violence*, trans. J. Herman, New York: Semiotext(e).

CLIA (Cruise Line International Association) (2000) Online. Available <http://www.cruising.com> (accessed 30 June 2000).

Clifford, J. (1997) *Routes: travel and translation in the late twentieth century*, Cambridge, MA: Harvard University Press.

——(1998) "Mixed feelings," in B. Robbins and P. Cheah (eds) *Cosmopolitics: thinking and feeling beyond the nation*, Minneapolis: University of Minnesota Press.

Cloke, P. and Perkins, H. (1998) "Cracking the canyon with the Awesome Foursome: representations of adventure tourism in New Zealand," *Society and Space: Environment and Planning D*, 16: 185–218.

Cloke, P., Philo, C., and Sadler, D. (1991) *Approaching Human Geography: an introduction to contemporary theoretical debates*, New York: Guilford Press.

Clunas, C. (1996) *Fruitful Sites: gardens in Ming dynasty China*, London: Reaktion Books.

Coast Staff (2002) "Old kitsch, new culture: Las Vegas," *Coast Magazine*. Online. Available <http://www.coastmagazine.com> (accessed 1 August 2003).

Cohen, E. (1974) "Who is a tourist? A conceptual clarification," *Sociological Review*, 22: 527–55.

——(1979a) "A phenomenology of tourist experiences," *Sociology*, 13: 179–201.

——(1979b) "Rethinking the sociology of tourism," *Annals of Tourism Research*, 6(1): 18–35.

——(1988) "Authenticity and commoditization in tourism," *Annals of Tourism Research*, 15(3): 371–86.

Cohen, M. (1989) "Walter Benjamin's phantasmagoria," *New German Critique*, 48: 87–107.

Cohen, S. and Taylor, L. (1992) *Escape Attempts: the theory and practice of resistance to/in everyday life*, 2nd edn., London: Routledge.

Commerson, P. (1769) *Mercure de France* (Paris), November.

Committee on Heritage (1988) *The Committee of Heritage Report*, Singapore: Advisory Council on Culture and the Arts.

Connell, R. (1987) *Gender and Power*, Stanford, CA: Stanford University Press.

Connell, R.W. (1998) "Masculinities and globalization," *Men and Masculinities*, 1(1): 3–23.

Conway, D. and Lorah, P. (1995) "Environmental protection policies in Caribbean small islands: some St. Lucian examples," *Caribbean Geography*, 6(1): 16–27.

Cook, S. (2000a) "Tourism futures—looking out to 2020," Presentation for 2000 Texas Travel Summit, 27 September.

——(2000b) "Outlook on domestic travel, 2000–2001," presentation for TIA's Marketing Outlook Forum, Anaheim, CA, 26 October.

Coram, R. (1993) *Caribbean Time Bomb: the United States' complicity in the corruption of Antigua*, New York: William Morrow.

Corbin, A. (1994) *The Lure of the Sea: the discovery of the seaside in the western world 1750–1840*, trans. J. Phelps, Cambridge: Polity Press and Berkeley, CA: University of California Press.

Corbin, C.J. and Singh, J.G. (1993) "Marine debris contamination of beaches in St. Lucia and Dominica," *Marine Pollution Bulletin*, 26(6): 325–8.

COSALC (Coast and Beach Stability in the Lesser Antilles) (1996) *Beach Erosion in Antigua and Barbuda Coast and Beach Stability in the Lesser Antilles*, San Juan, Puerto Rico: Sea Grant Printers.

Cosgrove, D.E. (1984a) *Social Formation and Symbolic Landscape*, London: Croom Helm.

——(1984b) *Social Formation and Symbolic Landscape*, Totowa, NJ: Barnes & Noble.

Cosgrove, D. and Daniels, S. (eds) (1988) *The Iconography of Landscape: essays on the symbolic representation, design, and use of past landscapes*, Cambridge: Cambridge University Press.

Costansó, M. (1911) *The Portola Expedition of 1769–1770: diary of Miguel Costansó*, trans. F.J. Teggert, Berkeley, CA: University of California, Publications of the Academy of Pacific Coast History.

Couldry, N. (2000) *The Place of Media Power*, London: Routledge.

Crang, M. (1997) "Picturing practices: research through the tourist gaze," *Progress in Human Geography*, 21(3): 359–73.

——(1999) "Globalization as conceived, perceived, and lived spaces," *Theory, Culture and Society*, 16(1): 167–77.

Crang, P. (1996a) "Popular geographies, guest editorial," *Environment and Planning D: Society and Space*, 14: 631–3.

——(1996b) "Displacement, consumption and identity," *Environment and Planning A*, 28: 47–67.

Crawshaw, C. and Urry, J. (1997) "Tourism and the photographic eye," in C. Rojek and J. Urry (eds) *Touring Cultures*, London: Routledge.

Crossley, N. (1995) "Merleau-Ponty, the elusive body and carnal sociology," *Body and Society*, 1(1): 43–61.

——(1996) *Intersubjectivity: the fabric of social becoming*, London: Sage.

Crouch, D. (ed.) (1999a) *Leisure/Tourism Geographies: practice and geographical knowledge*, Routledge London.

——(1999b) "The intimacy and expansion of space," in D. Crouch (ed.) *Leisure/Tourism Geographies: practice and geographical knowledge*, London: Routledge.

——(2000) "Tourism representations and non-representative geographies: making relationships between tourism and heritage active," in M. Robinson (ed.) *Tourism and Heritage Relationships: global, national and local perspectives*, Doxford International, Sunderland, UK: Business Education Publishers.

——(2001) "Spatialities and the feeling of doing," *Social and Cultural Geography*, 2(1): 61–75.

Crouch, D. and Matless, D. (1996) "Refiguring geography: the parish maps of common ground," *Transactions of the Institute of British Geographers*, 21(1): 236–55.

Crouch, D. and Ravenscroft, N. (1995) "Culture, social difference and the leisure experience: the example of consuming countryside," in G. McFee, W. Murphy, and G. Whannel (eds) *Leisure Cultures: values, genders, lifestyles*, Hove, UK: Leisure Studies Association.

Crozet, J. (1783) *Nouveau Voyage à la mer du Sud*, ed. A.M. de R. Barrois, Paris.

Csordas, T.J. (1990) "Embodiment as a paradigm for anthropology," *Ethos*, 18(1): 5–47.

——(1994) "The body as representation and being-in-the-world," in T.J. Csordas (ed.) *Embodiment and Experience: the existential ground of culture and self*, Cambridge: Cambridge University Press.

CTO (Caribbean Tourism Organization) (1994) *Dominica Cruise Passenger Survey*, Summer 1993, Barbados: CTO.

——(1996) *Caribbean Tourism Statistical Report*, 1996 edn., Barbados: CTO.

——(1997a) *Caribbean Tourism Investment Guide*, 1997 edn., Barbados: CTO.

——(1997b) "Proceedings of the Caribbean Tourism Organization's annual conference on sustainable tourism: developing a model ecotourism destination for the Caribbean," 21–24 May, Dominica, Barbados: CTO.

——(1998a) "Proceedings of the first Caribbean Hotel and Tourism Investment conference," Barbados: CTO.

——(1998b) *Keeping the Right Balance: proceedings of the Caribbean Tourism Organization second annual conference and trade show on sustainable tourism development*, Trinidad and Tobago, 15–19 April.

——(2000) *Caribbean Tourism Statistical Report*, 1999–2000 edn., Barbados: CTO.

——(2001) *Caribbean Tourism Organization Latest Statistics 2000*, Barbados: CTO.

CTRC (Caribbean Tourism Research Centre) (1978) "Revised Laws of Antigua (1951) Chapter 364 Hotels Aid," *Tourism Hotel Incentive Legislation: Volume I*, Christ Church, Barbados: CTRC.

Culler, J. (1981) "Semiotics of tourism," *American Journal of Semiotics*, 1(1–2): 127–40.

——(1988) *Framing the Sign: criticism and its institutions*, New York: Blackwell.

Cumming, G. (2000) "Paradise for a song," *New Zealand Herald*, Travel, 4 July.

Cupp, B.G. (1989) Untitled speech by Executive Vice President, Technologies and New Products Development, Presented to American Express employees, New York, 19 January, in Smith, G. (1991) "Tourism, Telecommunications, and Transnational Banking: a study in international interactions," PhD dissertation, The American University, Washington, DC.

Dagget, S. (1922) *Chapters on the History of the Southern Pacific*, New York: Ronald Press.

Dahles, H. (2002) "Gigolos and Rastamen: globalization, tourism and changing gender identities," in M.B. Swain and J.H. Momsen (eds) *Gender Tourism Fun?*, Elmsford, NY: Cognizant Communications.

Daily Observer (St. Johns, Antigua) (1997a) "Antigua and Barbuda win 1997 World Wide Award for Best Honeymoon Destination," 27 June.

——(1997b) "Guana is too good," 3 July.

——(1997c) "Conservation starts at home" 10 July.

Daniels, S. and Lee, R. (eds) (1996) *Exploring Human Geography: a reader*, London: Arnold.

Dann, G.M.S. (1996) *The Language of Tourism, A Sociological Perspective*, Wallingford: CAB International.

——(1997) "The green, green grass of home: nature and nurture in rural England," in S. Wahab and J.J. Pigram (eds) *Tourism, Development and Growth: the challenge of sustainability*, London: Routledge.

Davidson, L. and Gjerde, K. (1989) *An Evaluation of International Protection Offered to Caribbean Coral Reefs and Associated Ecosystems*, Washington, DC: Greenpeace International and Marine Policy Center, Woods Hole Oceanographic Institution.

Dawson, R. (1967) *The Chinese Chameleon: an analysis of European conceptions of Chinese civilization*, London: Oxford University Press.

——(1991) "Mythologizing Pacific Women: Cook's second voyage," Paper presented at the Pacific Science Congress, Honolulu, Unpublished.

DBEDT (2001) *Visitor Summary 2000*, Department of Business, Economic Development and Tourism, Honolulu. Online. Available <http://www.Hawaii.gov/dbedt/stats.html> (accessed 31 October 2002).

Debord, Guy (1994) *The Society of the Spectacle*, trans. D. Nicholson-Smith, New York: Zone Books.

de Certeau, M. (1984) *The Practice of Everyday Life*, Berkeley, CA: University of California Press.

Deere, C., Antrobus, P., Bolles, L., Melendry, E., Phillips, P., Riviera, M., and Safa, H. (1990) *The Shadow of the Sun: Caribbean development alternatives and US policy*, Boulder, CO: Westview Press.

Deleon, R.E. (1992) *Left Coast City: progressive politics in San Francisco, 1975–1991*, Lawrence, KS: University Press of Kansas.

——(2002) "Only in San Francisco? The city's political culture in comparative perspective," *SPUR Newsletter*, San Francisco Planning and Urban Research Association, November/December.

de Man, P. (1983) *Blindness and Insight: essays in the rhetoric of contemporary criticism*, Minneapolis: University of Minnesota Press.

Department of Taxation (1997) "Tax advisory on the application of the general excise tax to tourist wedding activities of churches," 21 April, Honolulu: State of Hawai'i.

Der Derian, J. (1996) "Speed pollution," *Wired*, 4(5): 120–1.

Derrida, J. (1982) *Margins of Philosophy*, trans. A. Bass, Chicago: University of Chicago Press.

——(1984) "No Apocalypse, Not Now (full speed ahead, seven missiles, seven missives)," *Diacritics*, 14(2): 20–31.

——(1993) *Aporias*, trans. T. Dutoit, Stanford, CA: Stanford University Press.

——(1994) *Specters of Marx: the state of the debt, the work of mourning and the new International*, trans. P. Kamuf, New York: Routledge.

——(1995) *The Gift of Death*, trans. D. Wills, Chicago: Chicago University Press.

Desforges, L. (2000) "Travelling the world: identity and travel biography," *Annals of Tourism Research*, 27(4): 926–45.

Desmond, J. (1999) *Staging Tourism: bodies on display from Waikīkī to Sea World*, Chicago: University of Chicago Press.

Dewsbury, J.D. (2000) Performativity and the event: enacting a philosophy of difference," *Society and Space: Environment and Planning D*, 18: 473–96.

d'Hauteserre, A.M. (1999) "A future for tourism in French Polynesia?," *Tourism Analysis*, 4(3/4): 201–11.

Dichter, E. (1960) *The Strategy of Desire*, Garden City, NY: Doubleday.

Dingeman, R. (1999) "Councilman to be cited for weddings at home," *Honolulu Advertiser*, 20 October.

——(2001) "Council member discloses wedding business earning," *Honolulu Advertiser*, 6 February.

Dirlik, A. (1994) *After the Revolution: waking to global capitalism*, Hanover, CT: Wesleyan University Press.

Domosh, M. (1998) "Those gorgeous incongruities: polite politics and public space on the streets of nineteenth-century New York," *Annals of the Association of American Geographers*, 88(2): 209–26.

Dorsett, E. (1997) "Bishop Dorsett speaks truth to power" (transcript of speech to Third UPP Convention, 26 June), *Antigua Outlet*, 1 July.

DOS (Directorate of Overseas Surveys) (1962) *Antigua Island, 1: 25,000*, London: Government of Great Britain.

——(1977) *Tourist Map of Antigua, West Indies, 1: 50,000*, London: Government of Great Britain.

Douglas, N. and Douglas, N. (1996) "Tourism in the Pacific: historical factors," in C. M. Hall and S. Page (eds) *Tourism in the Pacific: issues and cases*, London: International Thomson Business Press.

Duara, P. (1995) *Rescuing History from the Nation: questioning narratives of modern China*, Chicago: University of Chicago Press.

——(1997) "Nationalists among transnationals: overseas Chinese and the idea of China,

1900–1911," in N. Donald and A. Ong (eds) *Ungrounded Empires: the cultural politics of modern Chinese transnationalism*, New York: Routledge.

Dubinsky, K. (1999) *The Second Greatest Disappointment: honeymooning and tourism at Niagara Falls*, New Brunswick, NJ: Rutgers University Press.

du Halde, P. (1741) *The General History of China,* London: J. Watts.

Dumazedier, J. (1967) *Toward a Society of Leisure*, New York: Free Press.

Dumont d'Urville, J. (1832) "Notice sur les îles du Grand Océan et sur l'origine des peuples qui les habitent," *Bulletin de la Société de Géographie de Paris*, 17: 1–21.

Duncan, J. (1990) *The City as Text: the politics of landscape interpretation in the Kandyan kingdom*, Cambridge: Cambridge University Press.

Duncan, J. and Duncan, N. (1988a) "(Re)reading the landscape," *Environment and Planning D: Society and Space*, 6(2): 117–26.

——(1988b) "Ideology and Bliss: Roland Barthes and the secret histories of landscape," in T. Barnes and J. Duncan (eds) *Writing Worlds: discourse, text and metaphor in the representation of landscape*, New York: Routledge.

Duncan, J. and Ley, D. (1993) *Place/Culture/Representation*, London: Routledge.

Duncan, N. (1996) *BodySpace*, London: Routledge.

Dun Woo (1998) "What are the obvious distinctive characteristics of Guangzhou tourism?", *Proceedings of the Information of Guangdong Tourism Association*, Guangdong Tourism Association.

Durkheim, E. (1956) [1912] *The Elementary Forms of Religion*, trans. J. W. Swain, New York: Free Press.

Dwyer, L. and Forsyth, P. (1998) "Economic significance of cruise tourism," *Annals of Tourism Research*, 25(2): 393–415.

EAG'er (1997) "Guana Island Development: disaster for fisheries and endangered species," *The EAG'er*, St. Johns, Antigua: The Environmental Awareness Group, 18: 1–3.

East (1999) Singapore: Gnomadic Publishing.

Economists At Large (1997) *Grand Prixtensions: the economics of the magic pudding*, Prepared for the Save Albert Park Group, Melbourne: Economists At Large.

Edensor, T. and Kothari, U. (1994) "The masculinisation of Stirling's heritage," in V. Kinnaird and D. Hall (eds) *Tourism: a gender analysis*, Chichester: Wiley.

Edmond, R. (1997) *Representing the South Pacific*, Cambridge: Cambridge University Press.

Eliade, M. (1954) *The Myth of the Eternal Return,* trans. W. R. Trask, New York: Pantheon.

——(1959) *The Sacred and the Profane: the nature of religion*, New York: Harcourt Brace.

Elliot, J. and Pickersgill, R. (1984) *Captain Cook's Second Voyage: the journals of Lieutenant Eliot and Pickersgill*, ed. C. Holmes, London: Caliban Books.

Eltis, D. (1995) "The total product of Barbados, 1664–1701," *Journal of Economic History*, 55(2): 321–38.

Endo, Y. (2000) "Imaya kokunai nami no tehai ga kanou ni, Hawai uedingu saishin Jijou," *Crea*, 12(8): 112.

Evans, M. (1999) "Sydney linked to IOC crisis," *Sydney Morning Herald*, 22 January.

Faberon, J. Y. (ed.) (1996) *Le Statut du Territoire de la Polynésie Française, Bilan de dix ans d'application: 1984–1994*, Paris: Economica.

Feifer, M. (1985) *Going Places: tourism in history,* New York: Stein and Day.

Fitzgerald, C. P. (1935) *China: a short cultural history,* London: The Cresset Press.

FitzSimmons, M. and Goodman, D. (1998) "Environmental narratives and the reproduction of food," in Bruce Braun and Noel Castree (eds), *Remaking Reality: Nature at the Millennium*, New York: Routledge, 194–220.

Flaubert, G. (1992) *L'Éducation Sentimentale: les scénarios*, Paris: J. Corti.

Fletcher, J. (1986) "The Mongols: ecological and social perspectives," *Harvard Journal of Asiatic Studies*, 46(1): 11–50.

Flippo, H. (2002) *When in Germany, Do as the Germans Do*, New York: McGraw-Hill.

Forster, G. and Forster, J.R. (1968) *A Voyage round the World*, Berlin: Akademie-Verlag.

Foucault, M. (1980) *Power and Knowledge*, ed. C. Gordon, trans. C. Gordon, L. Marshall, J. Mepham, and K. Soper, New York: Pantheon.

——(1986) "Of other spaces," *Diacritics*, 16: 22–7.

Frank, A.G. (1998) *ReOrient: global economy in the Asian age*, Berkeley, CA: University of California Press.

French, H. (1990) "Winter in the sun: growing pains in the Caribbean," *The New York Times*, 9 December.

Frow, J. (1991) "Tourism and the semantics of nostalgia," *October*, 57: 121–51.

——(1997) *Time and Commodity Culture*, Oxford: Clarendon.

Fryer, J. (1975) *The Great Wall of China*, London: New English Library.

Game, A. (1991) *Undoing the Social: towards a deconstructive sociology*, Milton Keynes: Open University Press.

Garcia, K.J. (1995) "Joe was more than a football hero," *The San Francisco Chronicle*, 19 April.

Gauguin, P. (1949) *Paul Gauguin: letters to his wife and friends*, ed. M. Malingue, Cleveland, OH: World Publishing.

Geertz, C. (1973) *The Interpretation of Culture*, New York: Basic Books.

——(1988) *Work and Lives: the anthropologist as author*, Cambridge: Polity Press.

Giddens, A. (1979) *Central Problems in Social Theory: action, structure and contradiction in social analysis*, Berkeley, CA: University of California Press.

——(1990) *The Consequences of Modernity*, Stanford, CA: Stanford University Press.

——(1992) *The Transformation of Intimacy: sexuality, love and eroticism in modern societies*, Cambridge: Polity Press.

GIE Tahiti Tourisme, Service du Tourisme (1999) *Plan de développement stratégique du tourisme en Polynésie Française*, Papeété.

Gilardoni, L. (1995) *Storia di Bellagio,*. Bellagio: Amilcare Pizzi Editore.

Gillingham, P. (1998) "Calvary settles back taxes: Congregation questions whether church's mission includes weddings," *Pacific Business News*, 2 January.

Girard, R. (1977) *Violence and the Sacred*, trans. P. Gregory, Baltimore, MD: Johns Hopkins University Press.

Glennie, P.D. and Thrift, N. (1992) "Modernity, urbanism and modern consumption," *Environment & Planning D: Society and Space*, 10: 423–43.

Glovin, D. (2001) "Visa, MasterCard Suffer Legal Setback: judge rules firms can't block member banks from issuing rival credit cards," *Washington Post*, 9 October.

Goffman, E. (1959) *The Presentation of Self in Everyday Life*, Garden City, NJ: Doubleday.

——(1967) *Interaction Ritual: essays on face-to-face behavior*, Chicago: Aldine.

Gomes, L. (1997a) Unpublished speech to UPP rally, 19 June, St. Johns, Antigua.

——(1997b) Personal interview, 27 June, St. Johns, Antigua.

Goodall, B. (1990) "The dynamics of tourism place marketing," in G.J. Ashworth and B. Goodall (eds) *Marketing Tourism Places*, London: Routledge.

Goodell, J. (2000) *Sunnyvale: the rise and fall of a Silicon Valley family*, New York: Villard.

Goody, J. (1997) *Representations and Contradictions*, Oxford: Blackwell.

Goss, J. (1993a) "The 'Magic of the mall': an analysis of form, function, and meaning in the contemporary retail built environment," *Annals of the Association of American Geographers*, 83(1): 18–47.

——(1993b) "Placing the market and marketing the place: tourist advertising of the Hawaiian Islands, 1972–1992," *Environment and Planning D: Society and Space*, 11: 663–88.

——(1996) "Disquiet on the waterfront: nostalgia and utopia in the festival marketplace," *Urban Geography*, 17(3): 221–47.

——(1999) "Once-upon-a-time in the commodity world: an unofficial guide to Mall of America," *Annals of the Association of American Geographers,* 89(1): 45–75.

Graburn, N.H.H. (1976) "Introduction: the arts of the Fourth World," in N.H.H. Graburn (ed.) *Ethnic and Tourist Arts: cultural expressions from the fourth world*, Berkeley, CA: University of California Press.

——(1977, 2nd edn 1989) "Tourism, the sacred journey," in S. Valene (ed.) *Hosts and Guests: the anthropology of tourism*, 2nd edn., Philadelphia: University of Pennsylvania Press.

——(1983) "The anthropology of tourism," *Annals of Tourism Research*, 10(1): 9–33.

——(1989) "Tourism: the sacred journey," in V.L. Smith (ed.) *Hosts and Guests: the anthropology of tourism*, 2nd edn., Philadelphia: University of Pennsylvania Press.

Greenblatt, S.J. (1981) "Preface," in S.J. Greenblatt (ed.) *Allegory and Representation*, Baltimore, MD: Johns Hopkins University Press.

Greenspan, R. (2002) "China Pulls Ahead of Japan," *InternetNews.com – ISP News* (April 22). Online. Available <http://www.internetnews.com/isp-news/article.php/1013841> (accessed 7 June 2002).

Gregory, C.A. (1992) *Gifts and Commodities*, London: Academic Press.

Gregory, D. (1994) *Geographical Imaginations*, Cambridge, MA: Blackwell.

Greiner, N. (1994) "Inside running on Olympic bid," *The Australian*, 19 September.

Gren, M. (2001) "Time-geography matters," in N. Thrift and J. May (eds) *TimeSpace: geographies of temporality*, London: Routledge.

Griffiths, B. (1999) "Women's sexual objectification in the tourist industry in the United Kingdom," in "UNED-UK, 1999 Gender and Tourism: women's employment and participation in tourism," Report for the United Nations Commission on Sustainable Development, 7th Session, April, London: DFID.

GOCW (Guangdong Overseas Chinese Website) (2002). Online. Available: <http://www.gdover seaschn.com.cn/> (accessed 8 June 2002).

Green, E. and Donnelly, R. (2003) "Recreational Scuba Diving in Caribbean marine protected areas: do the users pay?", *AMBIO: A Journal of the Human Environment*, 32(2): 140–4.

Guangzhou Statistics Bureau (1999) *The Yearbook of Guangzhou Statistics 1998*, Beijing: China Statistical Press.

Guangdong Tourism Bureau (1999a) *The Statistics of Guangdong Tourism 1998*, Guangzhou: Guangdong Tourism Bureau.

——(1999b) *A Guide to Tourism of Guangzhou*, Beijing: China Travel and Tourism Press.

——(2001) *The Statistics of Guangdong Tourism*, Guangzhou: Guangdong Tourism Bureau.

Guardian Connections (1999) Personal ads, *San Francisco Bay Guardian*, various dates.

Guarrasi, V. (2001) "Paradoxes of modern and postmodern geography: heterotopia of landscape and cartographic logic," in C. Minca (ed.) *Postmodern Geography: theory and praxis*, Oxford: Blackwell.

Hage, G. (1997) "At home in the entrails of the west: multiculturalism, 'ethnic food' and migrant home building," in H. Grace, G. Hage, L. Johnson, J. Langsworth, and M. Symonds (eds) *Home/World: space, community and marginality in Sydney's west*, Annandale, NSW: Pluto Press.

Hall, C.M. (1992) *Hallmark Tourist Events: impacts, management and planning*, Chichester: Wiley.

——(1994) *Tourism and Politics: policy, power and place*, London: Bellhaven Press.

——(1996) "Hallmark events and urban reimaging strategies: coercion, community and the Sydney 2000 Olympics," in L.C. Harrison and W. Husbands (eds) *Practicing Responsible Tourism: international case studies in planning, policy and development*, New York: Wiley.

——(1997) "Geography, marketing and the selling of places," *Journal of Travel and Tourism Marketing*, 6(3/4): 61–84.

——(1998) "The politics of decision making and top-down planning: Darling Harbour, Sydney," in D. Tyler, M. Robertson, and Y. Guerrier (eds) *Tourism Management in Cities: policy, process and practice*, Chichester: Wiley.

Hall, C.M. and Hamon, C. (1996) "Casinos and urban redevelopment in Australia," *Journal of Travel Research*, 34(3): 30–6.

Hall, C.M. and Hodges, J. (1997) "Sharing the spirit of corporatism and cultural capital: the politics of place and identity in the Sydney 2000 Olympics," in M. Roche (ed.) *Sport, Popular Culture and Identity*, Chelsea School Research Centre Edition vol. 5, Aachen: Meyer & Meyer Verlag.

Hall, C.M. and Lew, A.A. (eds) (1998) *Sustainable Tourism: a geographical perspective*, New York: Longman.

Hall, D. (1991) "The ethnic revival," *Independent on Sunday*, 29 December.

Hall, M. (1995) *Tourism and Politics. Policy, Power and Place*, Chichester: Wiley.

Hall, S. (1988) "Brave new world," *Marxism Today*, October.

——(1995) "New Cultures for Old," in D. Massey and P. Jess (eds), *A Place in the World? Places, Cultures and Globalization*, Milton Keynes: Open University Press.

——(1991) "The local and the global: globalization and ethnicity," in A.D. King (ed.) *Culture, Globalization and the World System*, London: Macmillan.

Hammond, D. (ed.) (1970) *News from New Cythera, a Report of Bougainville's Voyage 1766–69*, Minneapolis: University of Minnesota Press.

Hanbury, R. (2002) "Impacts of 9/11 on the DC CVB," Guest lecture, The George Washington University, Washington, DC, 5 March.

Handler, R. and Linnekin, J. (1984) "Tradition, genuine or spurious?", *Journal of American Folklore*, 97(385): 273–90.

Hannerz, U. (1986) "Theory in anthropology: small is beautiful? The problem of complex societies," *Comparative Studies in Society and History*, 28(2): 362–7.

Hannigan, J.A. (1995) "Theme parks and urban fantasy-scapes," *Current Sociology*, 43(1): 183–91.

Hanssen, B. (1998) *Walter Benjamin's other History: of stones, animals, human beings and angels*, Berkeley, CA: University of California Press.

Haraway, D. (1991) *Simians, Cyborgs and Women: the reinvention of nature*, New York: Routledge.

Harkin, M. (1995) "Modernist anthropology and tourism of the authentic," *Annals of Tourism Research*, 22(4): 650–70.

Harré, R. and Gillet, G. (1994) *The Discursive Mind*, Thousand Oaks, CA: Sage.

Harris, M. (1998) "Masako's wedding," *Kansai Time Out*, No. 257, July.

Harris, N. (1991) "Urban tourism and the commercial city," in W.R. Taylor (ed.) *Inventing Times Square: commerce and culture at the crossroads of the world*, New York: Russell Sage Foundation.

Harris, R. (1999) *Lourdes: body and spirit in the secular age*, London: Allen Lane.

Harvey, D. (1989) *The Condition of Postmodernity*, Oxford: Basil Blackwell.

Haug, W.F. (1986) *Critique of Commodity Aesthetics: appearance, sexuality, and advertising in capitalist society*, trans. R. Buck, Minneapolis: University of Minnesota Press.

Hau'ofa, E. (1994) "Our sea of islands," *The Contemporary Pacific*, 6(1): 147–61.

Hawai'i Visitors and Convention Bureau (2001) "HVCB's 'Aloha Magic' Japan campaign enters 2nd year," Press release, 11 January.

Hawkins, J. and Roberts, C. (1993) "Effects of recreational SCUBA diving on coral reefs: trampling of reef flat communities," *Journal of Applied Ecology*, 30: 25–30.

He, Y. (1994) "Lunan minzu guanxi de lishi he xianzhuang (The history and status quo of nationality relations in Lunan)," *Yunnan Shihuikexue*, 6: 40–56.

Hechter, M. (1975) *Internal Colonialism: the Celtic fringe in British national development, 1536–1966*, Berkeley, CA: University of California Press.

Hector, T. (1997a) "William Cody Kelly – model foreign investor," 31 October. Online. Available <http://www.candw.ag/~jardinea/fanflame.97oct31.htm> (accessed 5 November 1997).

——(1997b) "As it was in the beginning, so it is now?" 31 October. Online. Available <http://www.candw.ag/~jardinea/fanflame.97oct31.htm> (accessed 5 November 1997).

——(1999) "Was the phantom election, illegitimate and illegal?" 7 May. Online. Available <http://www.candw.ag/~jardinea/f99may07.htm> (accessed 31 May 1999).

Heidegger, M. (1971) *Poetry, Language, Thought*, trans. A. Hofstadter, New York: Harper & Row.

Heller, A. (1999) *A Theory of Modernity*, Malden, MA: Blackwell.

Helmreich, S. (2000) *Silicon Second Nature: culturing artificial life in a digital world*, Berkeley, CA: University of California Press.

Henderson, D.J. (2000) "Wedding chapel destroys beauty of area," Letter to the editor, *Honolulu Advertiser*, 9 September.

Henningham, S. (1992) *France and the South Pacific*, Honolulu: University of Hawai'i Press.

The Herman Group (2002) "Herman Trend Alert: personal service on the rise," Strategic Business Management Consultants Online Newsletter, 19 March (www.herman.net).

Hewison, R. (1987) *The Heritage Industry: Britain in a climate of decline*, London: Methuen.

Hochschild, A.R. (1983) *The Managed Heart*, Berkeley, CA: University of California Press.

Honolulu Star-Bulletin (2000) "Wedding mills stir neighbors' opposition," Editorial, 25 July.

Horioka, C.Y. (1987) "Cost of marriages and marriage related savings," *Kyōto University Economic Review*, 57(1): 47–58.

Huang, R.F. (2000) *Taishan Past & Present 500 Years*, Macau: Macau Publishing. (In Chinese.)

Huang, S and Chang, T.C. (2003), "Selective disclosure: Romancing the Singapore River", in R.B.H. Goh and B.S.A Yeoh. (eds.) *Theorizing the Southeast Asian City as Text: Urban Landscapes, Cultural Documents and Interpretative Experiences*, Singapore: World Scientific, pp. 73-108.

Huang, S., Teo, P., Hend, P., and Heng, H. M. (1995) "Conserving the civic and cultural district: state policies and public opinion," in B.S.A. Yeoh and L. Kong (eds) *Portraits of Places: history, community and identity in Singapore*, Singapore: Times Editions.

Huaren.org (2002) "Chinese Diaspora." Online. Available <http://www.huaren.org/diaspora/> (accessed 23 May 2002).

Hubbard, P. (2000) "Desire/disgust: mapping the moral contours of heterosexuality," *Progress in Human Geography*, 24(2): 191–217.

Huffman, D. (2001) "Beyond the number crunching: we put the dot-economy on the psychologist's couch," *San Francisco Bay Guardian*, 30 May.

Hui, E.L.L. and McKercher, B. (2001) "Operational issues in marketing research: an example of the omnibus tourism survey," *Pacific Tourism Review*, 5(1/2): 5–14.

Hummon, D.M. (1988) "Tourist worlds: tourist advertising, ritual, and American culture," *Sociological Quarterly*, 29(2): 179–202.

Ibrahim, A. (1996) *The Asian Renaissance*, Singapore: Times Books International.

Illouz, E. (1997) *Consuming the Romantic Utopia: love and the cultural contradictions of capitalism*, Berkeley, CA: University of California Press.

Ing, E. (2000) "Don't let Felix hold ceremonies," Letter to the editor, *Honolulu Star Bulletin*, 15 January.

Inglis, F. (2000) *The Delicious History of the Holiday*, London: Routledge.

Intel Corporation. (2004) "Bumrungrad Hospital Transforms Healthcare Delivery with Integrated Information System on Intel Architecture" (Intel Business Center Case Study). Online. Available <http://www.intel.com/business/casestudies/bumrungrad.pdf> (accessed 26 July 2004).

Interbrand (1990) *Brands: An International Review*, London: Mercury Business Books, Gold Arrow Publications.

ISPA (2002) "Spa industry study reveals phenomenal growth," 26 September, Lexington, KY. Online. Available <http://www.experienceispa.com/media/2002_study.html> (accessed 30 October 2002).

Jackson, P. (1989) *Maps of Meaning: an introduction to cultural geography*, London: Unwin Hyman.

Jackson, P. and Taylor, J. (1996) "Geography and the cultural politics of advertising," *Progress in Human Geography*, 20(3): 356–71.

Jameson, F. (1999) "Marx's purloined letter," in M. Sprinker (ed.) *Ghostly Demarcations: a symposium on Jacques Derrida's Specters of Marx*, New York: Verso.

Japan Information Network (1997) "Toned down ceremonies: Japanese weddings enter age of tasteful restraint," *Trends in Japan*, 11 November. Online. Available <http://jin.jcic. or.jp/trends98/honbun/ntj971111.html> (accessed 22 January 2001).

Jenner, W.J.F. (1992) *The Tyranny of History, The Roots of China's Crisis*, London: Penguin.

Jolly, M. and Manderson, L. (1997) "Introduction," in M. Jolly and L. Manderson (eds) *Sites of Desire, Economies of Pleasure: sexualities in Asia and the Pacific*, Chicago: University of Chicago Press.

Jourdain, P. (1934) "Le yachting dans les Etablissements Français d'Océanie," *Bulletin de la Société d'Etudes Océaniennes*, 52: 397–400.

JTB (2000) Online. Available: <www.jtb.co.jp/soumu/english> (accessed 15 December 2000).

Judd, D.R. and Fainstein, S. (1999) *The Tourist City*, New Haven, CT: Yale University Press.

Kabbani, R. (1986) *Imperial Fictions: Europe's myths of orient*, London: Pandora.

Kahn, M. (2000) "Tahiti intertwined: ancestral land, tourist postcard, and nuclear test site," *American Anthropologist*, 102(1): 7–26.

Kaleikini, D. and Wong, L. (1997) "Headline was misleading in wedding chapel story," Letter to editor, *Honolulu Star Bulletin*, 17 April.

Kamata, S. (1985) "Wedding extravaganzas," *Japan Quarterly*, 32(2): 168–73.

Kanahele, G.H.S. (1986) *Ku Kanaka Stand Tall: a search for Hawaiian values*, Honolulu: University of Hawai'i Press.

Kaplan, C. (1996) *Questions of Travel: postmodern discourses of displacement*, Durham, NC: Duke University Press.

Kayal, M. (2000) "Japanese love more, spend less in islands," *Honolulu Advertiser*, 28 October.

Kearns, G. and Philo, C. (eds) (1993) *Selling Places: the city as cultural capital, past and present*, Oxford: Pergamon.

Keith, M. and Pile, S. (1993) "Introduction to part 2: the politics of place," in M. Keith and S. Pile (eds) *Place and the Politics of Identity*, London: Routledge.

Kellner, D. (1989) *Jean Baudrillard: from Marxism to postmodernism and beyond*, Stanford, CA: Stanford University Press.

Kelly, K. (1996) "What would McLuhan say?", *Wired*, 4(10): 148–9.

Kempadoo, K. (1999) *Sun, Sex and Gold: tourism and sex work in the Caribbean*, Lanham, MD: Rowman and Littlefield.

Keown, T. (1997) "Joe's mystical cool couldn't be duplicated," *The San Francisco Chronicle*, 15 December.

Khan, M. (1997) "Tourism development and dependency theory: mass tourism versus ecotourism," *Annals of Tourism Research*, 24(4): 988–92.

Kim, R. (2001) "Chinese lead Asian tally," *The San Francisco Chronicle*, May 16. Online. Available <http://www.sfgate.com/cgi-bin/article.cgi?file=/chronicle/archive/2001/05/16/MN101414.DTL> (accessed 23 May 2002).

Kincaid, J. (1988) *A Small Place*, New York: Plume.

King, A. (1990) *Urbanism, Colonialism, and the World Economy: cultural and spatial foundations of the world urban system*, London: Cambridge University Press.

King, B.E. (1997) *Creating Island Resorts*, London: Routledge.

King, R. (1993) "The geographical fascination of islands," in D.G. Lockhart, D.D. Smith, and J. Schembri (eds) *The Development Process in Small Island States*, London: Routledge.

Kinnaird, V. and Hall, D. (eds) *Tourism: a gender analysis*, New York: Wiley.

Kirby, K.M. (1996) *Indifferent Boundaries: spatial concepts of human subjectivity*, New York: Guilford.

Kirch, P. (2000) *On the Road of the Winds*, Berkeley, CA: University of California Press.

Kirshenblatt-Gimblett, B. (1998) *Destination Culture, Tourism, Museums, and Heritage*, Berkeley, CA: University of California Press.

Klyne, S. (1996) "The art of selling," *Asia Travel Trade*, March, 17–18.

Knight, F.W. (1990) *The Caribbean: genesis of a fragmented nationalism,* New York: Oxford University Press.

Knowles, T., Diamantis, D., and El-Mourhabi, J.B. (2001) *The Globalization of Tourism and Hospitality: a strategic perspective*, New York: Continuum.

Knox, P. (1991) "The restless urban landscape: economic and sociocultural change and the transformation of metropolitan Washington, DC," *Annals of the Association of American Geographers*, 81(2):181–209.

Kogawa, T. (1985) "The political economy of Japanese marriage," *Ampo*, 17(3): 48–53.

Koh, E. (1999) "Redical rising," *Men's Folio*, June–August, 92–4.

Kong, L. and Yeoh, B. (1994) "Urban conservation in Singapore: a survey of state policies and popular attitudes," *Urban Studies*, 31(2): 247–65.

——(1995) "The meanings and making of place: exploring history, community and identity," in B.S.A. Yeoh and L. Kong (eds) *Portraits of Places: history, community and identity in Singapore*, Singapore: Times Editions.

Kotkin, J. (2000) *The New Geography: how the digital revolution is reshaping the American landscape*, New York: Random House.

KPMG Peat Marwick (1993) *Sydney Olympics 2000 Economic Impact Study*, 2 vols, Sydney Olympics 2000 Bid Ltd, in association with Centre for South Australian Economic Studies, Sydney.

Kristeva, J. (1996) *The Portable Kristeva*, New York: Columbia University Press.

Kruger, R.E. (2001) "Facelift and a safari – scalpel tourism South African style," Deutsche Presse-Agentur, 4 May.

Kuliʻouʻou/Kalani Iki Neighborhood Board (2000) "Minutes of regular meeting January 6," Honolulu: City and County of Honolulu Neighborhood Commission Office.

Kurlansky, M. (1992) *A Continent of Islands: searching for the Caribbean destiny*, Reading, MA: Addison-Wesley.

Kymlicka, W. (1995) *The Rights of Minority Cultures,* Oxford: Oxford University Press.

La Capra, D. (1999) "Trauma, absence, loss," *Critical Inquiry*, 25(4): 667–727.

Lakoff, G. and Johnson, M. (1980) *Metaphors We Live By*, Chicago: University of Chicago Press.

——(1999) *Philosophy in the Flesh: the embodied mind and its challenge to western thought*, New York: Basic Books.

Landon, G.P. (1982) *Images of Crisis: literary iconography, 1750 to the present*, Boston: Routledge.

Lanfant, M.-F. (1995) "Introduction," in M.-F. Lanfant, J.B. Allcock, and E.M. Bruner (eds) *International Tourism: identity and change*, London: Sage.

La Provincia (1999) "E guerra contro la finta Bellagio," 23, 28 December.

Larbalestier, J. (1994) "Imagining the city: contradictory tales of space and place," in K. Gibson and S. Watson (eds) *Metropolis Now: planning and the urban in contemporary Australia*, Annandale, NSW: Pluto Press.

Larkham, P. (1994) "A new heritage for a new Europe: problems and potential," in G.J. Ashworth and L. Larkham (eds) *Building a New Heritage: tourism, culture and identity in the New Europe*, London: Routledge.

Lattimore, O. (1962) *Studies in Frontier History: collected papers 1928–1958*, London: Oxford University Press.

Lash, S. and Urry, J. (1994) *Economies of Signs and Space*, London: Sage.

Laurier, E. (1993) " 'Tackintosh': Glasgow's supplementary gloss," in C. Philo and G. Kearns (eds) *Selling Places: the city as cultural capital, past and present*, Oxford: Pergamon.

Law, C. (2002) *Urban Tourism: the visitor economy and the growth of large cities*, London: Continuum.

Law, M.C. (1993) *Urban Tourism: attracting visitors to large cities*, London: Mansell.

Lee, H.T. (1991) "The conservation dilemma", *Mirror*, 27: 1–4.

Lee, R.L. (1978) "Who owns boardwalk? The structure of control in the tourist industry of Yucatán," in M.D. Zamora, V.H. Sutlive and N. Altshuler (eds), *Tourism and Economic Change, Studies in Third World Society*, No. 6, Williamsburg, VA: Dept. of Anthropology, College of William and Mary, 19–36.

Leed, E.J. (1991) *The Mind of the Traveler: from Gilgamesh to global tourism*, New York: Basic Books.

Lefebvre, H. (1991) *The Production of Space*, trans. D. Nicholson-Smith, Oxford: Blackwell.

Lencek, L. and Bosker, G. (1998) *The Beach: a history of paradise on earth*, New York: Viking.

Lennon, J.J. and Foley, M. (1999) "Interpretation of the unimaginable: the U.S. Holocaust Museum, Washington, DC, and dark tourism," *Journal of Travel Research*, 38: 46–50.

Leoni, L. (2000) Personal interview.

Leung, M.W.H. (2003) "Notions of home among diaspora Chinese in Germany," in L.J.C. Ma and C. Cartier (eds) *The Chinese Diaspora: place, space, mobility and identity*, Lanham, MD: Rowman and Littlefield.

Levinas, E. (1961) *Totality and Infinity: an essay on exteriority*, trans. by A. Lingis, Pittsburgh: Duquesne University Press.

Levinson, J. (2000) "Redo Kapi'olani Park and Washington Place," Letter to the editor, *Honolulu Advertiser*, 8 October.

Lévi-Strauss, C. (1966) *The Savage Mind*, Chicago: University of Chicago Press.

Lew, A.A. and McKercher, B. (2002) "Trip destinations, gateways and itineraries: the example of Hong Kong," *Tourism Management*, 23 (6 December): 609–21.

Lew, A.A. and Wong, A. (2002) "Tourism and the Chinese diaspora," in C.M. Hall and A.M. Williams (eds) *Tourism and Migration: new relationships between production and consumption*, Dordrecht: Kluwer Academic.

——(2003) "News from the Motherland: a content analysis of existential tourism magazines in China," *Tourism Culture and Communication*, 4(2): 83–94.

——(2004) "Sojourners, *guangxi* and clan associations: social capital and overseas Chinese tourism to China," in D. Timothy and T. Coles (eds) pp. 202–214 *Diaspora and Tourism*, London: Routledge.

Lewis, D. and Bridger, D. (2000) *The Soul of the New Consumer: authenticity—what we buy and why in the new economy*, London: Nicholas Brealey.

Ley, D. (1981) "Behavioral geography and the philosophies of meaning," in K. Cox and R. Golledge (eds) *Behavioral Problems in Geography Revisited*, New York: Methuen.

Li, G. (ed.) (1960) *Ashima*, Kunming: Yunnan Renmin Chubanshi.

Li, L. (ed.) (1985) *Ashima Yuanshi Ziliao* (Ashima Epic Data), Kunming: Zhongguo Minjianwenyi Chubansi, Yunnan Ban.

Li, L. *et al.* (1993) "Maitou kugan zhong shixiao, shenggu kanhao Ashima (Engaged in a great effort, merchants are relying on Ashima)," *Yunnan Ribao*, 6 March.

Lim, P.P.H. (2000) "Genealogy and tradition among the Chinese of Malaysia and Singapore," in C. Huang, G. Zhuang, and T. Kyoko (eds) *New Studies on Chinese Overseas and China*, Leiden: International Institute for Asian Studies.

Lim, S.Y. (2000) "Fusion or confusion: the imaginative geographies of New Asia-Singapore," Unpublished honors thesis, National University of Singapore.

Lin, Y.S. (1979) *The Crisis of Chinese Consciousness*, Madison: University of Wisconsin Press.

Lind, I. (2000) "Wedding bells to ring in 'Āina Haina," *Honolulu Star Bulletin*, 2 November.

Linden, E. (1996) "Reimagining Polynesia," *Condé Nast Traveler*, June.

Linnekin, J. (1997) "Consuming cultures: tourism and the commoditization of cultural identity in the island Pacific," in M. Picard and R. Wood (eds) *Tourism, Ethnicity, and the State in Asian and Pacific Societies*, Honolulu: University of Hawai'i Press.

Lippard, L.R. (1997) *The Lure of the Local, Senses of Place in a Multicentered Society*, New York: The New Press.

——(1999) *On the Beaten Track: tourism, art and place*, New York: The New Press.

Liu, H. (1993) *Shilin Luyou Zhinan* (Stone Forest Directions in Tourism), Beijing: Zhongying Minzuxueyuan Chubanshi.

Liu, Q. (1992) "The full-length national dance drama 'Ashima'," *Women of China*, 7(92): 39–40.

Loercher, D. (1980) "Tourism in Jamaica: still an uneasy paradise," *Christian Science Monitor*, 17 June.

Lorah, P. (1995) "An unsustainable path: tourism's vulnerability to environmental decline in Antigua," *Caribbean Geography*, 6(1): 28–39.

Loti, Pierre (1878) *Le Mariage de Loti*, Paris: Calmann Lévy.

Lum, C. (2002) "Opponents fail to halt liquor sales at Makapu'u wedding chapel," *Honolulu Advertiser*, 31 May.

Lury, C. (1997) "The objects of travel," in C. Rojek and J. Urry (eds) *Touring Cultures: transformations of travel and theory*, New York: Routledge.

Lynch, R. (1998) "Japan's lovers sweet on isles," *Honolulu Star Bulletin*, 30 March.

——(2000) "Survey: Japan weddings in isles up 44% this fall," *Honolulu Star Bulletin*, 27 October.

Lyon, M.L. and Barbalet, J.M. (1994) "Society's body: emotion and the 'somatisation' of social theory," in T.J. Csordas (ed.) *Embodiment and Experience: the existential ground of culture and self*, Cambridge: Cambridge University Press.

Lyotard, J.-F. (1979) *La Condition Postmoderne: rapport sur les avoir*, Paris: Éditions de Minuit.

Ma, L.J.C. (2003) "Space, place and transnationalism in the Chinese Diaspora," in L.J.C. Ma and C. Cartier (eds) *The Chinese Diaspora: place, space, mobility and identity*, Lanham, MD: Rowman and Littlefield.

Macaulay, R. (1984) *Pleasure of Ruins*, London: Thames and Hudson.

MacCannell, D. (1973) "Staged authenticity: arrangements of social space in tourist settings," *American Journal of Sociology*, 79(3): 589–603.

——(1976) *The Tourist: a new theory of the leisure class*, New York: Schocken Books.

——(1984) "Reconstructed ethnicity: tourism and cultural identity in Third World communities," *Annals of Tourism Research*, 11(3): 375–91.

——(1989) "Introduction to special issue on the semiotics of tourism," *Annals of Tourism Research*, 16(1): 1–16.

——(1992) *Empty Meeting Grounds: the tourist papers*, London: Routledge.

——(1994) "Tradition's next step", in Scott Norris (ed.) *Discovered Country, Tourism and Survival in the American West*, Albuquerque, NM: Stone Ladder Press.

——(1999) *The Tourist: a new theory of the leisure class*, 2nd ed., Berkeley, CA: University of California Press.

MacCannell, D. and MacCannell, J. (1982) *The Time of the Sign*, Bloomington: Indiana University Press.

MacCannell, J. F. (2000) *The Hysteric's Guide to the Future Female Subject*, Minneapolis: University of Minnesota Press.

Macnaghten, P. and Urry, J. (1998) *Contested Natures*, London: Sage.

Macy, R. (2000) "Wynn finishes at Bellagio," *Las Vegas Sun*, 30 May.

Maffesoli, M. (1996) *The Time of Tribes*, London: Sage.

Magnay, J. (1999) "Games scandals hit Melbourne 2006 bid," *The Age*, 21 January.

Mahon, T. (1985) *Charged Bodies: people, power and paradox in Silicon Valley*, New York: New American Library.

Malbon, P. (1998) "Clubbing," in T. Skelton and G. Valentine (eds) *Cool Places*, London: Routledge.

Malpas, J.E. (1999) *Place and Experience: a philosophical topography*, Cambridge: Cambridge University Press.

Mancall, M. (1984) *China at the Center: 300 years of foreign policy*, New York: Free Press.

Manderson, L. and Jolly, M. (eds) (1997) *Sites of Desire, Economies of Pleasure: sexualities in Asia and the Pacific*, Chicago: University of Chicago Press.

Margueron, D. (1987) "Permanence d'un mythe," *Bulletin de la Société d'Etudes Océanienne*, 238(3): 1–10.

Mariñez, P.A. (1996) "Las relaciones de México con el Caribe, Un enfoque sobre sus estudios," *Revista Mexicana del Caribe Ano*, 1.

Marra, J. (1967) *Journal of the Resolution's Voyage in 1771–1775,* New York: Da Capo Press.

Marshment, M. (1998) "Gender takes a holiday: representation in holiday brochures," in M.T. Sinclair (ed.) *Gender, Work and Tourism*, London: Routledge.

Martin-Allanic, J.E. (1964) *Bougainville: navigateur et les découvertes de son temps*, 2 vols, Paris.

Martini, M. (1665) *Novus Atlas Sinensis*. Amsterdam: Blau.

Martis, J. (1999) "The muchachas of Orange Walk Town, Belize," in K. Kempadoo *Sun, Sex and Gold: tourism and sex work in the Caribbean*, Lanham, MD: Rowman and Littlefield.

Marx, K. (1977) *Capital: a critique of political economy, Volume 1*, trans. B. Fowkes, New York: Vintage.

Mason, M.G. (1939) *Western Concepts of China and the Chinese*, New York: Seeman.

Masser, T., Sviden, O., and Wegener, M. (1994) "What new heritage for which new Europe? Some contextual considerations," in G.J. Ashworth and L. Larkham (eds) *Building a New Heritage: tourism, culture and identity in the New Europe*, London: Routledge.

Massey, D. (1988) "Uneven development: social change and spatial divisions of labor," in D. Massey and J. Allen (eds) *Uneven Re-development: cities and regions in transition*, London: Hodder and Stoughton.

——(1992) "A place called home?", *New Formations*, 17: 3–17.

——(1993) "Power geometry and a progressive sense of place," in J. Bird, B. Curtis, T. Putnam, G. Robertson, and L. Tickner (eds) *Mapping the Futures: Local Cultures, Global Change*, New York: Routledge.

——(1994) *Space, Place, and Gender*, Minneapolis: University of Minnesota Press.

——(1995) "The conceptualization of place," in D. Massey and P. Jess (eds) *A Place in the World? Places, cultures and globalization*, Milton Keynes: Open University Press.

Mathews, G. (2000) *Global culture/individual identity: searching for home in the cultural supermarket*, London: Routledge.

Matsuo, S. (2001) Director of Sales for Japan Market, Hawai'i Visitors and Convention Bureau. Personal interview in Honolulu, Hawai'i, 16 January.

McBride, B. (1999) "The (post)colonial landscape of Cathedral Square: urban redevelopment and representation in the 'cathedral city'," *New Zealand Geographer*, 55(1): 3–11.

McCauley, S. (1998) "Television: the yellow brick road to 70's San Francisco," *The New York Times*, 7 June.

McCole, J. (1993) *Walter Benjamin and the Antinomies of Tradition*, Ithaca, NY: Cornell University Press.

McDowell, L. (1999) *Gender, Identity and Place: understanding feminist geographies*, Minneapolis: University of Minnesota Press.

McElroy, J. and Albuquerque, de K. (1988) "Migration transition in small northern and eastern Caribbean states," *International Migration Review*, 22(3): 30–58.

——(1991) "Tourism styles and policy responses in the open economy-closed environment context," in N.P. Girvan and D.A. Simmons *Caribbean Ecology and Economics*, (eds) Barbados: Caribbean Conservation Association.

McGinn, F. (1986) "Ersatz Place," *Canadian Heritage*, 12: 25–9.

McKillop, P. (1999) "Letter from Japan: For richer, for poorer," *Time* (Asia) 24 November. Online. Available <http://www.time.com/time/asia/asiabuzz/9911/24/> (accessed 12 January 2001).

Mecker, D.L. and Tisdel, C. (1990) *Development Issues and Small Island Economics*, New York: Praeger.

Meinig, D.W. (ed.) (1979) *The Interpretation of Ordinary Landscapes*, Oxford: Oxford University Press.

Merleau-Ponty, M. (1962) *The Phenomenology of Perception*, London: Routledge.

Michelet, Claude (1980) *Des Grèves aux Loups*, Paris: Hachette.

Midgett, D.K. (1984) "Distorted development: the resuscitation of the Antiguan sugar industry," *Studies in Comparative International Development*, 19(2): 33–58.

Miller, J. (1991) *Seductions: readings in meanings and culture*, Cambridge, MA: Harvard University Press.

Miller, R. (2002) "The future of travel and tourism: in light of September 11, 2001," Speech, NY Office, WTTC, 23 February.

Minca, C. (2001) "Postmodern temptations," in C. Minca (ed.) *Postmodern Geography: Theory and Praxis*, Oxford: Blackwell.

Ministère du Tourisme (1999) *Statistiques Touristiques*, Papeété: Service du Tourisme.

Missac, P. (1995) *Walter Benjamin's Passages*, trans. S.W. Nicholsen, Cambridge, MA: MIT Press.

Mission (2000) *Rapport de Prospective*, Papeété: Présidence du Gouvernement de la Polynésie Française.

Mitchell, A. (1989) *The Fragile South Pacific, An Ecological Odyssey*, Austin: University of Texas Press.

Mitchell, D. (2000) *Cultural Geography: a critical introduction*, Oxford: Blackwell.

Mitchell, K. (2000) "Networks of ethnicity," in E. Sheppard and T. Barnes (eds) *A Companion to Economic Geography*, London: Blackwell.

Mitchell, T. (1988) *Colonizing Egypt*, Cambridge: Cambridge University Press.

Mitchell, W.T.J. (ed.) (2002) *Landscape and Power*, 2nd edn, Chicago: University of Chicago Press.

Moerenhout, J. A. (1837) *Voyage aux Iles du Grand Océan*, Paris: Arthur Bertrand.

Momsen, J. and Lewis, H. (1990) "Erosion rates from Barbados," Discussion paper, Department of Geography, University of Newcastle upon Tyne.

Momsen, J.D. (1977) "Second homes in the Caribbean," in J.T. Coppock (ed.) *Second Homes: curse or blessing?*, Oxford: Pergamon.

Momsen, J.H. (1994) "Tourism, gender and development in the Caribbean," in V. Kinnaird and D. Hall (eds) *Tourism: a gender analysis*, New York: Wiley.

Morse, H. (1997) "Wedding chapel plan draws friends, foe," *Honolulu Star Bulletin*, 14 October. Online. Available <http://www.star-bulletin.com/97/10/14/news/story5.html> (accessed 2 January 2001).

Mowforth, M. and Munt, I. (1998) *Tourism and Sustainability: new tourism in the Third World*, London: Routledge.

MTI (Ministry of Trade and Industry) (1984) *Report of the Tourism Task Force*, Singapore: Ministry of Trade and Industry.

——(1998) *Committee on Singapore's Competitiveness*, Singapore: Ministry of Trade and Industry.

Multer, H.G., Weiss, M.P. and Nicholson, D.V. (1986) *Antigua: reefs, rocks and highroads of history*, St. John's, Antigua: Leeward Islands Science Associates.

Murphy, P. (1985) *Tourism: a community approach*, New York: Methuen.

Murphy, P. and Watson, S. (1997) *Surface City: Sydney at the millennium*, Annandale, NSW: Pluto Press.

Mydans, S. (2002) "The perfect Thai vacation: sun, sea and surgery," *The New York Times*, 9 September.

Nägele, R. (1991) *Theater, Theory, Speculation: Walter Benjamin and the scenes of modernity*, Baltimore, MD: Johns Hopkins University Press.

Nair, S. (1996) "Expressive countercultures and postmodern Utopia: a Caribbean context," *Research in African Literatures*, 27(4): 71–87.

Nash, C. (1994) "Remapping the body/land: new cartographies of identity, gender and landscape in Ireland," in A. Blunt and G. Rose (eds) *Writing Women and Space: colonial and postcolonial geographies*, New York: Guilford Press.

——(1996) "Reclaiming vision: looking at landscape and the body," *Gender, Place and Culture*, 3: 149–70.

Nash, R. (1982) *Wilderness and the American Mind*, New Haven, CT: Yale University Press.

Nast, H.J. and Pile, S. (1998) *Places through the Body*, London: Routledge.

National Informer (St. Johns, Antigua) (1997) "Regional planners say Antigua is on right track," 5 July.

Needham, J. (1954) *Science and Civilization in China*, Cambridge: Cambridge University Press.

Nelson, B. (1988) *Workers on the Waterfront: seamen, longshoremen, and unionism in the 1930s*, Urbana: University of Illinois Press.

Newbury, C. (1980) *Tahiti Nui, Change and Survival in French Polynesia, 1767–1945*, Honolulu: University of Hawai'i Press.

Nielsen, N. K. (1995) "The stadium in the city," in J. Bale (ed.) *The Stadium and the City*, Keele: Keele University Press.

Nolte, C. (2002) "Bay Area Voters are state's contrarians: some winning measures lose big here," *The San Francisco Chronicle*, 10 March.

Norse, E. (1993) *Global Marine Biological Diversity*, Washington, DC: Island Press.

Nuryanti, W. (1996) "Heritage and postmodern tourism," *Annals of Tourism Research*, 23(2): 249–60.

Oakes, T. (1997) "Place and the paradox of modernity," *Annals of the Association of American Geographers*, 87(3): 509–31.

——(1998) *Tourism and Modernity in China*, London: Routledge.

——(1999) "Bathing in the far village: globalization, transnational capital, and the cultural politics of modernity in China," *Positions: East Asia Cultures Critique*, 7(2): 307–42.

Office of Health Status Monitoring (2001) "Table 89, Marriages by Geographic Area of Marriage and Residency of Bride and Groom, 2000," Honolulu: State of Hawai'i.

Ogden, J. and Gladfelter, E. (1983) *Coral Reefs, Sea Grass Beds and Mangroves: their interaction in the coastal zones of the Caribbean*, Montevideo: UNESCO.

Ogden, J.C. and Zieman, J.C. (1977) "Ecological aspects of coral reef – sea grass bed contacts in the Caribbean," *Proceedings of the International Coral Reef Symposium*, 3: 377–82, Miami.

O'Harrow, R. Jr. (2002) "Intricate screening of fliers in works," *Washington Post*, 1 February.

Ohmae, K. (2000) *The Invisible Continent: four strategic imperatives of the new economy*, New York: Harper Business.

Oliver, D. (1989) *Oceania, the Native Cultures of Australia and the Pacific Islands*, Honolulu: University of Hawai'i Press.

Olson, S.C. (1999) *Hollywood Planet*, Mahwah, NJ: Lawrence Erlbaum.

Olwing, K. (2002) *Landscape, Nature and the Body Politic: from Britain's Renaissance to America's New World*, Madison: University of Wisconsin Press.

Ong, A. (1999) *Flexible Citizenship: the cultural logics of transnationality*, Durham, NC: Duke University Press.

Outlet (St. Johns, Antigua) (1997a) "Antigua's off-shore islands: the last bastion of our environmental legacy," 10 June.

——(1997b) "Asian Village does not justify cost to Antigua – Bar," 20 June.

——(1997c) "Lester Bird concedes he can't govern in local interest," 24 June.

——(1997d) "National Redemption Day: the struggle heightens," 27 June.

——(1997e) "The old must make way for the new," 27 June.

——(1997f) "We must fight the good fight," 1 July.

——(1997g) "Better days are coming," 4 July.

Overend, W. (2003) "The Gold Coast of California," *Los Angeles Times*, 14 June.

Oxfeld, E. (2001) "Imaginary Homecomings: Chinese villagers, the overseas Chinese relations, and social capital," *Journal of Socio-Economics*, 30(2): 181–6.

Page, S. (1995) *Urban Tourism*, London: Routledge.

Pan, L. (1990) *Sons of the Yellow Emperor: a history of the Chinese diaspora*, Boston: Little, Brown.

Parker, A. and Sedgwick, E.K. (1995) "Introduction: performativity and performance," in A. Parker and E. Sedgwick (eds) *Performativity and Performance*, New York: Routledge.

Parry, R. L. (1996) "Japan's teens giddy over virtual idol," *San Francisco Examiner*, 15 August.

Pattullo, P. (1996) *Last Resorts: the cost of tourism, in the Caribbean*, London: Cassell.

Pawson, E. (1997) "Branding strategies and languages of consumption," *New Zealand Geographer*, 53(2): 16–21.

PCC (1995) *All the Spirit of the Islands*, Souvenir Edition, La'ie, HI: Polynesian Cultural Center.

Pearce, P.L. (1988) *The Ulysses Factor: evaluating visitors in tourist settings*, New York: Springer Verlag.

Pearson, W.H. (1969) "European intimidation and the myth of Tahiti," *Journal of Pacific History*, 4: 199–217.

Percy, W. (1975) *The Message in the Bottle*, New York: Farrar, Straus, Giroux.

Perry, M., Kong, L., and Yeoh, B.S.A. (1997) *Singapore: a developmental city-state*, Chichester: Wiley.

Petersen, Y.Y. (1995) "The Chinese landscape as a tourist attraction: image and reality," in A.A. Lew and L. Yu (eds) *Tourism in China: geographical, political and economic perspectives*, Boulder, CO: Westview Press.

Phillips, J.L. (1999) "Tourism and the sex trade in St. Maarten and Curaçao, the Netherlands Antilles," in K. Kempadoo *Sun, Sex and Gold: tourism and sex work in the Caribbean*, Lanham, MD: Rowman and Littlefield.

Philo, C. and Kearns, G. (1993) "Culture, history, capital; a critical introduction to the selling of places," in C. Philo and G. Kearns, (eds) *Selling Places: the city as cultural capital, past and present*, Pergamon Press, Oxford.

Pier 39 (2002) "Pier 39 media." Online. Available <http://www.pier39.com/media/media.cfm/category/retrospective> (accessed 1 May 2002).

Pieterse, J.N. (1995) "Globalisation as hybridization," in M. Featherstone, S. Lash, and R. Robertson (eds) *Global Modernities*, London: Sage.

Pietz, W. (1993) "Fetishism and materialism: the limits of theory in Marx," in E. Apter and W. Pietz (eds) *Fetishism as Cultural Discourse*, Ithaca, NY: Cornell University Press.

Pile, S. (1996) *The Body and the City: psychoanalysis, space and subjectivity*, London: Routledge.

Pile, S. and Thrift, N. (1995) *Mapping the Subject: geographies of cultural transformation*, London: Routledge.

Poston, D.L. Jr. and Yu, M.-Y. (1990) "The distribution of the overseas Chinese in the contemporary world," *International Migration Review*, 24(3): 480–508.

Power, B. (1999) "$350m tax jolt just 'academic' says Egan," *Sydney Morning Herald*, 22 January.

Pragg, S. (1999) "Rights-Jamaica: beach access reveals deep divisions in society" *Inter Press Service*, 26 May. Online. Available <http://web.lexis-nexis.com/universe> (accessed 30 June 2001).

Prakash, G. (1992) "Postcolonial criticism and Indian historiography," *Social Text*, 10(31–2): 8–19.

Pratt, M. L. (1992) *Imperial Eyes*, London: Routledge.

Pred, A. (1986) *Place, Practice, and Structure: social and spatial transformation in southern Sweden: 1750–1850*, Totowa, NJ: Barnes and Noble.

Pruitt, D. and LaFont, S. (1995) "For love and money: romance tourism in Jamaica," *Annals of Tourism Research*, 22(2): 422–40.

Raban, J. (1974) *Soft City*, London: Hamilton.

Rabinow, P. (1989) *French Modern: norms and forms of the social environment*, Cambridge, MA: MIT Press.

Rach, L. (2002) "Cruise Ship Condos," *I-DIGEST*, (3)2: 1–7.

Radley, A. (1990) "Artefacts, memory and a sense of the past," in D. Middleton and D. Edwards (eds) *Collective Remembering*, London: Sage.

——(1995) "The elusory body and social constructionist theory," *Body and Society*, 1(2): 3–23.

——(1996) "Displays and fragments: embodiment and the configuration of social worlds," *Theory and Psychology*, 6(4): 559–76.

Ramrattan, P. (1997) Personal interview, 1 July, Five Islands, Antigua.

Ratto, R. (1997) "San Francisco's unbelievable hero," *The San Francisco Examiner*, 15 December.

Ravenscroft, N. (1999) "Hyper-reality in the official (re)construction of leisure sites: the case of rambling," in D. Crouch (ed.) *Leisure/Tourism Geographies: practices and geographical knowledge*, London: Routledge.

Redmond, T. (2001) "Nobody's winning: the boom, the bust, and the really stupid politics," *San Francisco Bay Guardian*, 30 May.

Res & Co. (1999) *Jamaica: master plan for sustainable tourism development. Diagnostic & strategic options*, Draft Report prepared for the Office of the Prime Minister, Tourism. Kingston, Jamaica.

Richter, L. (1989) *The Politics of Tourism in Asia*, Honolulu: University of Hawai'i Press.

Robbins, B. (1998) "Actually existing cosmopolitanism," in B. Robbins and P. Cheah (eds) *Cosmopolitics: thinking and feeling beyond the nation*, Minneapolis: University of Minnesota Press.

Robinson, J. (1991) "The other Jamaica; Caribbean adventures," *The Boston Globe*, 27 January.

Robischon, N. (1996) "Digital recording for the analogue soul," *Wired*, 4(11): 78.

Roche, M. (1992) "Mega-events and micro-modernization: on the sociology of the new urban tourism," *British Journal of Sociology*, 43(4): 563–600.

——(1994) "Mega-events and urban policy," *Annals of Tourism Research*, 21(1): 1–19.

Roig, S. (2000) "Sea Life Park's chapel plan challenged," *Honolulu Advertiser*, 10 May.

——(2002) 'Āina Haina church arrival worries residents,' *Honolulu Advertiser*, 1 July.

Rojek, C. (1993a) *Ways of Escape: modern transformations in leisure and travel*, London: Macmillan.

——(1993b) *Ways of Escape: modern transformations in leisure and travel*, New York: Routledge.

——(1995) *Deconstructing Leisure*, London: Sage.

Rolett, B. (1996) Colonisation and cultural change in the Marquesas, in J. Davidson, G. Irwin, F. Leach, A. Pawley, and D. Brown (eds) *Oceanic Culture History, Essays in Honour of Roger Green*, Dunedin: New Zealand Journal of Archaeology Special Publication.

Roof, W.C. (1999) *Spiritual Marketplace*, Princeton, NJ: Princeton University Press.

Rosaldo, R. (1989) *Culture and Truth: the remaking of social analysis,* Boston: Beacon Press.

Rose, G. (1993) *Feminism and Geography: the limits of geographical knowledge*, Minneapolis: University of Minnesota Press.

Rothman, H.K. and Holder, D.J. (2001) "Administrative history of the Golden Gate National Recreation Area," Online. Available <http://www.nps.gov/goga/history/index.html> (accessed 15 May 2001).

Rowe, N.A. (1949) *Voyage to the Amorous Islands*, Fairlawn, NJ: Andre Deutsch.

Rowling, M. (1971) *Everyday Life of Medieval Travelers*, London, New York: B.T. Batsford, G.P. Life Putnanam's Sons.

Rushdie, S. (1991) *Imaginary Homelands*, London: Granta.

Sack, R. (1992) *Place, Modernity and the Consumer's World: a relational framework for geographical analysis,* Baltimore, MD: Johns Hopkins University Press.

Safire, W. (2000) "The way we live now: 10–01–00; on language," *The New York Times*, 1 October.

——(2001) "Essay: California power failure," *The New York Times*, 11 January.

Sahlins, M. (1985) *Islands of History*, Chicago: University of Chicago Press.

Said, E.W. (1978) *Orientalism,* New York: Pantheon.

Sandals (2002) *Sandals Resorts Homepage*. Online. Available <http://www.allinclusivesavings.com/main/Antigua> (accessed 31 July 2002).

St. Lucia Tourist Board (2000) Unpublished Tourism Statistics.

Saura, B. (1998) *Des Tahitiens, des Français, leurs Representations Réciproques Aujourd'hui*, Papeété: Les Essais.

Sayer, A. (1997) "The dialectic of culture and economy," in R. Lee and J. Wills (eds) *Geographies of Economies*, London: Arnold.

Scemla, J.J. (1994) *Le Voyage en Polynésie. Anthologie des voyageurs occidentaux de Cook à Segalen*, Paris: Laffont.

Schama, S. (1995) *Landscape and Memory*, New York: A.A. Knopf.

Schein, L. (1997) "Gender and internal orientalism in China," *Modern China*, 23(1): 69–98.

——(2000) *Minority Rules: the Miao and the feminine in China's cultural politics*, Durham, NC: Duke University Press.

Schwartz, S. (1998) *From West to East: California and the making of the American mind*, New York: Free Press.

Seabrook, J. (1988) *The Leisure Society*, New York: Basil Blackwell.

Segalen, V. (1956) *Les Immémoriaux*, Paris: Librairie Pion.

Selwyn, T. (1996) *The Tourist Image: myth and myth-making in tourism*, Chichester: Wiley.

SFCVB (San Francisco Convention and Visitors Bureau) (1999) *Summary Report: San Francisco hotel guest survey year end, 1999*, San Francisco: San Francisco Convention and Visitors Bureau.

——(2000) *Brand Audit Final Report: San Francisco as a travel destination*, San Francisco: San Francisco Convention and Visitors Bureau.

Sharkey, D. A. and Momsen, J.H. (1995) "Tourism in Dominica: problems and prospects," *Caribbean Geography*, 6(1): 40–51.

Sharpe, S. (1999) "Bodily speaking: spaces and experiences of childbirth," in E.K. Teather (ed.) *Embodied Geographies*, London: Routledge.

Shaughnessy, D. (1989) "On baseball: sox can't coast in West," *The Boston Globe*, 11 May.

Shields, R. (1991) *Places on the Margin: alternative geographies of modernity*, London; New York: Routledge.

Shotter, J. (1993) *The Cultural Politics of Everyday Life: social constructionism, rhetoric and knowledge of the third kind*, Milton Keynes: Open University Press.

Sigogne, L. (1922) "Le tourisme en Océanie," *Bulletin de la Société d'Etudes Océaniennes*, 6: 52–4.

Simmons, F., Goodman, M., and Goodman, D. (1998) "Environmental narratives and the reproduction of food," in B. Braun and N. Castree (eds) *Remaking Reality: nature at the millennium*, New York: Routledge.

Sims, K. (2000) "San Francisco Economy: Implications for Public Policy," Report prepared for the San Francisco Planning and Urban Research Association, July 10. Online. Available : <http://www.spur.org/documents/sims150.pdf> (accessed 27 July 2004).

Simson, V. and Jennings, A. (1992) *The Lords of the Rings: power, money and drugs in the modern Olympics*, London: Simon & Schuster.

Slusser, S. (1998) "Athletes topic of show: ESPN program includes segment on how young, others deal with rumors," *The San Francisco Chronicle*, 16 December.

Smith, A. (1986) *The Ethnic Origins of Nations*, New York: Basil Blackwell.

Smith, G. (1991a) "Tourism, Telecommunications, and Transnational Banking: a study in international interactions," PhD dissertation, The American University, Washington, DC.

——(1991b) "International tourism and political instability: the 'collateral damage' of the Persian Gulf War," Unpublished essay.

Smith, K.B. and Smith, F.C. (1986) *To Shoot Hard Labor: the life and times of Samuel Smith, an Antiguan working man 1877–1982*, Scarborough, Ontario: Edan's.

Smith, N. (1984) *Uneven Development*, Oxford: Basil Blackwell.

——(1998) "Nature at the millennium: production and re-enchantment," in B. Braun and N. Castree (eds) *Remaking Reality: nature at the millennium*, New York: Routledge.

Smith, V. (1998) *Literary Culture and the Pacific*, Cambridge: Cambridge University Press.

Sobers, A. (1999) Personal communication, Caribbean Tourism Organization.

Sofield, T.H.B. and Li, F.S.M. (1996) "Rural tourism in China," in S. Page and D. Getz (eds) *The Business of Rural Tourism*, London: International Thomson Business Press.

——(1998) "China: tourism development and cultural policies," *Annals of Tourism Research*, 25(2).

Sombart, W. (1967) *Luxury and Capitalism*, trans. W.R. Dittmar, Ann Arbor: University of Michigan Press.

Sontag, S. (1977) *On Photography,* New York: Anchor Books.

Spate, O.H.K. (1979) *The Spanish Lake: The Pacific since Magellan*, vol. 1, Canberra: Australian National University Press.

——(1988) *Paradise Found and Lost*, vol. 3 of *The Pacific since Magellan,* Canberra: Australian National University Press.

Spence, J. (1992) "Western perceptions of China from the late sixteenth century to the present," in P.S. Ropp (ed.) *Heritage of China: contemporary perspectives on China*, Berkeley, CA: University of California Press.

Spivack, S. and Chernish, W. (1999) "Assessing the need for mutual understanding: health education for tourism educators and tourism education for health educators," *Asia Pacific Journal of Tourism Research*, 3(1): 45–54.

State of California (2003) "Travel industry: research & statistics," California tourism, California's Top Attractions. Online. Available <http://gocalif.ca.gov/state/tourism/tour_htmldisplay.jsp?iOID=29041&sFilePat=tourism/htdocs/research_stats/RS_TopAttractions.html> (accessed June 2003).

STB (Singapore Tourism Board) (1996) *Tourism 21: vision of a tourism capital*, Singapore: STB.

——(1998a) "Promoting the New Asia-Singapore branding task force report," *Singapore Tourism Conference '98*, 1–17, Singapore: STB.

——(1998b) *STB Media Release*, 28 January.

——(1999) *Singapore Annual Report on Tourism Statistics 1998*, Singapore: STB.

——(2001) *The Singapore Tourism Board Year Book 2000–2001*, Singapore: Singapore Tourism Board.

Stephenson, N. (1996) "The hacker tourist travels the world to bring back the epic story of wiring the planet," *Wired*, 4(12): 97–140, 144–5, 148, 152, 156, 158, 160.

Stevens, M. (1998) "Bellagio shopping offers a variety of riches," *Las Vegas Sun*, 16 October.

Stevens, M. and Lehmann, J. (1999) "IOC purge starts race for reform," *The Australian*, 26 December.

Stewart, L. (1995) "Bodies, visions, and spatial politics: a review essay on Henri Lefebvre's *The Production of Space*," *Environment and Planning D: Society and Space*, 13(5): 609–18.

Stewart, S. (1993) *On Longing: narratives of the miniature, the gigantic, the souvenir, the collection*, Durham, NC: Duke University Press.

Stinchcombe, A.L. (1995) *Sugar Island Slavery in the Age of Enlightenment: the political economy of the Caribbean world*, Princeton, NJ: Princeton University Press.

Stock, B. (1993) "Reading, community and a sense of place," in J. Duncan and D. Ley (eds) *Place/Culture/Representation*, London: Routledge.

The Straits Times, various editions, Singapore: Singapore Press Holdings.

Strow, D. (2001) "Bellagio is shattering industry profit records," *Las Vegas Sun*, 27 April.

Stryker, S. and Van Buskirk, J. (1996) *Gay by the Bay: a history of queer culture in the San Francisco Bay Area*, San Francisco: Chronicle Books.

Stumpy, C. (1996) "This country," *Calypso Talk '96*, St. John's, Antigua: Wadadli Productions.

Sturma, M. (1999) "Packaging Polynesia's image," *Annals of Tourism Research*, 26(3): 712–15.

Sun, W. (2002) *Leaving China: media, migration, and transnational imagination*, New York: Rowman and Littlefield.

Sun, Y.S. (1922) *Sun-wen hsueh-shuo* (Sun Yat Sen Collected Papers), Taipei: Yuang-tung.

The Sunday Times, 8 August 1999, Singapore: Singapore Press Holdings.

Swain, M.B. (1995a) "Gender in tourism," *Annals of Tourism Research*, 22(2): 247–67.

——(1995b) "Pre Paul Vial and the Gni-Pa: orientalist scholarship and the Christian Project," in S. Harrell (ed.) *Cultural Encounters on China's Ethnic Frontiers*, Seattle: University of Washington Press.

——(2001) "Cosmopolitan tourism and minority politics in the Stone Forest," in T.C. Beng, S. Cheung, and Y. Hui (eds) *Tourism, Anthropology and Chinese Society: in memory of Professor Wang Zhusheng* (English Volume), Bangkok: White Lotus Press.

——(2002) "Gender/tourism/fun(?): an introduction," in M. Swain and J. Momsen (eds) *Gender/Tourism/Fun(?)*, Elmsford, NY: Cognizant Communication Corporation.

Sweet, J. (1989) "Burlesquing 'the Other' in Pueblo performance," *Annals of Tourism Research*, 16(1): 62–75.

Symonds, G., Black, K.O., and Young, I.R. (1995) "Wave driven flow over shallow reefs," *Journal of Geophysical Research*, 100: 2639–48.

Taillemite, E. (1978) *Bougainville et ses Compagnons autour du Monde*, Paris: Imprimerie Nationale.

Tang, H. (1999) "HVCB hopes to find love with national promotions," *Honolulu Star Bulletin*, 30 July.

Taylor, C. (1989) *Sources of the Self: the making of modern identity*, Cambridge, MA: Harvard University Press.

——(1995) "Gates, journalist settle suit over wedding arrest," *Seattle Times*, 13 April.

Taylor, F.F. (1993) *To Hell With Paradise: a history of the Jamaican tourism industry*, Pittsburgh: University of Pittsburgh Press.

Teiawa, T.K. (2001) "L(o)osing the edge," *The Contemporary Pacific*, 13(2): 343–57.

Tempest, R. (2002) " 'Left Coast' stereotype faces test," *The Los Angeles Times*, 12 December.

TenBruggencate, J. (2000) "Grace Guslander, visionary hotelier, dies," *Honolulu Advertiser*, 6 April.

Teo, M. (1999) "Cuisine on the Cutting Edge," *Elite*, February: 34–7.

Teo, P. and Chang, T.C. (2000) "Singapore: tourism development in a planned context," in S. Hall and M. Page (eds) *Tourism in South and Southeast Asia: issues and cases*, Oxford: Butterworth–Heinemann.

Teo, P. and Huang, S. (1995) "Tourism and heritage conservation in Singapore," *Annals of Tourism Research*, 22(3): 589–615.

Teo, P. and Yeoh, B.S.A. (1997) "Remaking local heritage for tourism," *Annals of Tourism Research*, 24(1): 192–213.

Tester, K. (ed.) (1994) *The Flâneur*, London: Routledge.

Thomas, A., Crow, B., Frenz, P., Hewitt, T., Kassam, S., and Tregust, S. (1994) *Third World Atlas*, Washington, DC: Taylor & Francis.

Thomas, N. (1991) *Entangled Objects: exchange, material culture and colonialism in the Pacific*, Cambridge, MA: Harvard University Press.

Thomas-Hope, E. (ed.) (1998) *Solid Waste Management: critical issues for developing countries*, Jamaica: University of the West Indies, Canoe Press.

Thompson, G. (1998) "Bellagio's art collection is a treasure for Nevada," *Las Vegas Sun*, 16 October.

Thrift, N. (1996) *Spatial Formations*, London: Sage.

——(1997) "The still point: resistance, expressive embodiment and dance," in M. Keith and K. Pile (eds) *Geographies of Resistance*, London: Routledge.

Thrift, N. and Dewsbury, J.D. (2000) "Dead geographies—and how to make them live," *Environment and Planning D: Society and Space*, 18: 411–32.

TIA (Travel Industry Association of America, Inc.) (2001) "TIA Travel Forecasting Model," 11 November.

Timothy, D. and Coles, T. (eds) (2003) *Diaspora and Tourism*, London: Routledge.

Timothy, D.J. (2001) *Tourism and Political Boundaries*, New York: Routledge.

Tobin, J.J. (ed.) (1992) *Re-made in Japan: everyday life and consumer taste in a changing society*, New Haven, CT: Yale University Press.

Toullelan, P.Y. (1991) *Tahiti et ses Archipels*, Paris: Karthala.

Touraine, A. (1995) *Critique of Modernity*, Oxford: Blackwell.

Tourism Victoria (1997) *Annual Report 1996–97*, Melbourne: Tourism Victoria.

Tourtellot, J. (2002) Sustainable Tourism Resource Center, National Geographic Society. Online. Available <http://test.nationalgeographic.com/travel/sustainable/index.html> (accessed 29 November 2001).

Towner, J. (1996) *An Historical Geography of Recreation and Tourism in the Western World: 1540–1940*, New York: Wiley.

Trask, H.K. (1991/92) "Lovely hula hands: corporate tourism and the prostitution of Hawaiian culture," *Borderlines*, 23: 22–9.

——(1993) *From a Native Daughter*, Monroe, ME: Common Courage Press.

Tu, W.M. (ed.) (1994) "Cultural China: the periphery at the center," in W.M. Tu (ed.) *The Living Tree: the changing meaning of being Chinese today*, Stanford, CA: Stanford University Press.

Tuan, Y. F. (1979) *Landscapes of Fear,* New York: Pantheon.

Tunteng, P.-K. (1975) "Reflections on labour and governing in Antigua," *Caribbean Studies*, 15(2): 36–56.

Turner, B.S. (1994) *Orientalism, Postmodernism and Globalism*, London: Routledge.

Turner, V. (1967) "Betwixt and between: the liminal period in rites of passage," in V. Turner (ed.) *The Forest of Symbols*, Ithaca, NY: Cornell University Press.

——(1969) *The Ritual Process: structure and anti-structure*, London: Routledge & Kegan Paul.

Turner, V. and Turner, E. (1978) *Image and Pilgrimage in Christian Culture: anthropological perspectives*, New York: Columbia University Press.

TWSF (*Tourism Works for San Francisco*) (2001) "Report based on research conducted by Economic Research Associates and the San Francisco Convention and Visitors Bureau," San Francisco. Online. Available <http://www.sfvisitor.org/research/download/tourb&w.pdf> (accessed 15 May 2001).

UCEA (University Continuing Education Association) (2001/2002) "Indepth Trends – Internet Use Democratizing," December/January.

Ullman, J. and Dinhofer, A. (1968) *Caribbean Here and Now*, New York: Macmillan.

UNED-UK (1999) "Gender and tourism: women's employment and participation in tourism," Report for the United Nations Commission on Sustainable Development, 7th Session, April, London: DFID.

UPR (University of Puerto Rico) (1994a) *Caribbean Newsletter*, April/June.

——(1994b) *Caribbean Newsletter*, October/December.

URA (Urban Redevelopment Authority) (1988) *Historic Districts in the Central Area: a manual for Chinatown conservation area,* Singapore: URA.

Urry, John (1990) *The Tourist Gaze: leisure and travel in contemporary societies*, London: Sage.

——(1995) *Consuming Places*, London: Routledge.

——(2000) *Sociology beyond Societies: mobilities for the twenty-first century*, London: Sage.

——(2002) *The Tourist Gaze*, 2nd edn, London: Sage.

Uzzell, D. (ed.) (1989) *Heritage Interpretation: the natural and built environment,* London: Belhaven Press.

Van Gennep, A. (1960) [1909] *The Rites of Passage*, trans M. B. Vizedom and G. L. Caffee, Chicago: University of Chicago Press.

Verbiest, F. (1685) *Voyages de L'Emperor de la Chine dans la Tartarie,* Paris: Estienne Michallet.

Vermeer, E.B. (1977) *Water Conservancy and Irrigation in China,* Leiden: Leiden University Press.

Vial, P. (1898) *Les Lolos: historie, religion, moeure, langue, ecriture*, Changhai: Imprimerie de la Mission Catholique.

Von Strokirch, K. (2000) "Polynesia in review: issues and events," *The Contemporary Pacific*, 12(1): 221–7.

Vorsino, M. (2002) "Chapel liquor OK raises concern," *Honolulu Star Bulletin*, 13 June.

Wagner, V. (1998) "Western addition residents," *The San Francisco Examiner*, 21 April.

Waimānalo Neighborhood Board (2000a) "Regular meeting minutes September 11," Honolulu: City and County of Honolulu Neighborhood Commission Office.

——(2000b) "Regular meeting minutes November 13," Honolulu: City and County of Honolulu Neighborhood Commission Office.

Wakeman, F. (1975) *The Fall of Imperial China*, New York: Free Press.

Waldron, A.N. (1990) *The Great Wall of China: from history to myth*, Cambridge: Cambridge University Press.

Walker, R.A. (1998) "An appetite for the city," in J. Brook, C. Carlsson, and N.J. Peters (eds) *Reclaiming San Francisco: history, politics, culture*, San Francisco: City Lights Books.

——(2001) "California's golden road to riches: natural resources and regional capitalism, 1848–1940," *Annals of the Association of American Geographers*, 91(1): 167–99.

Walter, S. (1997a) "Of Perception and reality," *Daily Observer*, 20 June.

——(1997b) "Sowing the seeds of strife and discord," *Antigua Daily Observer*, 5 July.

Wang, G. (2000) *The Chinese Overseas: from earthbound China to the quest for autonomy*, Cambridge, MA: Harvard University Press.

Wang, N. (2000) *Tourism and Modernity: a sociological analysis*, Amsterdam: Elsevier and Oxford: Pergamon.

Washington, S. (1999) "IOC allegations may affect funding," *Australian Financial Review*, 19 January.

The Washington Post (1979) "Violence in Jamaica kills 7, threatens good tourist season," 12 January.

——(1997) "S.F. stadium consultant apologizes for bawdy party," 10 May.

Watabe Wedding (2001a) *Hawai: Kaigai Uedingu Monogatari*, Kyōto: Watabe Uedingu Kabushiki Kaisha.

——(2001b) Online. Available <http://www.watabe-wedding.com/./main/profile.html> (accessed 1 July 2001 in Japanese), Kyōto: Watabe Uedingu Kabushiki Kaisha.

Watanabe, C.C. (1998) "Japan nuptials lose frill," *Detroit News*, 21 October.

Watanabe, J. (2000) "Kokua Line: Waimānalo chapel to be 'low impact'," *Honolulu Star Bulletin*, 5 September.

Watts, D. (1987) *The West Indies, Patterns of Development, Culture and Environmental Change since 1492*, Cambridge: Cambridge University Press.

——(1995) "Environmental degradation, the water resource and sustainable development in the Eastern Caribbean," *Caribbean Geography*, 6(1): 2–15.

Wayne, S. (2002a) "WTTC's position on the General Agreement on Trade in Services," Washington, DC: SW Associates.

——(2002b) "Sustainable tourism, sustainable livelihoods," Washington, DC: SW Associates.

Wearing, B. and Wearing, S. (1996) "Refocussing the tourist experience: the flâneur and the choraster,". *Leisure Studies*, 1: 229–43.

Weaver, D.B. (1988) "The evolution of a 'plantation' tourism landscape on the Caribbean island of Antigua," *Tijdschrift voor Economische en Sociale Geografie*, 79: 319–31.

——(1998) "Peripheries of the periphery: tourism in Tobago and Barbuda," *Annals of Tourism Research*, 25(2): 292–313.

Weber, M. (1946) *From Max Weber: essays in sociology*, trans. H. H. Gerth and C. Wright Mills (ed.), New York: Oxford University Press.

Wee, L.A. (2001) "Fewer foreign patients in Singapore," *The Straits Times*, 5 June.

Weiss, M. (1999) "Commentary: '70s lifestyle vs. dot-com: Silicon Valley is zapping S.F.'s cultural revolution," *The San Francisco Chronicle*, 9 December.

Weiss, M. and Multer, H.G. (1988) "Modern reefs and sediments of Antigua, W.I.," Department of Geology, Northern Illinois University.

West Indian Commission (1992) "Time for Action: overview of the report of the West Indian Commission," Barbados.

Whatley, T. (1770) *Observations on Modern Gardening*, London.

Wilbert, C. (2000) "Anti-this-Anti-that: resistances along a human-non-human axis," in J. Sharp, P.

Routledge, C. Philo, and R. Paddison (eds) *Entanglements of Power: geographies of domination/resistance*, New York: Routledge.

Wilkinson, P. F. (1994) "Tourism and small island states: problems of resource analysis, management and development," in A.V. Seaton, C.L. Jenkins, P.U.C. Dieke, M.M. Bennett, L.R. MacLellan, and R. Smith (eds) *Tourism: the state of the art*, Chichester: Wiley.

Williams, J.G. (1991) *The Bible, Violence, and the Sacred: liberation from the myth of sanctioned violence*, New York: HarperCollins.

Williams, R. (1980) *Problems of Materialism and Culture*, New York: Columbia University Press.

——(1983) *Keywords*, London: Fontana.

Wilson, M. (1993) "Bird backs bulldozers," *Caribbean Week*, 6 February.

Wired (1995) Advertisement, 3(3): 77.

——(1996) Advertisement, 4(10): 23.

Wolch, J., West, K., and Gaines, T. (1995) "Transspecies urban theory," *Environment and Planning D: Society and Space*, 13: 735–60.

Wollen, P. (1982) *Walter Benjamin: an aesthetic of redemption*, New York: Columbia University Press.

Wong, H. (1997) "The North American (US) Chinese experience," in G. Zhang (ed.) *Ethnic Chinese at the Turn of the Centuries*, vol. 2, Fuzhou, China: Fujian People's Publishing Co.

Wood, R. (1997) "Tourism and the state: ethnic options and constructions of otherness," in M. Picard and R. Wood (eds) *Tourism, Ethnicity, and the State in Asian and Pacific Societies*, Honolulu: University of Hawai'i Press.

Woods, G. (1995) "Fantasy islands: popular topographies of marooned masculinity," in D. Bell and G. Valentine (eds) *Mapping Desire: Geographies of sexualities*, London: Routledge.

Woods, M. (1998) "Rethinking elites: networks, space, and local politics," *Environment and Planning A*, 30: 2101–19.

Woon, Y.F. (1989) "Social change and continuity in South China: overseas Chinese and the Guan lineage of Kaiping County, 1949–87," *The China Quarterly*, 118: 324–44.

Wozniak, M. (2001) "Where the sidewalk ends," *San Jose Mercury News*, 31 July.

Wright, H. and Rapport, S. (eds) (1957) *The Great Explorers*, New York: Harper & Row.

Wright, W. (1999) "Champagne ban at chapel will stand," *Honolulu Advertiser*, 29 October.

WTO (World Tourism Organization) (2000) *Basic Reference on Tourism Statistics*, Madrid: WTO. Online. Available <http://www.world-tourism.org/statistics/tsa_project/basic_references/index-en.htm> (accessed 1 August 2002).

——(2001a) *Tourism Highlights, 2001*, Madrid: WTO. Online. Available <http://www.world-tourism.org/market_research/facts&figures/latest_data/Highlightsupdatedengl.pdf> (accessed 1 August 2002).

——(2001b) Tourism market trends. Online. Available <http://www.world-tourism.org/market_research/facts&figures/market_trends/ita.htm (accessed 1 August 2002).

——(2001c) *Tourism Highlights 2001*, Madrid: WTO.

——(2001d) "Overview 2001." Online. Available <http://www.world-tourism.org/market_research/facts&figures/menu.htm> (accessed 1 May 2002).

——(2001e) "International tourism receipts by (sub)region." Online. Available <http://www.world-tourism.org/market_research/facts&figures/lateSt_data/titr01_07–02.pdf> (accessed 31 July 2002).

WTTC (World Travel and Tourism Council) (2001) "WTTC/WEFA Year 2000 Tourism Satellite Accounting Research Estimates and Forecasts for Governments and Industry, London: WTTC."

Wyllie, R.W. (1998) "Not in our backyard: opposition to tourism development in a Hawaiian community," *Tourism Recreation Research*, 23(1): 55–64.

Yamamoto, D. and Gill, A.M. (2002) "Issues of globalization and reflexivity in the Japanese tourism production system: the case of Whistler, British Columbia," *The Professional Geographer*, 54(1): 83–93.

Yang, G. (1981) *Ashima*, Beijing: Foreign Languages Press.

Yang, Z. (ed.) (1980) *Ashima*, Beijing: Zhongguo Qingnian Chubanshi.

Yeoh, B.S.A. and Kong, L. (1994) "Reading landscape meanings: state constructions and lived experiences in Singapore's Chinatown," *Habitat International*, 18(4): 17–35.

Yeoh, B.S.A. and Lau, P. (1995) "Historic district, contemporary meanings: urban conservation and the creation and consumption of landscape spectacle in Tanjong Pagar," in B.S.A. Yeoh and L. Kong (eds) *Portraits of Places: history, community and identity in Singapore*, Singapore: Times Editions.

Yeoh, B.S.A. and Teo, P. (1996) "From Tiger Balm Gardens to Dragon World: philanthropy and profit in the making of Singapore's first cultural theme park," *Geografiska Annaler (series B)*, 78B: 27–42.

Yim, K. (2001) Manager, Watabe Wedding Honolulu. Personal interview in Honolulu, Hawai'i, 8 February.

York, P. and Jennings, C. (1995) *Peter York's 80's*, London: BBC Books.

Young, I.M. (1990) *Throwing Like a Girl and other Essays in Feminist Philosophy and Social Theory*, Bloomington: Indiana University Press.

Zachariah, E. (1997a) "What is making the beaches in Antigua erode? Part 1," *Antigua Sun*, 13 June.

——(1997b) "What is making the beaches in Antigua erode? Part 2," *Antigua Sun*, 20 June.

Zhong, X. (1983) *Yunnan Travelogue – 100 Days in Southwest China*, Beijing: New World Press.

Zizek, S. (1997) *The Plague of Fantasies*, New York: Verso.

Zuercher, M. (2000) "Mission workers are always looking for opportunities," *Mennonite Weekly Review*, Internet Edition, 23 May. Online. Available <http://www.mennoweekly.org/index_20000523.html> (accessed 15 January 2001).

Zukin, S. (1991) *Landscapes of Power: from Detroit to Disney World*, Berkeley, CA: University of California Press.

——(1996) "Space and symbols in an age of decline," in A. King (ed.) *Re-Presenting the City: ethnicity, capital and culture in the twenty-first century metropolis*, Houndmills: Macmillan.

Zukovic, B. (2000) "Edward Teller and the mach I car," *Journal of Culture and the Unconscious*, 1: 121–33.

Zurick, D. (1995) *Errant Journeys*, Austin: University of Texas Press.

Index